How Users Matter

Inside Technology
edited by Wiebe E. Bijker, W. Bernard Carlson, and Trevor Pinch

How Users Matter
The Co-Construction of Users and Technology

edited by Nelly Oudshoorn and Trevor Pinch

The MIT Press
Cambridge, Massachusetts
London, England

First MIT Press paperback edition, 2005

© 2003 Massachusetts Institute of Technology

Set in New Baskerville by The MIT Press.

Library of Congress Cataloging-in-Publication Data

How users matter : the co-construction of users and technology / edited by Nelly Oudshoorn and Trevor Pinch.
p. cm. — (Inside technology)
Includes bibliographical references and index.
ISBN 987-0-262-15107-8 (hc. : alk. paper)—978-0-262-65109-7 (pb.)
1. Technology—Social aspects. I. Oudshoorn, Nelly, 1950– II. Pinch, T. J. (Trevor J.) III. Series.
T14.5.H69   2003
303.48'3—dc21   2003046410

Contents

II
Multiple Spokespersons: States and Social Movements as Representatives of Users

5
Citizens as Users of Technology: An Exploratory Study of Vaccines and Vaccination   103
Dale Rose and Stuart Blume

6
Knowledge Is Power: Genetic Testing for Breast Cancer and Patient Activism in the United States and Britain   133
Shobita Parthasarathy

7
Who Represents the Users? Critical Encounters between Women's Health Advocates and Scientists in Contraceptive R&D   151
Jessika van Kammen

8
Inclusion, Diversity, and Biomedical Knowledge Making: The Multiple Politics of Representation   173
Steven Epstein

III
Multiplicity in Locations: Configuring the User during the Design, the Testing, and the Selling of Technologies

9
Materialized Gender: How Shavers Configure the Users' Femininity and Masculinity   193
Ellen van Oost

10
Clinical Trials as a Cultural Niche in Which to Configure the Gender Identities of Users: The Case of Male Contraceptive Development   209
Nelly Oudshoorn

11
The Mediated Design of Products, Consumption, and Consumers in the Twentieth Century 229

Johan Schot and Adri Albert de la Bruheze

12
Giving Birth to New Users: How the Minimoog Was Sold to Rock and Roll 247

Trevor Pinch

Preface

Users are everywhere gaining prominence. In fields such as information technology they are currently the center of much research. Collectives that speak for users, such as patient and consumer organizations, have successfully claimed to have a say in the development of medical therapies and digital products. This volume presents reflections on users and on a variety of technologies from the perspective of social studies of science and technology. It also presents case studies selected from recent scholarship. We do not claim to offer a complete treatment—the topic is too vast for that. But we do think it is representative of some of the most interesting work on users currently being conducted in Science and Technology Studies in Europe and North America.

The majority of the papers collected in this volume were first presented at sessions on The Co-construction of Users and Technologies at the Annual Meeting of the Society for Social Studies of Science held in San Diego in October 1999. The collection as a whole has benefited enormously from the critical scrutiny the papers received then and subsequently from the referees who graciously agreed to examine each paper in more detail. We thank the following referees: Olga Amsterdamska, Stuart Blume, Adele Clarke, Ken Green, Ron Kline, Merete Lie, Hugh Mackay, David Nye, Ann Rudinow Saetnan, Knut Sørenson, Lucy Suchman, Jane Summerton, Steve Woolgar, and Sally Wyatt. The volume as a whole was read by Wiebe Bijker. The authors took our comments and those of the referees with good grace and in a timely manner. We thank Hilde Meijer-Meijer for helping us with the preparation of the manuscript. Larry Cohen of The MIT Press helped give birth to this volume with his usual patience and sensible advice.

Nelly Oudshoorn
Trevor Pinch

How Users Matter

Introduction

How Users and Non-Users Matter
Nelly Oudshoorn and Trevor Pinch

New uses are always being found for familiar technologies. Sometimes these changes in use are dramatic and unexpected. Before September 11, 2001, no one foresaw that an airliner could be turned by a small number of its occupants into a giant Molotov cocktail. After the Gulf War of 1991, it was discovered that an effective way to put out oil-rig fires was to strap down captured Mig jet fighters and blow out the fires using their exhaust. Such examples remind us that we can never take the use of a technology for granted.

Susan Douglas (1987) has pointed out how amateur operators discovered new uses to which the emerging technology of radio could be put, and how commercial operators soon followed the amateurs' lead. Claud Fischer (1992) and Michele Martin (1991) have drawn attention to the use of the telephone by rural women to overcome their isolation—a use not foreseen by telephone companies, which conceived of the telephone mainly as a business instrument.

Our concern in this book is with the role of users in the development of technology in general. We are interested in how users consume, modify, domesticate, design, reconfigure, and resist technologies. In short, our interest is in whatever users do with technology.

There is no one correct use for a technology. "What is an alarm clock for?" we might ask. "To wake us up in the morning," we might answer. But just begin to list all the uses to which an alarm clock can be put and you see the problem. An alarm clock can be worn as a political statement by a rapper; it can be used to make a sound on a Pink Floyd recording; it can be used to evoke laughter, as Mr. Bean does in one of his comic sketches as he tries to drown his alarm clock in his bedside water pitcher; it can be used to trigger a bomb; and, yes, it can be used to wake us up. No doubt there are many more uses. Of course, there may be one dominant use of a technology, or a prescribed use, or a use that confirms the manufacturer's warranty, but there is no one essential use that can be deduced from the artifact itself. This is an axiomatic assumption for the scholars whose work we collect here. All the contributors follow the research path of studying technologies in their "context of use"—the society and the web of other artifacts within which technologies are always embedded. In short, we look at how technologies are actually used in practice.

In addition to studying what users do with technology, we are interested in what technologies do to users. Users of technologies do not arrive de novo. Think of the camera. When George Eastman developed his revolutionary new technology of roll film and a cheap camera, he had one outstanding problem: There were as yet no users for it. Photography was seen as a high-end activity practiced by a small group of skilled professionals. Eastman had to define explicitly who the new users might be, and he had to figure out how to recruit them to his new technology. He had to redefine photography and the camera. After he did, photography became something that anyone could participate in, and cameras became usable by all (Jenkins 1975). Working out who the new users are and how they will actually interact with a new technology is a problem familiar to many innovators of new technologies. Some fields, including information technology, are particularly cognizant of the problem of users. It has long been recognized that the most sophisticated and complex computer hardware and software will come to naught if users don't known how to use them. Studies of human-computer interaction, of work practices, and of user interfaces are often carried out by the computer industry, and they have become important not only for that industry but also for developing new ideas of how the user-technology nexus should be conceptualized (Suchman 1994; Woolgar 1991).

One important research question addressed in this book is how users are defined and by whom. For instance, are users to be conceived of as isolated autonomous consumers, or as self-conscious groups? How do designers think of users? Who speaks for them, and how? Are users an important new political group, or a new form of social movement? In short, what general lessons are to be drawn from a renewed focus on users in today's technologically mediated societies?

Different Approaches to Users

Users and technology are too often viewed as separate objects of research. This book looks for connections between the two spheres.

Users and technology are seen as two sides of the same problem—as co-constructed. The aim is to present studies of the co-construction of users and technologies that go beyond technological determinist views of technology and essentialist views of users' identities.

In this introduction we discuss several influential approaches to user-technology relations,[1] focusing in particular on the conceptual vocabulary developed within the different approaches and on the similarities and differences between them.

The SCOT Approach: Users as Agents of Technological Change

In the 1980s and the 1990s, the old view of users as passive consumers of technology was largely replaced in some areas of technology studies, and along with it the linear model of technological innovation and diffusion. One of the first approaches to draw attention to users was the social construction of technology (SCOT) approach.

Pinch and Bijker (1984), in defining the SCOT approach, conceived of users as a social group that played a part in the construction of a technology. Different social groups, they noted, could construct radically different meanings of a technology. This came to be known as a technology's interpretive flexibility. In a well-known study of the development of the bicycle, it was argued that elderly men and women gave a new meaning to the high-wheeled bicycle as the "unsafe" bicycle, and that this helped pave the way for the development of the safety bicycle. The SCOT approach specifies a number of closure mechanisms—social processes whereby interpretative flexibility is curtailed. Eventually, a technology stabilizes, interpretative flexibility vanishes, and a predominant meaning and a predominant use emerge (Bijker and Pinch 1987; Bijker 1995). The connection between designers and users was made more explicit with the notion of a technological frame (Bijker 1995). Users and designers could be said to share a technological frame associated with a particular technology.

Many of the classic SCOT studies were of the early stages of technologies. For example, there were studies of how the bicycle, fluorescent lighting, and Bakelite moved from interpretative flexibility to stability. Early on, social groups were seen as the shaping agents. Not until later, with notions such as that of sociotechnical ensembles, did SCOT fully embrace the idea of the co-construction or mutual shaping of social groups and technologies (Bijker 1995b). The SCOT approach was rightly criticized for its rather cavalier attitude toward users—it closed down the problem of users too early, and it did not show how users could actively modify stable technologies (Mackay and Gillespie 1992). Kline and Pinch (1996) remedied this with their study of how a stable technology, the Model T automobile, could be appropriated and redesigned by groups such as farmers who used cars as stationary power sources. Kline and Pinch referred to such users as "agents of technological change." Also attempting to correct SCOT's neglect of gender, Kline and Pinch argued that users should be studied as a crucial location where often-contradictory gender identities and power relationships were woven around technologies. Bijker (1995) argued for a semiotic conception of power whereby power is embedded and mediated by artifacts as well as by frames and social groups. However, this semiotic notion of power (like most semiotic approaches within technology studies) seems inevitably to leave out invisible actors and social groups, which in the SCOT approach might be termed "non-relevant social groups."

Feminist Approaches: Diversity and Power
Feminist scholars have played a leading role in drawing attention to users. Their interest in users reflects concerns about the potential problematic consequences of technologies for women and about the absence of women in historical accounts of technology. Since the mid 1980s, feminist historians have pointed to the neglect of women's role in the development of technology. Because women were historically underrepresented as innovators of technology, and because historians of technology often focused exclusively on the design and production of technologies, the history of technology came to be dominated by stories about men and their machines. Moreover, these stories represented a discourse in which gender was invisible. Historians did not consider it relevant in settings where women were absent, thus reinforcing the view that men had no gender.[2] Feminist historians suggested that focusing on users and use rather than on engineers and design would enable historians to go beyond histories of men inventing and mastering technology (Wajcman 1991; Lerman et al. 1997). In response to this criticism, users were gradually included in the research agenda of historians of technology.[3] This "turn to the users" can be traced back to Ruth Schwartz Cowan's exemplary research on user-technology relations. In the late 1970s, Cowan brought the fields of history of technology and women's history together, emphasizing that women as users of technology perceive technological change in significantly different ways from men (Pursell 2001). Cowan's notion of "the consumption junction," defined as "the place and time at which the consumer makes choices between competing technologies" (Cowan 1987:

263), was a landmark. Cowan argued that focusing on the consumer and on the network relations in which the consumer is embedded enables historians and sociologists of technology to improve their understanding of the unintended consequences of technologies in the hands of users. Focusing on users would enrich the history of technology with a better understanding of the successes and failures of technologies (ibid.: 279). In contrast to actor-network theory (which we will discuss below), Cowan urged historians and sociologists of technology to choose the user, rather than the artifact or the technologist, as a point of departure in network analyses of technology, and to look at networks from the consumer's point of view (ibid.: 262). The scholarship that Cowan inspired rejects the idea that science and technology begin or end with the actions of scientists and engineers. Scholars in the field of Science and Technology Studies (STS) were urged to follow technologies all the way to the users (Rapp 1998: 48). An exemplary study is Cynthia Cockburn and Susan Ormrod's 1993 book on the microwave oven in the United Kingdom, which analyzes the design, the production, and the marketing as well as the use of a new technology.

Gender studies, like technology studies in general, reflect a shift in the conceptualization of users from passive recipients to active participants. In the early feminist literature, women's relation to technology had been conceptualized predominantly in terms of victims of technology. The scholarship of the last two decades, however, has emphasized women's active role in the appropriation of technology. This shift in emphasis was explicitly articulated in the first feminist collection of historical research on technology, *Dynamos and Virgins Revisited* (Trescott 1979), which included a section on "women as active participants in technological change" (Lerman et al. 1997: 11). The authors of the essays in that section argued that feminists should go beyond representations of women as essentially passive with respect to technology. Having accepted that challenge, feminist historians, anthropologists, and sociologists have published numerous accounts of how women shape and negotiate meanings and practices in technology, including studies of the relationship between reproductive technologies and women's health and autonomy,[4] of the gendered medicalization of bodies,[5] of women's relations to computers and the impact of computer technologies on women's work,[6] of the consequences of household technologies for women's lives,[7] and of the exclusion of women from technologies.[8] Granting agency to users, particularly women, can thus be considered central to the feminist approach to user-technology relations.

Another important concept in feminist studies of technology is diversity. As Cowan (1987) suggested, users come in many different shapes and sizes. Medical technologies, for example, have a wide variety of users, including patients, health professionals, hospital administrators, nurses, and patients' families. "Who is the user?" is far from a trivial question. The very act of identifying specific individuals or groups as users may facilitate or constrain the actual roles of specific groups of users in shaping the development and use of technologies. Different groups involved in the design of technologies may have different views of who the user might or should be, and these different groups may mobilize different resources to inscribe their views in the design of technical objects (Oudshoorn et al., forthcoming). And these different type of users don't necessarily imply homogeneous categories. Gender, age, socio-economic, and ethnic differences among users may all be relevant. Because of this heterogeneity, not all users will have the same position in relation to a specific technology. For some users, the room for maneuvering will be great; for others, it will be very slight. Feminist sociologists thus emphasize the diversity of users and encourage scholars to pay attention to differences in power relations among the actors involved in the development of technology.

To capture the diversity of users[9] and the power relations between users and other actors in technological development, feminist sociologists have differentiated "end users," "lay end users," and "implicated actors." End users are "those individuals and groups who are affected downstream by products of technological innovation" (Casper and Clarke 1998). The term "lay end users" was introduced to highlight some end users' relative exclusion from expert discourse (Saetnan et al. 2000: 16). Implicated actors are "those silent or not present but affected by the action" (Clarke 1998: 267). And there are two categories of implicated actors: "those not physically present but who are discursively constructed and targeted by others" and "those who are physically present but who are generally silenced/ignored/made invisible by those in power" (Clarke, forthcoming).[10] All three terms reflect the long-standing feminist concern with the potential problematic consequences of technologies for women and include an explicit political agenda: the aim of feminist studies is to increase women's autonomy and their influence on technological development. A detailed understanding of how women as "end users" or "implicated actors" matter in technological development may provide information that will be useful in the empowerment of women or of spokespersons for them, such as social movements and consumer groups.

The concept of the implicated actor also reflects a critical departure from actor-network approaches in technology studies. Feminists have criticized the sociology of technology, particularly actor-network theory, for the almost exclusive attention it gives to experts and producers and for the preference it gives to design and innovation in understanding socio-technical change.[11] This "executive approach" pays less attention to non-standard positions, including women's voices (Star 1991; Clarke and Montini 1993: 45; Clarke 1998: 267).[12] Moreover, the "executive approach" implicitly assumes a specific type of power relations between users and designers in which designers are represented as powerful and users as disempowered relative to the experts. Feminist sociologists suggest that the distribution of power among the multiple actors involved in socio-technical networks should be approached as an empirical question (Lie and Sørensen 1996: 4, 5; Clarke 1998: 267; Oudshoorn et al., forthcoming). Thus, the notion of the implicated actor was introduced to avoid silencing invisible actors and actants and to include power relations explicitly in the analysis of user-expert relations.[13]

Another important word in the feminist vocabulary is "cyborg." Donna Haraway was the first to use this word to describe how by the late twentieth century humans had become so thoroughly and radically merged and fused with technologies that the boundaries between the human and the technological are no longer impermeable. The cyborg implies a very specific configuration of user-technology relations in which the user emerges as a hybrid of machine and organisms in fiction and as lived experience. Most important, Haraway introduced the cyborg figure as a politicized entity. Cyborg analyses aim to go beyond the deconstruction of technological discourses. On page 149 of her "Cyborg Manifesto" (1985), Haraway invites us to "question that which is taken as 'natural' and 'normal' in hierarchic social relations." Haraway writes of cyborgs not to celebrate the fusion of humans and technology, but to subvert and displace meanings in order to create alternative views, languages, and practices of technosciences and hybrid subjects.[14] In the 1990s, the concept of the cyborg resulted in an extensive body of literature that described the constitution and transformation of physical bodies and identities through technological practices.[15]

Semiotic Approaches to Users: Configuration and Script
An important new approach to user-technology relations was introduced by STS scholars who extended semiotics, the study of how meanings are built, from signs to things. The concept of "configuring the

user" is central to this approach. Exploring the metaphor of machine as text, Steve Woolgar (1991: 60) introduced the notion of the user as reader to emphasize the interpretive flexibility of technological objects and the processes that delimit this flexibility. Although the interpretative flexibility of technologies and questions concerning the closure or stabilization of technology had already been addressed in the SCOT approach, Woolgar focused on the design processes that delimit the flexibility of machines rather than on the negotiations between relevant social groups. He suggested that how users "read" machines is constrained because the design and the production of machines entails a process of configuring the user. For Woolgar, "configuring" is the process of "defining the identity of putative users, and setting constraints upon their likely future actions" (ibid.: 59). He describes the testing of a new range of microcomputers as "a struggle to configure (that is to define, enable, and constrain) the user," a struggle that results in "a machine that encourages only specific forms of access and use" (ibid.: 69, 89). In this approach, the testing phase of a technology is portrayed as an important location in which to study the co-construction of technologies and users. In contrast to the approaches discussed thus far, this semiotic approach draws attention to users as represented by designers rather than to users as individuals or groups involved or implicated in technological innovation.

In recent debates, the notion of the configuration of users by designers has been extended to better capture the complexities of designer-user relations. Several authors criticized Woolgar for describing configuration as a one-way process in which the power to shape technological development is attributed only to experts in design organizations. For example, Mackay et al. (2000: 752) suggested that "designers configure users, but designers in turn, are configured by both users and their own organizations," and that this is increasingly the case in situations where designer-user relations are formalized by contractual arrangements (ibid.: 744). The capacity of designers to configure users can be further constrained by powerful groups within organizations that direct design projects. In large organizations, designers usually have to follow specific organizational methods or procedures that constrain design practices (ibid.: 741, 742, 744; Oudshoorn et al. 2003). In many companies in the information and communication technologies sector, for example, designers are allowed to test prototypes of new products only among people who work in the organization. In this highly competitive sector, companies are reluctant to test new products among wider groups of users for fear that other firms will become aware of their plans at an early phase of product development (European Commission 1998: 22; Oudshoorn et al. 2003).

Another criticism and extension of the configuration approach was introduced by scholars who questioned who was doing the configuration work. In Woolgar's studies, configuration work was restricted to the activities of actors within the company who produced the computers. Several authors broadened this rather narrow view of configuration to include other actors and to draw attention to the configuration work carried out by journalists, public-sector agencies, policy makers, and social movements acting as spokespersons for users (van Kammen 2000a; van Kammen, this volume; Epstein, this volume; Parthasarathy, this volume; Oudshoorn 1999; Rommes 2002). Other scholars attempted to broaden the scope of the analysis by including the agency of users. Whereas Woolgar explored the metaphor of machine and text to highlight "encoding," thus focusing attention on the work performed by the producers of texts and machines, a more symmetrical use of the metaphor requires that we also focus on the processes of "decoding," the work done by readers and users to interpret texts and machines (Mackay et al. 2000: 739, 750, 752). A similar criticism of the asymmetry of Woolgar's work was voiced by scholars who had adopted domestication approaches to technology.

A second central notion in the semiotic approaches to user-technology relations is the concept of script. Madeleine Akrich and Bruno Latour, in theorizing relationships between users and technology, use this term to describe the obduracy of objects. The concept of script tries to capture how technological objects enable or constrain human relations as well as relationships between people and things. Comparing technologies to film, Akrich (1992: 208) suggested that "like a film script, technical objects define a framework of action together with the actors and the space in which they are supposed to act." To explain how scripts of technological objects emerge, she drew attention to the design of technologies. Akrich suggested that in the design phase technologists anticipate the interests, skills, motives, and behavior of future users. Subsequently, these representations of users become materialized into the design of the new product. As a result, technologies contain a script (or scenario): they attribute and delegate specific competencies, actions, and responsibilities to users and technological artifacts. Technological objects may thus create new "geographies of responsibilities" or transform or reinforce existing ones (ibid.: 207, 208). Rooted in actor-network theory, Akrich and Latour's work challenges social constructivist approaches in which only people are given the status of actors.[16] The script approach aims to describe how technical objects "participate in building heterogeneous networks that bring together actants of all types and sizes, whether humans or nonhumans" (ibid.: 206).

In the 1990s, feminist scholars extended the script approach to include the gender aspects of technological innovation. Adopting the view that technological innovation requires a renegotiation of gender relations and an articulation and performance of gender identities, Dutch and Norwegian feminists introduced the concept of genderscript to capture all the work involved in the inscription and de-inscription of representations of masculinities and femininities in technological artifacts (Berg and Lie 1993; Hubak 1996; van Oost 1995; van Oost, this volume; Oudshoorn 1996; Oudshoorn et al. 2003; Oudshoorn et al., forthcoming; Rommes et al. 1999; Spilkner and Sørensen 2000). This scholarship emphasizes the importance of studying the inscription of gender into artifacts to improve our understanding of how technologies invite or inhibit specific performances of gender identities and relations. Technologies are represented as objects of identity projects—objects that may stabilize or de-stabilize hegemonic representations of gender (Oudshoorn, this volume; Saetnan et al. 2000). Equally important, the genderscript approach drastically redefines the exclusion of specific groups of people from technological domains and activities. Whereas policy makers and researchers have defined the problem largely in terms of deficiencies of users, genderscript studies draw attention to the design of technologies (Oudshoorn 1996; Oudshoorn et al., forthcoming; Rommes et al. 1999; Rommes 2002). These studies make visible how specific practices of configuring the user may lead to the exclusion of specific users.[17]

At first glance, the script approach seems to be very similar to Woolgar's approach of configuring the user, since both approaches are concerned with understanding how designers inscribe their views of users and use in technological objects. A closer look, however, reveals important differences. Although both approaches deal with technological objects and designers, the script approach makes users more visible as active participants in technological development. Akrich in particular is very much aware that a focus on how technological objects constrain the ways in which people relate to things and to one another can be easily misunderstood as a technological determinist view that represents designers as active and users as passive. To avoid this misreading, she emphasizes the reciprocal relationship between objects and subjects and explicitly addresses the question of the agency of users (Akrich 1992:

207). Although technological objects can define the relationships between human and nonhuman actors, Akrich suggests that "this geography is open to question and may be resisted" (ibid.). To avoid technological determinism, Akrich urges us to analyze the negotiations between designers and users and concludes that "we cannot be satisfied methodologically with the designer's or user's point of view alone. Instead we have to go back and forth continually between the designer and the user, between the designer's projected users and the real users, between the world inscribed in the object and the world described by its displacement" (ibid.: 209).

To further capture the active role of users in shaping their relationships to technical objects, Akrich and Latour have introduced the concepts of subscription, de-inscription, and antiprogram. "Antiprogram" refers to the users' program of action that is in conflict with the designers' program (or vice versa). "Subscription" or "de-inscription" is used to describe the reactions of human (and nonhuman) actors to "what is prescribed and proscribed to them" and refers respectively to the extent to which they underwrite or reject and renegotiate the prescriptions (Akrich and Latour 1992: 261). In contrast to Woolgar's work on configuring the user, script analyses thus conceptualize both designers and users as active agents in the development of technology. However, compared to domestication theory, the script approach gives more weight to the world of designers and technological objects. The world of users, particularly the cultural and social processes that facilitate or constrain the emergence of users' antiprograms, remains largely unexplored by actor-network approaches. More recently, this imbalance has been repaired to an extent by the work of scholars who have extended actor-network theory to include the study of "subject networks." These studies aim to understand the "attachment" between people and things, particularly but not exclusively between disabled people and assistive technologies, and to explore how technologies work to articulate subjectivities (Callon and Rabehariso 1999; Moser 2000; Moser and Law 1998, 2001).[18] This scholarship conceptualizes subjects in the same way as actor-network theorists previously approached objects. Subject positions such as disability and ability are constituted as effects of actor networks and hybrid collectives.[19]

Cultural and Media Studies: Consumption and Domestication
In contrast to the approaches to user-technology relations we have discussed thus far, scholars in the fields of cultural and media studies acknowledged the importance of studying users from the very beginning.

Whereas historians and sociologists of technology have chosen technology as their major topic of analysis, those who do cultural and media studies have focused primarily on users and consumers. Their central thesis is that technologies must be culturally appropriated to become functional.[20] This scholarship draws inspiration from Bourdieu's (1984) suggestion that consumption has become more important in the political economy of late modernity. Consequently, human relations and identities are increasingly defined in relation to consumption rather than production. In his study of differences in consumption patterns among social classes, Bourdieu defined consumption as a cultural and material activity and argued that the cultural appropriation of consumer goods depends on the "cultural capital" of people (ibid.).[21] This view can be traced back to the tradition of the anthropological study of material culture, most notably the work of Mary Douglas and Baron Isherwood (1979). Among the first to criticize the view (then dominant among consumption theorists) that consumption is merely an economic activity, they suggested that consumption is always a cultural as well as an economic phenomenon (Lury 1996: 10). Describing the use of consumer goods in ritual processes, they defined consumer culture as a specific form of material culture, and they conceptualized the circulation of material things as a system of symbolic exchange. This scholarship articulates the importance of the sign value rather than the utility value of things. From this perspective, material things can act as sources and markers of social relations and can shape and create social identities (Lury 1996: 10, 12, 14; Douglas and Isherwood 1979; McKracken 1988; Appadurai 1986).

Feminist historians have also been important actors in signaling the relevance of studying consumption rather than production (McGaw 1982). Feminists have long been aware of the conventional association and structural relations of women with consumption as a consequence of their role in the household and as objects in the commodity-exchange system (de Grazia 1996: 7).[22] Whereas early feminist studies focused on the (negative) consequences of mass consumption for women, more recent studies address the question of whether women have been empowered by access to consumer goods. They conceptualize consumption as a site for the performance of gender and other identities.[23] The notion of consumption as a status and identity project was elaborated further by Jean Baudrillard (1988), who criticizes the view that the needs of consumers are dictated, manipulated, and fully controlled by the modern capitalist marketplace and by producers. Theodor

Adorno, Herbert Marcuse, and Max Horkheimer of the Frankfurt School had argued that the expansion of the production of consumer goods throughout the twentieth century had resulted in an increase in ideological control and manipulation by the "culture industries" (Adorno 1991; Horkheimer and Adorno 1979; Marcuse 1964).[24] Since the 1970s, this view of consumption as manipulation had resulted in a literature dominated by studies oriented toward production and marketing—studies that highlighted big companies and advertising agencies as the forces driving consumption. In these studies, consumption was characterized as a passive and adaptive process and consumers are represented as the anonymous buyers and victims of mass production.[25] In contrast, Baudrillard emphasized the mutual dependencies between production and consumption and suggested that consumers are not passive victims but active agents in shaping consumption, social relations, and identities.

Cultural and media studies also emphasize the creative freedom of users to "make culture" in the practice of consumption as well as their dependence on the cultural industries, not because they control consumers but because they provide the means and the conditions of cultural creativity (Storey 1999: xi). This scholarship portrays consumers as "cultural experts" who appropriate consumer goods to perform identities, which may transgress established social divisions (du Gay et al. 1997: 104; Chambers 1985).

Semiotic approaches to analyzing user-technology relations also came to the fore in cultural and media studies. Stuart Hall, one of the leading scholars in this field, introduced the "encoding/decoding" model of media consumption (Hall 1973), which aims to capture both the structuring role of the media in "setting agendas and providing cultural categories and frameworks" and the notion of the "active viewer, who makes meaning from signs and symbols that the media provide" (Morley 1995: 300). Since the 1980s, the symbolic and communicative character of consumption has been studied extensively by scholars in the fields of cultural and media studies. Consumption fulfills a wide range of social and personal aims and serves to articulate who we are or who we would like to be; it may provide symbolic means of creating and establishing friendship and celebrating success; it may serve to produce certain lifestyles; it may provide the material for daydreams; it may be used to articulate social difference and social distinctions (Bocock 1993; du Gay et al. 1997; Lie and Sørensen 1996; Mackay 1997; Miller 1995; Storey 1999). Cultural and media studies thus articulate a perspective on user-technology relations that emphasizes the role of technological objects in creating and shaping social identities, social life, and culture at large.[26]

Roger Silverstone coined the term "domestication" to describe how the integration of technological objects into daily life involves "a taming of the wild and a cultivation of the tame." New technologies have to be transformed from "unfamiliar, exciting, and possible threatening things" into familiar objects embedded in the culture of society and the practices and routines of everyday life (Silverstone and Hirsch 1992; Lie and Sørensen 1996). Domestication processes include symbolic work, in which people create symbolic meanings of artifacts and adopt or trans-form the meanings inscribed in the technology; practical work, in which users develop a pattern of use to integrate artifacts into their daily rou-tines; and cognitive work, which includes learning about artifacts (Lie and Sørensen 1996: 10; Sørensen et al. 2000). In this approach, domesti-cation is defined as a dual process in which both technical objects and people may change. The use of technological objects may change the form and the practical and symbolic functions of artifacts, and it may enable or constrain performances of identities and negotiations of status and social positions (Silverstone et al. 1989; Lie and Sørensen 1996).[27] The notion of domestication also reflects a preference for studying the use of technology in a specific location: the home. British scholars in this tradition have largely restricted their analyses to the household and the politics of family life (Silverstone 1989, 1992). In their work, processes of domestication are understood in terms of the "dynamics of the house-hold's moral economy" (Silverstone, Hirsch, and Morley 1992). More recently, Norwegian scholars have extended the scope of research to other domains. Merete Lie and Knut Sørensen (1996: 13, 17) argue that the domestication of technical objects has been too easily associated with the "private sector" (meaning the home).[28] Various chapters in the vol-ume edited by Lie and Sørensen show how similar processes are taking place in work, in leisure, and within subcultures.

Domestication approaches have enriched our understanding of user-technology relations by elaborating the processes involved in consump-tion. In *Consuming Technologies*, Roger Silverstone and his colleagues specify four phases of domestication: appropriation, objectification, incorporation, and conversion. Appropriation occurs when a technical product or service is sold and individuals or households become its own-ers (Silverstone et al. 1992: 21). In objectification, processes of display reveal the norms and principles of the "household's sense of itself and its place in the world" (ibid.: 22). Incorporation occurs when technological objects are used in and incorporated into the routines of daily life. "Conversion" is used to describe the processes in which the use of technological objects shape relationships between users and people outside the household (ibid.: 25). In this process, artifacts become tools for making status claims and for expressing a specific lifestyle to neighbors, colleagues, family, and friends (Silverstone and Haddon 1996: 46).

Although at first sight "domestication" and "decoding" or "de-inscription" may seem synonymous, there is an important difference. By specifying the processes involved in the diffusion and the use of technology, domestication approaches take the dynamics of the world of users as their point of departure. The concepts of decoding and de-inscription, on the other hand, give priority to the design context in order to understand the emergence of user-technology relations. Domestication approaches thus emphasize the complex cultural dynamics in which users appropriate technologies (ibid.: 52). This contrasts with semiotic approaches that tend to define the user as an isolated individual whose relationship to technology is restricted to technical interactions with artifacts.[29] As Silverstone and Haddon suggest, a focus on how designers configure the user runs the risk of reifying the innovator's conceptions of users. In contrast, domestication approaches conceptualize the user as a part of a much broader set of relations than user-machine interactions, including social, cultural, and economic aspects. By employing cultural approaches to understand user-technology relations, this scholarship aims to go beyond a rhetoric of designers' being in control. Semiotic approaches tend to reinforce the view that technological innovation and diffusion are successful only if designers are able to control the future actions of users. Although semiotic approaches have introduced notions that are useful in understanding the worlds of designers and users, "script" and "configuring the user" conceptualize the successes and failures of technologies mainly in terms of the extent to which designers adequately anticipate users' skills and behavior. In this view, users tend to be degraded to objects of innovators' strategies. The semiotic approaches have therefore been criticized for staying too close to the old linear model of technological innovation[30] and diffusion, which prioritizes the agency of designers and producers over the agency of users[31] and other actors involved in technological innovation (Oudshoorn 1999). Even the concept of antiprogram, introduced by Akrich and Latour to describe how users may try to counter the original intentions of the design of the artifact, remains within the rhetoric of designer's control (Sørensen 1994: 5). The only option available to the user seems to be to adopt or to reject the designers' intended use and meaning of technological objects. These approaches are inadequate to understand the full dynamics of technological innovation where users invent completely new uses and meanings of technologies or where users are actively involved in the design of technologies.

Most important, cultural and media studies inspire us to transcend the artificial divide between design and use. This scholarship has drastically reconceptualized the traditional distinction between production and consumption by re-introducing Karl Marx's claim that the process of production is not complete until users have defined the uses, meanings, and significance of the technology: "Consumption is production."[32] They describe design and domestication as "the two sides of the innovation coin" (Lie and Sørensen 1996: 10).

An Overview of the Book

One of the aims of this volume is to bridge the approaches to users that have been developed in technology studies, in feminist scholarship, and in cultural and media studies. The scholarship presented in this book acknowledges the creative capacity of users to shape technological development in all phases of technological innovation. The authors are interested in and sensitive to the multiplicity and diversity of users, spokespersons for users, and other actors involved in socio-technical change. This approach makes visible how the co-construction of users and technologies may involve tensions, conflicts, and disparities in power and resources among the different actors involved. By doing this, we aim to avoid the pitfall of what David Morley (1992) has called the "don't worry, be happy" approach. A neglect of differences among and between producers and users may result in a romantic voluntarism that celebrates the creative agency of users, leaving no room for any form of critical understanding of the social and cultural constraints on user-technology relations.

Part I focuses on the active role of users and non-users in shaping socio-technical change during the domestication of technologies. Christina Lindsay tells the story of the TRS-80 personal computer, a technology that is kept alive and fully functional by users almost 25 years after its introduction and long after the original designers, producers, and marketers moved on. She describes the changing roles of users during the TRS-80's life history. The users in this story begin as somewhat stereotypically gendered representations constructed by the designers of the computer and end up as designers, producers, retailers, and technical support for the technology. They take responsibility for the further development of the TRS-80, and in the process they rework their own identities as computer users in relation to this technology. Lindsay shows how the co-construction of users, user representations, and technology was not a static, one-time exercise by the designers of the TRS-80, but was a dynamic ongoing process through the whole life history of the technology in which many different groups, including the users, participated. The important insight to be gained from this chapter is that users can have multiple identities. In addition to being users, they can perform activities and identities traditionally ascribed to designers.

The other three chapters in part I highlight two aspects of the agency of potential users of technology that have largely remained unexplored in domestication approaches: resistance and non-use. Ronald Kline challenges common perceptions and theoretical understandings of resistance, which view resistance to technology as irrational or heroic. Instead, he suggests resistance can be considered as a common feature of the processes underlying socio-technical change. Acts considered as resistance by promoters, mediators, and users are crucial aspects of the creation of new technologies and social relations. Adopting SCOT as an analytical framework, Kline describes how farm people domesticated the telephone and electrification into their daily life in the early twentieth century. Most important, this detailed and fascinating account of the domestication of these technologies is not restricted to the users. Kline shows the usefulness of a methodology that does not focus only on use but which also includes the interplay of the actions and reactions of both producers and users. Producers responded to users' resistance and created new techniques, hardware, and mediating organizations to adapt the new technologies to fit the social patterns of rural life. This chapter is not only a story about the contested aspects of a modernization process. Inspired by feminist approaches to user-technology relations, Kline elaborates the theme of diversity of users by showing how the production, use, and interpretation of new technologies can only be understood in the context of a gendered system of social relations.

Sally Wyatt provides an interesting new understanding of the non-use of technologies. Beginning with an analysis of the use of Internet, she questions the assumptions—dominant in many policy documents and in much of the academic literature—that non-use of a technology always and necessarily involves inequality and deprivation. Producers and policy makers usually promote the Internet as a universal medium whose users will have a better socio-economic position than its non-users. In this modernist discourse, non-use is portrayed as a deficiency and an involuntary act. Challenging this view, Wyatt introduces a reconceptualization of the category of non-use that includes the voluntary and the involuntary aspects. This preliminary taxonomy identifies four different types of non-users: "resisters" (people who have never used the technology because they do not want to), "rejectors" (people who no longer use the technology, because they find it boring or expensive or because they have alternatives), "the excluded" (people who have never used the technology, because they cannot get access for a variety of reasons), and "the expelled" (people who have stopped using the technology involuntarily because of cost or loss of institutional access). Wyatt's study warns us to avoid the pitfalls of implicitly accepting the rhetoric of technological progress, including a worldview in which adoption of new technologies is the norm. She urges us to take non-users and former users seriously as relevant social groups in shaping socio-technical change.

Anne Sofie Laegran explores patterns of use and non-use in a comparative study of the appropriation of the Internet and automobiles among young people in rural Norway. Adopting a revised concept of domestication that extends the analysis to settings other than the household, she focuses on differences between two youth cultures to understand how both technologies are reinterpreted and gain different symbolic and utility values. In contrast with the preceding two chapters, which follow a SCOT approach and emphasize the importance of use and non-use in understanding the design of technologies, this chapter relates use and non-use to the construction of symbolic meanings of the technology and the articulation of identities among users. Laegran describes how the youths who construct their identity in relation to the urban culture interpret the Internet as a medium enabling communication in a global context. In contrast, the rural youths totally reject this symbolic meaning of the Internet. They use the automobile, interpreted as a local means of transportation and as an icon, for building their identities. This construction of symbolic meanings and identities has important consequences for the adoption or rejection of both technologies in the two youth cultures. Laegran's study nicely illustrates how a focus on the construction of identities enables us to understand how people eventually become users or non-users of a technology.

In summary, all the chapters in part I show how users and non-users matter in the stabilization and de-stabilization of technologies. Whereas the first chapter demonstrates the multiple identities and roles of users, the other chapters show how an adequate understanding of socio-technical change should include an analysis of resistance and non-use. Most important, the chapters in this part introduce a new conceptualization of these phenomena. Instead of representing resistance and non-use as irrational, heroic, or involuntary, they argue that these reactions to technology should be considered as rational choices shaping the design and (de)stabilization of technologies. Moreover, Kline and Laegran in particular provide important insights into how people eventually become non-users or resisters of technologies. They suggest that resistance and non-use are most likely to occur in situations in which the prescribed uses and the symbolic meanings attached to the technology by its producers and its promoters do not correspond to the gender relations, the cultural values, and the identities of specific groups of people.

Part II further elaborates the themes of agency, multiplicity, and diversity by focusing on the multiple collectives who speak for users and the ways in which they represent the diversity of users. The chapters in this part develop a perspective that goes beyond a conceptualization of user-technology relations in which the configuration work is solely in the hands of experts and users are categorized as a singular group. Dale Rose and Stuart Blume explore the theme of multiple spokespersons by focusing on the state. Their chapter encourages us to rethink the ways in which we conceptualize users. Instead of looking at users merely as consumers, Rose and Blume extend the analysis to include users as citizens of states. Most important, they address the theoretical problem of how to conceptually link the notion of individuals as users of technologies, that of individuals as consumers of commodities, and that of individuals as citizens of states. Based on an analysis of the development and provision of vaccines against human infectious diseases, they reject the consumer/citizen distinction. They describe how, in most Western industrialized nations, the state configures two types of vaccine users: the consumer of a commodity and a more passive public citizen whose actions as a user of these technologies defines that person as fulfilling a civic responsibility to be a good citizen. This configuration of users has important consequences for citizens. Citizens who reject technologies developed by the state for a common public good (in this case, vaccines for the prevention of diseases), or who fail to use them in the prescribed way, not only become inappropriate users of technologies but also fail in their civic responsibilities and eventually are deemed "bad" citizens. The structure in which vaccines are developed thus erases the distinction between users as consumers and users as citizens. States configure consumers ultimately as passive citizens.

Shobita Parthasarathy also focuses on the state. Instead of analyzing the state as a spokesperson for users, she addresses the question of how national political cultures shape the multiple and differing representation strategies of advocacy groups who speak for users. Her comparative study of the development of genetic testing for breast cancer in the United States and Britain shows how cultural norms and values influence the identities of and negotiations among a variety of different actors who want to influence the testing. Parthasarathy, like Rose and Blume, argues that the current conceptualization of individuals as users is inadequate if we want to take into account the structural constraints of state regulations. She uses the term "civic individual" instead of "user" in order to capture the broader political and cultural dimensions suggested by the normative language of rights and responsibilities. The chapter thus adds an important cultural dimension to our understanding of the role of user-technology relations. In exploring the different political cultures, Parthasarathy problematizes the relationship between activists and the civic individual. She describes how, while many advocacy groups claim to speak for and empower the individual, this relationship is made much more complex by the cultural contingencies embedded in activist identities and definitions of the "empowered" civic individual. Patient advocacy groups, which appear superficially similar, construct very different identities and definitions of the "empowered" civic individual. These divergent politics lead to the co-production of unique technologies and "empowered" civic individuals in the United States and Britain.

Jessika van Kammen similarly problematizes the relationship between advocacy groups who speak for users and the actual users. She analyzes the configuration work of experts as well as political representatives who speak on behalf of users. This chapter reveals the complexities that emerge when representations of users articulated by user interest groups do not correspond with the user representations of scientists. Based on a study of the development of anti-fertility vaccines, van Kammen shows how the attempts by scientists to align the multiple user representations and incorporate them into the artifact eventually resulted in a "technological monster," a sophisticated artifact that was unable to attract any users. The multiplicity and diversity of user representations not only constrained the design of the new contraceptive, it also shaped the strategies of the advocacy groups. Instead of trying to represent the enormous social, cultural, and individual diversity of women using contraceptives and to speak on behalf of the needs of women, the women's health advocates gave voice to "users' perspectives." They profiled themselves as political representatives of users and mobilized their experiences of working on women's health and rights issues in the political arena to acquire credibility in the eyes of scientists. Crucially, this strategy enabled them to relate to contraceptive technologies in capacities other than that of potential future users: as researchers or as advocates. Room was created for women's health advocates to introduce different frames of meaning, such as the kind of social relations that one or another technology might constitute. Speaking from users' perspectives also reinforced their position as partners in a dialogue with the contraceptive developers.

Steven Epstein explores the role of patient advocacy groups and other heterogeneous sets of actors in representing the user by examining the reform in US policies that included women, minorities and children in biomedical research and in drug development. Focusing on what he calls the "multiple politics of representation," Epstein analyzes the different representational strategies of women's health advocates, women politicians and professionals, scientists, clinicians, and representatives of the pharmaceutical industry. In order to call for the greater representation of previously underrepresented groups as subjects in biomedical research, various actors had to position themselves successfully as legitimate representatives of social interests and collectivities, invoking the needs, wishes, and interests of groups such as "women" and "African Americans." At the same time, these representatives speaking for the group had to frame their demands by making claims about the nature of the group—that is, they had to claim to offer a symbolic depiction of fundamental group characteristics. This chapter makes an important contribution to our understanding of how categories of ethnicity, gender, and age are used to depict users in ensuring "fair representation" in clinical research. Epstein reveals the complex configurations of power and knowledge that are involved in configuring user identities. The heterogeneous set of actors involved in this case competed and collaborated to speak on behalf of socio-demographic categories that do not speak in a single voice.

The chapters in part II nicely illustrate how a methodology that focuses on the multiple groups who try to represent the user, including both experts and advocacy groups, reveals the cultural contingencies and the politics involved in the co-construction of users and technologies. It enriches our understanding of how the politics of users become manifest in today's technologically mediated state.

In part III, the focus shifts to the multiple locations that are important in understanding the configuring of users in the development of technology. In line with semiotic approaches to user-technology relations, Ellen van Oost draws attention to the design phase of technology. Inspired by feminist scholarship, she carefully avoids the trap of analyzing users as a monolithic category by showing how designers differentiate between male and female users and eventually between "male" and "female" artifacts. Van Oost presents a fascinating account of the history of the development of Philips electric shavers. Adopting a genderscript approach, she describes how a single artifact, first designed to be used by both women and men, gradually developed into two different design trajectories and separate products: the Philishave for men and the Ladyshave for women. She analyzes the design strategies used to construct a "female" shaver as distinct from a "male" one and shows how the design trajectory of the Ladyshave was characterized by masking technology. This resulted in a product that users could not open and repair. In contrast, the shavers for men were designed to display and emphasize the technology inside, and they could be opened and repaired. Van Oost concludes that Philips produced not only shavers but also gender. Whereas the genderscript of the Ladyshave "told" women that they ought to dislike technology, the script of the Philishave invited men to see themselves as technologically competent users. The designs of these shavers thus constructed and reinforced dominant views of gender identities that emphasized the bond between masculinity and technology. The chapter illustrates the power of the genderscript approach as a tool to account for the diversity of users.

Nelly Oudshoorn further explores how gender matters in configuring the user and shifts the analysis to the testing phase of technologies. Whereas the previous chapter described a technology that created strong links between artifacts and dominant notions of masculinity, this chapter describes a technology with a weak alignment with male identities that constitutes a major barrier for technological innovation. Oudshoorn challenges conceptualizations of users underlying semiotic approaches by arguing that the narrow focus on users' competence fails to take into account the articulation and performance of subject identities as crucial aspects of technological innovation. Based on an analysis of the clinical testing of hormonal contraceptives for men, she describes how technological innovation requires the mutual adjustment of technologies and gender identities. Innovation in contraceptives for men involved a de-stabilization of conventionalized performances of gender identities in which contraceptive use was excluded from hegemonic forms of masculinity. Oudshoorn adds a new aspect to our understanding of the co-

construction of users and technologies by conceptualizing clinical trials as a cultural niche in which experts and potential users articulate and perform alternative gender identities to create and produce the cultural feasibility of the technology.

Johan Schot and Adri Albert de la Bruheze introduce yet another set of locations that are relevant to study the co-construction of users and technologies. Inspired by Schumpeterian studies of technological innovation, they draw attention to the mediation process between production and consumption. They characterize this mediation process as a process of mutual articulation and alignment of product characteristics and user requirements. They don't restrict their analysis to the work of producers and users; they also include the work of mediators, such as consumer organizations and women's collectives, who claim to represent the user. This chapter introduces the concept of the mediation junction, "a series of forums and arenas where mediators, consumers, and producers meet and negotiate." Like the chapters in part I, this chapter portrays users and representatives of users as co-designers of new products. Analyzing the mediation work involved in two Dutch consumer products, the disposable milk carton and snacks, Schot and Albert de la Bruheze reveal two different mediation patterns: a mediation process fully controlled by producers and a mediation process not fully controlled by producers in which various mediators play an important role. They suggest that a mediation junction that is located outside the firm, and not fully controlled by producers, seems to create more favorable conditions for user representatives to shape the mediation process. Compared to a mediation junction that is located inside the firm, an "out-house" mediation junction facilitates the matching of projected, represented, and real users. This type of mediation process thus may contribute to the type of technological development that incorporates the interests of producers as well as those of (representatives of) users. Schot and Albert de la Bruheze conclude by suggesting that mediation processes have been constitutive for the shaping of the twentieth-century Dutch consumer society.

In the final chapter, Trevor Pinch also addresses the role of intermediaries in the development of technology. He urges us to pay attention to salespeople, whom he describes as the "true missing masses" in technology studies. Pinch argues that, because salespeople occupy a strategic position between users and designers, studying selling strategies is important to understanding the co-construction of user and technologies. Based on an analysis of the development of the electronic music synthesizer, the chapter shows how frequent interaction with users enables salespeople to see how users improve technologies and even invent new uses. They often communicate this information to the designers and manufacturers, thus providing important feedback for design. Pinch tells the fascinating story of how David Van Koevering identified a new use and a new group of users for the Minimoog synthesizer. Whereas earlier synthesizers had been designed as studio instruments for composers and elite rock musicians, Van Koevering marketed the Minimoog as a synthesizer that could be used on stage by young rock musicians. Pinch challenges the linear model of production and selling as sequential and distinct activities. He shows how Van Koevering's own experiences as a user played a crucial role in his activities. The chapter thus further contributes to the book's perspective of multiple identities. Whereas Lindsay's chapter describes the multiple identities of users who acted as users, designers, and producers, Pinch shows the conflating identities of users and sellers, thus illustrating the fluidity of boundaries between sales and use.

In summary, our authors argue that a thorough understanding of the role of users in technological development requires a methodology that takes into account the multiplicity and diversity of users, spokespersons for users, and locations where the co-construction of users and technologies takes place. From this perspective, technological development emerges as a culturally contested zone where users, patient advocacy groups, consumer organizations, designers, producers, salespeople, policy makers, and intermediary groups create, negotiate, and give differing and sometimes conflicting forms, meanings, and uses to technologies. The focus on multiplicity and diversity shows how users not only matter once a technology is in use, but also play an important role in the design, the production, and the selling of technologies. Most important, the chapters in this book challenge any a priori distinction between users and technologists. They emphasize the multiple and conflating identities of users, producers, and salespeople.

The focus on multiplicity and diversity also reveals how the work involved in configuring and representing the user is not restricted to technologists but includes the activities of many other groups of actors such as states (Rose and Blume), patient advocacy groups (Epstein, van Kammen, and Parthasarathy), and consumer organizations (Schot and Albert de la Bruheze). In the picture that emerges, states and national political cultures construct differing and often conflicting representations of users that shape and constrain the agency of users as citizens as well as the representation strategies of advocacy groups. The diversity of users further complicates the work of these advocacy groups. Speaking on behalf of the user is a complicated endeavor now that users no longer speak in a single voice.

The authors note the multiple ways in which identities of users are articulated, performed, and transformed during the development and use of technologies. User identities, including gender, age, race, and ethnicity, become materialized in the design of technological artifacts (van Oost) and biomedical discourses (Epstein). They are articulated and performed during the testing of technologies (Oudshoorn, Schot and Albert de la Bruheze). They play a crucial role in the domestication of technologies (Kline, Laegran). Consumer and medical technologies thus emerge as identity projects with a twofold function: they facilitate and constrain the daily lives of people as well as the design and the (de)stabilization of technologies.

Finally, our authors present stories that go beyond a voluntaristic view of the agency of users. Although they show the creative agency of users in shaping socio-technical change, they also reveal constraints induced by state regulations and national political cultures (Rose and Blume, Parthasarathy), by hegemonic gender relations and youth cultures (Kline, Oudshoorn, van Oost, Laegran), by the boundary work of scientists and technologists (van Kammen, Schot and Albert de la Bruheze), and by costs and skills (Wyatt, Lindsay). Our focus on the agency of users has led us to important insights in the role of non-users and resisters of technologies (Wyatt, Kline, Laegran). Non-users and people who resist technologies can be identified as important actors in shaping technological development. *How Users and Non-Users Matter* therefore might have been a more appropriate title for this collection.

I

Users and Non-Users as Active Agents in the (De-) Stabilization of Technologies

1

From the Shadows: Users as Designers, Producers, Marketers, Distributors, and Technical Support

Christina Lindsay

The TRS-80 was introduced to the market by Radio Shack in August 1977. Although its name probably does not immediately, if at all, spring to mind today when thinking about personal computers, for 7 years the TRS-80 was at the forefront of home computing power. However, in 1984 Radio Shack ceased to upgrade the TRS-80 and changed it into a clone of the IBM personal computer, which had reached the market 3 years earlier. The TRS-80 vanished from the personal computer scene and, for some, became just one of the fond memories of the early days of home computing. The original intention of my research was to focus on the entire life history, albeit a short one, of this personal computer. The aim of the project was to examine how different ideas about the users were constructed by the developers and producers of this computer, and how these ideas both shaped and were shaped by the technology.

The twist in this tale came when I found that some people are still using this supposedly obsolete technology. Contrary to its perceived disappearance from mainstream personal computing, the TRS-80 has moved beyond obsolescence, emerging alive and well with a new lease on life. The designers, engineers, producers, marketers, advertisers, support staff, and software developers have long since moved on, leaving just some users and the TRS-80 itself. Almost 25 years after its first introduction, the TRS-80 is being kept alive and fully functional by some remaining users, who not only are further developing the technology, but are also defining their identities and constructing new ideas of what it means to be a user of the TRS-80.

Bringing this story full circle is the complex relationship of the current TRS-80 and its users to earlier TRS-80 users. Contemporary TRS-80 users define themselves as being in resonance with the computer hobbyists who first used the TRS-80 in the 1970s and in contrast to users of current personal computers. Ironically, this contemporary community is being kept alive primarily through e-mail communication and the Internet, neither of which can be accessed fully through the TRS-80.

This then is an account of the changing roles of users during the life history of one particular technology. The users in this particular story begin as somewhat stereotypically gendered representations constructed by the designers of the computer and end by becoming designers, producers, and retailers providing technical support for the technology and taking responsibility for its further development. In the process they also rework their own identities as computer users in relation to this technology. Throughout the life of the TRS-80, different representations of users have been constructed and negotiated, and have influenced its design and its use. This chapter shows that the co-construction of users, user representations, and technology is not a static, one-time exercise by the designers of the TRS-80, but is a part of a dynamic ongoing process in which many different groups, including the users themselves, participate.

This chapter is thus a story in two parts. In the first part, I examine the "what, by whom, and how" of the co-construction of user representations and technology in the design, development, and marketing of the TRS-80 in the late 1970s. I have relied mainly on secondary data for this part of the case study, using resources both written and online about the history of the TRS-80 and early personal computers. The second part jumps ahead to the late 1990s and asks the same questions. Here, I used information from web sites run by current TRS-80 users. I also conducted extensive online interviews with 40 such users, some of whom have been using this computer since its introduction. The enthusiasm of these people for my interest in the TRS-80 was a rich source of stories and experiences.

User Representations

The biography of the TRS-80 shows that it is not just the actual, real-life users who matter, but that ideas about the user—user representations—are just as important in the relationships between users and technology. The argument that the users are designed along with the technology is crucial to this discussion. Three frameworks have informed my research.

In their work looking at the creators of the personal computer, Thierry Bardini and August Horvath introduced the idea of the reflexive user (Bardini and Horvath 1995). By looking at the linkage between technical development and cultural representations, particularly those of the user, Bardini and Horvath asked how the creators of the personal computer at the Stanford Research Institute and at Xerox PARC in the 1970s envisioned the user, and how this influenced subsequent decisions made about technical options. They suggested that the innovators were the first users of the PC technology, and consequently defined the future users in their own image. The reflexive user is therefore the future "real" user in the minds of the developers, and this enables them to anticipate potential uses of the technology. This reflexive user, while influential in the early development of the technology, is shown to be a static, one-time view that is bound to disappear and to be replaced by the real user.

Steve Woolgar's work also looks at the development of a particular computer. He shows how the design and production of a new technological entity amounts to a process of configuring its user. This act of configuration involves defining the identity of putative users and the setting of constraints upon their likely future actions through the functional design of the physical artifact with a focus on how it can be used (Woolgar 1991). In effect, the new machine becomes its relationship with its configured users. Woolgar claims that, whereas insiders know the machine, users have a configured relationship with it, such that only certain forms of access and use are encouraged. For example, insiders or experts, unlike regular users, are able to take the back off the computer box and play around with the electronics inside. Woolgar's configured user is inextricably intertwined with the development, especially the testing phase, of the technology.

Madeleine Akrich (1992) introduced the idea of the projected user, created by the designers of technology. The projected users are defined with specific tastes, competencies, motives, aspirations, and political prejudices. The innovators then inscribe this vision or script about the world and about the users into the technical content of the object, and thus attempt to predetermine, or prescribe, the settings the users are asked to imagine for a particular piece of technology. The prediction about the user is thus built into, or scripted into, the technology. Akrich recognizes that the projected user is an imaginary user, and asks what happens when the projected user does not correspond to the actual user.

Each of these three frameworks presents the representation of the user as a one-time static view constructed by the developers of a technology in the design phase of its life cycle. Bardini and Horvath's framework of the reflexive user fails to recognize that the technology, its uses, and its users may change after introduction to the marketplace. Their ideas present a stationary picture in which the reflexive user is constructed by the developers and is then replaced by "real" users, who act as mere consumers of the technology with no role in shaping either the technology or their role as users. While his model is equally static, Woolgar, in saying that the new machine becomes its relationship with the configured users, draws attention to the importance of interactions between the users (in the testing phase) and the technology. However, he stops much too soon. While he argues that the design and production of a new entity amounts to a process of configuring the user, he does not study the actual use of the technology by these people, to understand whether the configuration process still continues. Like Bardini, Woolgar does not seem to consider that the user may play a role in shaping the uses of the technology. In both of these models, limited groups of people create these reflexive or configured users, and neither Bardini nor Woolgar consider that the reflexive or configured users may be created throughout the life history of a technology. Akrich's projected user is also constructed only by the designers of the technology.

I argue that there is much more to these imagined users than a static image constructed by one group sometime during the development phase of the technology's life history. I propose that "user representations" encompass many other imagined users, and that these user constructions are not built, and do not exist, in isolation. Each of the social groups involved with a technology throughout its life history, even those that are not directly involved, will have its own ideas about who and what the user is. My approach is to study the interplay of these various user representations, the ways in which the social groups use these constructions to reinforce or challenge their own ideas of the user, and how individuals' relationships to the technology are mediated by these ideas about the user.

The Introduction of the TRS-80

In the 1970s, though the main sites of computer power were still in the mainframes and minicomputers found in corporations, in government, in universities, and in the military, computer power had begun to enter the home. Computers came into the household through the basements and the garages of computer hobbyists, usually young men, who frequently had knowledge of and experience with computers through their work sites. The early personal computers, such as the Altair 8800, were sold as kits and marketed as minicomputers for people who wanted their own computing power. It was possible for individuals to have their own personal computers if they were able to assemble the many pieces of electronics that came in the computer kits and make them all work. Doing this required some specific hobbyist skills, and much persistence and perseverance. It was into this "hacker" culture (in the original sense of the word as "hobbyist" or "enthusiast") that Radio Shack introduced the TRS-80.

Under the umbrella of the Tandy Corporation, Radio Shack, with its origins in "ham" radio, was a major retailer of consumer electronics. Between June 1975 and June 1976, Radio Shack opened 1,200 outlets in the United States and Canada, bringing the number of North American outlets to 4,599, and the number worldwide to 5,154. Radio Shack did not sell off-the-shelf consumer electronics such as televisions and radios, but instead served as a hobbyist store.

In this period, one of Radio Shack's biggest markets was for Citizens' Band (CB) radio (Farman 1992). Although Radio Shack brought the very first 40-channel CB radio to the marketplace in January 1977, the company feared that the CB radio boom might be coming to an end. Many other companies were flooding the market with similar gear. To stay competitive, Radio Shack decided to develop new products. To this end, the company concentrated on developing a new line of calculators, because "they were kind of the latest thing" (ibid.: 400). Some of the people on the development teams enjoyed the hobbyist activity of putting together "computer parts jigsaw puzzles." At the instigation of these people, Radio Shack initiated a half-hearted project to look at developing its own computer kit. Steve Leininger, recruited from National Semiconductor to lead the project, decided to take a risk and worked to develop a ready-wired complete system, a computer that would not need much putting together by the user. This was a new idea. At the time, home computers were available only in kit form. The computer that was developed was called the TRS-80—T for Tandy, R for Radio Shack, and 80 for the Zilog Z-80 microprocessor.

The TRS-80 was developed without a long-range plan. Charles Tandy, the CEO of the company, was somewhat reluctant to support the project. "A computer!" Tandy blared. "Who needs a computer?" (Farman 1992: 404) One of the reasons for Tandy's skepticism was that the development group envisioned the new computer selling for $500–$600, and Radio Shack had never sold anything that expensive. Indeed, Radio Shack's median sales ticket at this time was just $29.95. The members of the development group admitted that there were no existing customers for the new device and that it was virtually impossible to identify buyers. They had designed the computer for hobbyists like themselves.

Engineers' designing for themselves, in effect considering themselves to be representatives of the future users, is not uncommon. Such engineers are reflexive users (Bardini and Horvath 1995). In this method of constructing user representations, which Akrich named the "I-methodology," the personal experiences of the designers (in this case, the engineers) are used to make statements on behalf of the future users. Already, however, these ideas about the users do not directly correspond to the developers' ideas. The TRS-80 was designed so that its few components would require less skill to put together than the computers in kit form already available and used by the design engineers.

The TRS-80 was formally introduced at a press conference in August 1977 in New York. For $599.95 the buyer got a 1.77-megahertz processor with 4 kilobytes of storage, a black-and-white visual display unit, a keyboard, and a cassette tape player, which was used to store the programs and the data.[1] There was also an operating system and a very limited selection of software, including games, educational programs for teaching multiplication and subtraction, and conversion tables. Basic was the programming language used; it had been licensed from Bill Gates for a one-time fee. For people who could not afford the complete computer, the processor alone was available for $399.

After developing the TRS-80 for users who were conceived of as being similar to the developers, Radio Shack sought to explain its computer to two existing groups they felt might be interested. The first group, targeted through post and computer mailings, consisted of existing Radio Shack customers, described as "the guy that's gonna put up his own antenna, the guy that's gonna repair his own telephone, the guy that's gonna fix his own hi-fi" (Farman 1992: 143). A spokesperson at the time stated: "We thought we had a good product that was of interest to the traditional Radio Shack customer, and at $600 we hoped we could sell it." (ibid.: 410) Even though one of the main features of the TRS-80 was that it was not in kit form, the user representation was linked to the consumer electronics do-it-yourselfers who were already Radio Shack customers. The second group consisted of current computer enthusiasts and hobbyists. These were reached through advertisements in *Byte* magazine, with its focus on the hobbyist computer market in which the users built personal computers from kits. *Byte*'s articles were "written by individuals who are applying personal systems or who have knowledge which will prove useful to our readers" (*Byte* 2, 1977, October). It was anticipated that members of this second group would transfer their interest in computers, along with their skills in building them, to the TRS-80.

This process of recruiting new user groups can be likened to Woolgar's concept of replacing the configured user with the actual user. The original representation of the users had been expanded. In this way, knowledge of users of existing consumer electronics technologies were added to ideas about who the users of the TRS-80 would be. However, the skills and interests of the people in the two new groups still closely matched those of the engineers.

The envisioned users of the TRS-80 would then be members of the early hacker culture—people who had not only an existing interest in electronics, but also the skills and knowledge to put together their own machines. These computer hobbyists were usually young and male, with an extensive knowledge of electronics or computers obtained from work experience or from previous hobbies. Such users wanted their own computing power. They were in general not the businessmen anticipated by Charles Tandy as the initial users of the TRS-80. In its 1977 annual report, the Tandy Corporation had stated: "The market we foresee is businesses, schools, services and hobbyists. The market others foresee is 'in-home'— computers used for recipes, income tax, games, etc. We think home use is a later generation happening." The original aim had been to make computers available to a lot of people, "especially businessmen who have never had a chance, that could not afford a computer" (Farman 1992: 419).

The contradictions in the user representations are more obvious in the initial print advertisement for the TRS-80, which appeared in October 1977. This advertisement makes the representation and the role of the users more apparent by depicting the TRS-80 in its intended environment of use. The computer had been moved both literally and figuratively out of the basement or garage of the computer hobbyists and into the home, specifically the kitchen.

The advertisement shows a man seated at a kitchen table using the computer while a woman looks on, smiling. These are the projected users (Akrich 1992) made visible, with their tastes, competencies, and motives described through pictures and text. While the user is still conceived to be a man, he is neither a young hobbyist nor working in a business environment. The gendered messages of this ad are complex. Bringing the computer into the household suggests that it is now a domestic technology. However, placing it in the kitchen, traditionally the woman's domain raises the question of whether it was to be seen as a "white goods," like the stove and the refrigerator, or as a leisure appliance such as the television and stereo (Cockburn and Ormrod 1993). Was the personal computer to be just another home appliance, such as the toaster or the stove, to be used by women? Apparently not. The ad shows a man, presumably the man of the house, happily using the computer while the woman stands at the counter, smiling at him. Bringing the computer into the kitchen does not necessarily position it as a technology that would, or should, be used by women. The users of such computers would still be men, but now the computer's use would fit comfortably into the household chores.

In the move from the basement and the garage into the living space of the home, the personal computer began to be marketed as a tool for household management tasks, as well as a hobbyist machine for the writing of programs. However, the constructed user of the new technology was always male, and implicitly knowledgeable concerning the technology's construction and use. Although most subsequent TRS-80 advertisements were not explicitly gendered (in fact, there were no people in the adverts at all), the extensive computer skills and knowledge required to be a TRS-80 user were still possessed mainly by men. The TRS-80 was advertised as "a system ready for you to plug in and use." "Program it to handle . . . " was the opening phrase of the next sentence. In fact, the TRS-80 was not as easy to put together and keep running as the advertisement suggested, and some hands-on craft skills, other than programming, were needed. The machine was introduced into a climate of hobbyist enthusiasms and skills. Although it did not come in kit form, the various components had to be linked together, presuming at least a familiarity with consumer electronics and the requisite assembly skills.

The user representations displayed in the ad were textually supported by a list of possible uses for the TRS-80. Most of the uses concerned control of the household, but some related directly to tasks associated with the kitchen and food preparation. The uses suggested in the original press release concerned household tasks, including personal financial management, learning, games, and (with the addition of an external device controller) future control of the operation of appliances, security systems, and the monitoring of a home weather station.[2] There were quite a few suggested uses given for the kitchen, including storing recipes, keeping a running inventory of groceries, menu planning, and adapting recipes for larger or smaller servings.[3]

The design and development of the TRS-80 is thus a story of "technology push" in which engineers and businesspeople searched for a new market for a new technology. In developing the TRS-80 explicitly for people like themselves, the engineers provide a wonderful illustration of Bardini and Horvath's "reflexive users" and of Akrich's "I-methodology" of product development. But soon other user representations were developed. Charles Tandy envisioned a business user. Other ideas about who the user might be came from Radio Shack's experience and knowledge with other technologies. Current Radio Shack consumer electronics customers would have the knowledge and skills needed to assemble the computer, and readers of *Byte* would have an interest in home computing and also an interest in, if not the skill to build, home computers. The different uses listed for the computer, such as business software and household applications, also contributed to the different representations of the user.

Mediators

After the introduction of the TRS-80, other groups became involved in constructing who the user was or could be. Of particular interest are the groups that served as mediators between the technological and the social realms—those who, in effect, explained the technology to the user.

One such group consisted of computer magazine writers who reviewed and evaluated the TRS-80 for readers. User representations were developed in two ways, the first of which was again a reflection of their knowledge about users of related technologies, such as kit computers. In this case, users would apply their considerable interests and skills acquired from kit computers to the more ready-to-use TRS-80. User representations were further developed by a comparison of TRS-80 users to those of a similar personal computer, the Apple II. By focusing on the differences between the two computers, writers were able to reflect on the interests and the skills of the relevant users. Many of these product reviews compared the TRS-80 with the Apple II. Although both computers were considered "high-end," the reviewers established a distinction between the users of the two computers built around the skills needed to use each machine. The Apple II was an all-in-one machine that did not have to be assembled, and it was marketed as a home computer for which the user did not need prior specialized knowledge. The relatively high price of the Apple II served to further distinguish the two kinds of users. One of the current TRS-80 users interviewed (Bathory-Kitsz) expressed the view that the "TRS-80 was always kind of the people's computer and the Apple was the elite's computer."

Ideas about who the users could be, or would be, began to change with the success of the TRS-80. Another important group of mediators were the writers of technology columns and articles in computer magazines. While not directly evaluating the TRS-80, they often provided help and offered information about how to deal with problems with the computer.

They published the source code of programs for various kinds of new applications, and they effectively served as co-producers of the technology.

These writers not only began to develop a new role for themselves as TRS-80 users in providing help as part of their job, but also were instrumental in changing ideas about who the users could be. Their work provided a new avenue of assistance for TRS-80 users, who now did not have to be individually knowledgeable about all aspects of the computer and did not have to be limited to the few programs that were available. Of course, a user still had to be able to program the computer! The representation was of a user who could draw on other resources. This was very important to the success a person might have in using the TRS-80. The Tandy Corporation did not provide much service support for its computer, and the knowledge of the salespeople in the Radio Shack stores was very limited. The user as originally envisioned by Radio Shack was someone who had all the knowledge necessary to build and use the computer, and also to trouble-shoot any problems that might arise. Technical articles in computer magazines were the first step to reducing the amount of knowledge an individual user needed.

The TRS-80 users themselves were instrumental in continuing to change the idea of who could be a successful user, and the formation of support groups turned out to be an important influence. Special-interest groups formed to provide support and advice for TRS-80 users formed all over the United States. Most of these groups held monthly meetings and produced newsletters.

To talk about users as co-producers of the TRS-80 may seem an overstatement. But to become a successful user of this early personal computer, one had to actively put together the machine and to type already-written programs into it—no easy task when just one incorrect letter or number could cause havoc. Thus a dichotomy between developers and users is not useful here, as there is a third group, the expert users. To use the TRS-80 successfully, the user had to open it up and then wire in the selected peripherals.

If co-production is considered as not just using the technology, but as taking the development of the technology further, then the TRS-80 users became very active co-producers. They did this by developing software programs that they then exchanged among the support groups. In addition, some attempts were made, primarily through advertisements in computer magazines, to set up software exchanges of user-written programs. This fits in with the prevailing "homebrew" ideology of the hacker culture into which the TRS-80 was introduced.

By November 1978, a further new user representation had developed. Some of the early users, in addition to becoming co-producers of the TRS-80 technology through their hobbyist activities, had begun to make a commercial business of it. Both individuals and businesses began to advertise software for sale. For example, a software company called Micro Systems Service advertised "dial-a-program." Using a toll-free number, the caller could have programs for the TRS-80 transmitted over the phone line. The data was received using a standard home phone pickup onto audiocassette tape. While inventive, this method was not successful because the programs could not be transmitted reliably (*Byte* 2, 1977, September). Another software development company advertised a TRS-80 programming contest. Thus a picture of a different user emerged: one who used the TRS-80 for the commercial business of producing and selling TRS-80 materials.

More details can be added to the changing picture of the user by studying the classified ads in *Byte*. In the June 1978 issue, the first advertisement appeared for the sale of a TRS-80. The computer was still in its wrapping, and its owner was moving up to a more powerful system. Numerous such advertisements appeared over the next few years in *Byte* and other computing magazines. The TRS-80 user was now perceived as no longer being satisfied with the performance of the original Model I.

Different types of users were identified, differentiated by what they would want to use the TRS-80 for and by the power it would have. The TRS-80 would now have a role in the business world as well as at home. The Radio Shack developers came back into the picture, introducing different levels of Tandy machines for different types of user needs. These included the Model II, a "small-business computer for people who like to pay less than the 'going price'" (*Byte* 4, 1979, October), and also a new level of Basic with advanced features, including full editing and integer arithmetic. The developers were now explicitly distinguishing between home users and businesspeople.

In tracing the path of the TRS-80 from the developers to the user, I have illustrated different representations of who the users were or could be. The initial representations of the user were changed by the different groups who became involved with the TRS-80 and with changes to the technology itself—whether those changes were made by Radio Shack in producing new models or by the users who wrote software for the computer. Representations of the users were developed in various ways, including the initial I-methodology of the developers, the anticipated transference of interests and skills from users of similar or related technologies, and the intervention of the users who developed new roles for and expectations of future users. One of the most significant changes in the user representation was from the TRS-80 user as a individual knowledgeable about, and skilled in, the DIY aspects of consumer electronics to a user who could be less knowledgeable and could obtain support and knowledge from a TRS-80 community, either through reading computer magazines or through joining a computer club. The mediating groups and the users became co-producers both of the technology of the TRS-80 and of the user representations that were constructed.

The TRS-80 had a shorter commercial life span than the Apple II or the IBM personal computer (introduced in 1981). For a few years, the TRS-80 was at the forefront of the home computing market and was considered to be a sales leader (Levering, Katz, and Moskowitz 1984). The TRS-80 Color Computer, one of the first computers capable of displaying color, was introduced in 1980. It had nothing in common with the original TRS-80. It provided color capability at a low price, and there was nothing else like it on the market at the time. The Welsh Dragon computer, introduced in 1982, was the first clone of the TRS-80. In 1983, Radio Shack introduced a portable version of the TRS-80.

By 1984, Radio Shack had stopped introducing new models and had turned the TRS-80 into a clone of the IBM PC. The TRS-80, in its original form, subsequently vanished from mainstream personal computing, and the Apple, the IBM PC, and the Macintosh became the personal computers of choice.

Obsolescence

The story now turns to the late 1990s (23 years after the TRS-80 was introduced to the public) and moves beyond the standard technological life trajectory of "design to use" to consider obsolescence. It is in this period of perceived obsolescence that the biography of the TRS-80 takes an interesting turn.

Some people today are still using their old TRS-80 computers. The producers, advertisers, support staff, commercial TRS-80 software developers and salespeople have long since vanished, leaving behind the users and the artifacts of the TRS-80 itself. The focus of this part of the research is thus only on the current users of the TRS-80. As there are only a small number of such users, it is possible to focus not only on user representations but also on the actual users themselves. As the only group left in the story, the users are active in constructing their own identities and in maintaining and further developing the technology. For this part of the research, I interviewed 40 of these current users, all men. In what turned out to be a paradoxical element to the story, and one that I'll address in detail later, I found these people by searching for "TRS-80" on the World Wide Web.

I was unable to find a female user of the TRS-80. This is more disappointing than surprising, as technology, in general, has been shown by researchers to be a masculine culture.[4] Men's "love affair with technology" (Oldenziel 1997), both in childhood and in adulthood, reflects an experience of technology as fun. Computer culture continues this masculine association (Haddon 1992; Turkle 1984). The cultural context of computers has been gendered masculine with respect to intellectual strength and abstract thinking, two skills associated with men rather than with women (Lie 1996). There may also be an element of fear: women may be reluctant to become involved with a tool they see as threatening (Turkle 1988). This is amplified within the computer hacker culture, with its image of hackers as young technology-focused men without social skills (Haddon 1988; Hafner and Markoff 1991; Turkle 1984). While there are women hackers on the World Wide Web (Gilboa 1996), the hacking skills needed for the TRS-80 are not only the skills to write programs but also the hands-on ability to manipulate the hardware components of the computer.

The co-construction of users and technology in current TRS-80 culture is very clear, the difference being that now it is only the users who are involved. This raises the question of whether it is felicitous to continue to talk about user representations when it is now possible to talk to the users themselves and to obtain their self-descriptions. Can these self-descriptions or subjective identities (Cockburn and Ormrod 1993) be considered user representations? I would argue that, insofar as these ideas about the current users still continue to influence who can or should be using the TRS-80 and how it is to be used, they have a useful purpose in a public forum as user representations. In addition, the relationship of current user representations to those of the past became an important element in the contemporary relationships of users with the TRS-80.

The TRS-80 Users Today
The first step in examining the current co-construction of users and the TRS-80 is to introduce the users and to examine why and for what they are continuing to use the TRS-80. These people's constructions of their

own identities as computer users are tied inextricably to the TRS-80 and to current personal computers and also to past ideas about computer users as hackers. By examining these identities, or representations, we can understand how the users construct their own identities with respect to the technology and how they are changing the technology in relationship to those identities.

Many of the TRS-80 users I interviewed were either engineers or had obtained technological skills and knowledge from childhood, whether at home or at school. Several of the interviewees cited working with other members of their family as their introduction to the world of building things. The gendered relationships of children with technology have been well researched. From model cars (Oldenziel 1997) to robots (Faulkner 2001), building things as a hobby has been shown to be primarily a masculine pursuit. In general, parents tend to encourage boys more than girls to play with technology, and so to acquire manual and problem-solving skills (Millard 1997). This early exposure to technology often leads to a familiarity with technology that many boys carry forward to later life, whether in the careers they choose or in their leisure pursuits (Berner and Mellström 1997).

The initial involvement of the interviewees with the TRS-80 was often through exposure to computers at high school, university or graduate school. In addition, many of the users worked in computers already, or were professionals in the computing industry. Two of the respondents specifically worked on and used arcade game machines and were able to transfer their skills and interests to the TRS-80. Several of the users worked for Radio Shack or Tandy, and learned of the TRS-80 when it came into the Radio Shack retail stores. Several people mentioned linking childhood pursuits to adult experiences. "Toys for boys" was a theme throughout some of these early exposures to the TRS-80, with one interviewee stating that his father gave him the choice of getting a speedboat or a computer.

Many users continue to use their TRS-80s because they meet their current needs. One user stated that his TRS-80 boots up faster than his three well-equipped PCs with Windows. TRS-80s are still sometimes used for word processing and text manipulation. For one user, these computers are the only ones he uses. He has one at home and another in his office and uses them for all his home and business word processing needs. Another user wrote his doctoral dissertation (on technology ethics and public policy) using his TRS-80. Some other users are still writing game programs for the TRS-80 and distributing them as freeware. One enthusiastic radio amateur uses his computer to send and receive Morse code on the amateur bands.

Time and the New Users
Central to the relationships of the current users and the TRS-80 of today is the concept of time—more accurately, the two periods of time detailed in this case study. The identities of the current TRS-80 users were constructed in relation to the identities of computer users not only in the past but also in the present. Likewise, the relationships of these people with the TRS-80 were linked both to the TRS-80 of the past and to current personal computer technologies. Thus, this is not a straightforward story of the co-construction of users and technology, it is made complex by linkages across time.

The current identities of the TRS-80 users were constructed through processes of aligning or identifying themselves with certain groups and differentiating themselves from others. Both of these processes were to be found within the explicit links made to past TRS-80 user representations.

The current TRS-80 users who responded to my questions identified and labeled themselves in various ways. The term "tinkerer" was used by several people, referring to either or both of hardware or software tinkering. This label resonates with the hacker culture into which the TRS-80 was first introduced. One user alluded to the specialist knowledge of the TRS-80 users when he stated that "the TRS-80s are perhaps remembered in a kinder light because the community that used these machines were not your typical home user awash in myths."

A process of differentiation concerning skills and knowledge was used by these current users to distinguish themselves from the developers of the TRS-80. Many of the users identified themselves as critics of Radio Shack's technical, marketing, and service support. They claimed to have more knowledge and skills than that available from the company. Even people who had worked in the technology support area of Radio Shack acknowledged the lack of adequate technical help.

The self-identities of the current TRS-80 users in relation to computer users of today were established primarily through a process of differentiation from other computer groups. TRS-80 users did not consider themselves to be the same as the run-of-the-mill computer users of today. Echoing Woolgar's idea that configuring the user is an act of drawing boundaries between the inside and the outside of the machine, one user labeled himself a programmer and not an end user, thereby aligning himself with those who knew the workings of the computer and were active in using and expanding it rather than with the users who "just" used the machine. Along these lines, another user stated that he was a marketer of software previously written by himself, and two others were publishers of TRS-80 magazines.

These people also differentiated themselves from the programmers of today's personal computers. Some of the TRS-80 users were there when you had to do "more with less" of the technology, and learned to use these same skills on today's computers. One interviewee stated:

Today's programmers are not the same caliber of people we had in my day. Today's software is large and fat and wastes huge amounts of memory, CPU and disks for very little benefit. . . . You sit there saying "I've got a machine 100 times faster [than the TRS-80], with 100 times more memory, and 10,000 times more disk and IT'S SLOWER." It makes you wonder what today's programmers know.

This interviewee was using his knowledge of the TRS-80 to construct an identity of a "real programmer" that set him apart from today's programmers, and even further from today's end users. He was not alone in this. Many of the respondents stated that using the TRS-80 allows them to get at the "guts" of the machine and the code, reinforcing their self-identity as tinkerers, hardware hackers, and hobbyists.

In all of the boundary drawing being performed here by the TRS-80 users—between themselves, end users, Tandy, and today's programmers—their construction of their own identities as computer users was invariably attached to the artifact of the TRS-80, and they were also aligning themselves with and declaring their affinity for past periods of computing. They were qualifying their labels as "users" and "programmers" by attaching the label "TRS-80," and bringing along all that is associated with that computer. They were resisting being associated with the users and programmers of current PC technologies, and in doing so were trying to keep alive their virtuosity as technically oriented people. This user identity implies that greater skills and knowledge are needed to use the TRS-80 computer than are needed for today's machines. In effect, it implies that programmers and users from the past can easily use the current technology, but that the programmers and users of today would not have survived in the personal computing world of the late 1970s.

The user identities constructed by the current TRS-80 users strongly echo the user representations that developed along with the technology. Even the terms "tinkerer" and "hacker" link them directly back to the "homebrew" computer culture of the late 1970s. Indeed, many of the contemporary TRS-80 enthusiasts I interviewed were there at the beginning. However, their identities as computer users were linked more closely with the past than just possessing a sense of history.

Simplicity and Reliability of Use

Why are the identities articulated by users important to the present TRS-80 community? I posed this question explicitly to my respondents. Many of the answers I received concerned the effectiveness of the TRS-80 as a piece of technology, especially when compared with present-day personal computers, particularly with respect to simplicity, reliability, and cost.

One TRS-80 user identified himself as a "neo-Luddite" and stated that he used his old computer as it was simpler than those of today. Many other users echoed this sentiment. However, this is not the limited simplicity of ease of use; rather, it is a simplicity that can only be appreciated by people who have the skills and knowledge to work within the computer's limitations and the rudimentary programming that it allowed. Some comments emphasized this point:

No finicky complex OS, no bloated applications, no oddball device drivers, NO SEGMENTED MEMORY ARCHITECTURE, no pointy-clicky, no icons, no GUI— just get up and go computing. Anywhere.

The OS [operating system] was simple. . . . It didn't have all that security crap.

The lack of CPU power is actually a blessing since there is no time for kludges.

In the language used here ("pointy-clicky," "GUI" (graphical user interface), "security crap"), the desire for simplicity served as a rejection of the user-friendly additions to personal computer systems associated in the main with Macintosh computers and with making the computer accessible to all levels of users. Simplicity can also mean speed. One user remarked that he uses his TRS-80 "to run very simple programs that aren't worth firing up the PC for."

Along with this simplicity comes reliability:

[The TRS-80 is] simple, extremely reliable, runs on regular AA batteries and is quite quiet in the library.

One user wanted a TRS-80 Model I to do his business mailing labels:

. . . as long as you feed the labels to the old dot matrix printer, the old computer and the old printer will "chug along" all day and all night with little or no assistance from individuals who can as well be doing something better.

In addition to simplicity and reliability, several users cited cost as a factor. One said:

This is important because I am a neo-Luddite and do not wish to contribute to the useless and wasteful faddishness that has been going on for some time now.

Another expressed a view held by several other users:

All of the ports . . . come with the motherboard—[you] don't have to spend thousands of dollars buying boards and some Mickey Mouse software to run it.

Nostalgia

The most-cited reason for continuing to use the TRS-80 once again linked the past and the present; it also served to underscore the relationships between users, user representations, and the technology. Practicalities of use, reliability, and cost aside, one of the primary reasons that people were still using their TRS-80 machines was nostalgia. This was more than just wishing for a time of (depending upon one's viewpoint) less or more complex computing. However, there was an element of fond remembering of people's introduction to the computing world. "These were our first computers," said one user. "We loved them." Another user was rebuilding his system "just for old times' sake." But this nostalgia also included an element of exploration. Many of the users I interviewed portrayed the TRS-80 user as a hacker who had been at the frontier of personal computing in the 1970s and who was now at the frontier of the physical boundary of the artifact of the personal computer. Unlike users of current PC technologies, these current TRS-80 users were able to cross this physical boundary and to manipulate their machines:

These days computers have been reduced to pre-packaged consumer tools, and that aspect of exploration is out of reach of many potential hobbyists.

Some of us stick with it (in addition to the modern powerful machines) because it is fun to experiment with, or just out of nostalgia, like pipe organs or pinball machines.

The nostalgia also included a sense of fun:

It's fun to see if I can still remember how to program the old dinosaurs. . . . I am reminded that, once upon a time, 64K of memory was a lot of space, and you could have an entire operating system on one 360K floppy disk.

A major part of the fun was in being able to cross successfully the physical boundary between the user and the computer:

They're fun. Today's machines have taken all the fun out of hardware and software hacking; like "tinkering"; taking it apart and putting new pieces in to see if I can get it to run better.

You can get inside and tinker.

The pleasure that men take in technology, from the hands-on tinkering with the hardware or the cognitive analytical problem solving of programming has been noted for other technologies, such as software engineering and robot building (Faulkner 2001).

One particular result of this nostalgia was to explicitly and purposefully link the past with the present. There was also a historical and educational aspect to this nostalgia—a desire to show people who were not involved in early personal computing what it was all about. Some of the current users took their TRS-80s to computer shows, especially vintage computer shows. This was not done to show people how small and slow the machines were; some users treated these forums as opportunities to instruct the public on how programmers were able to "do more with less"—less computer memory, less speed, less "user-friendly" computing languages, less help from computer companies and less experience. This served to reinforce the identity of such TRS-80 users as tinkerers, both in the heyday of the TRS-80 and now in their continued ability to use the machine:

A very very high percentage of all owners learned to program to at least a small degree. Practically no one who buys a PC ever learns to write programs on it.

Resistance

For some users, nostalgia for the good old days in which TRS-80 users did "more with less" went beyond identity construction and beyond merely differentiating themselves as TRS-80 programmers as distinct from "end users" and from the programmers of today's computers. Instructing people in how the TRS-80 was used signified a form of explicit resistance. One user, who was engaged in a personal project of documenting the development of the operating systems of the TRS-80, stated:

What I want to record for future generations is what REAL programming looks like. . . . I want to tell people that just because Microsoft does things one way, that doesn't mean that it's the ONLY way to do things. Here's another way. Here's why this way is BETTER. Perhaps you should question why you do things their way.

Thus, the TRS-80 users constructed their identities through relationships with the users representations and technologies of both the past and the present, forging links between the two times and the different computer cultures. These users' representations of their own identities as computer users were similar to the original conception of the users when the TRS-80 was introduced. The past and present ideas about the users are centered around the TRS-80 and are linked by an emphasis on skills, knowledge, and technical competence. However, as the only group left in relationship to the TRS-80, the current users not only constructed their own identities as computer users, but also moved from co-producers of the technology to assuming fully the responsibility of maintaining and developing the TRS-80. The role of the user has become one of designer, developer, marketer and technical support.

One link between the time periods is the desire of the present users to improve the artifacts of the past. The users have improved on the original technology by actively changing the TRS-80. In light of the poor technical support from Tandy, TRS-80 users have always had a history of "tweaking" their computers and solving their own problems. However, in later years, after Tandy dropped all support, the users gradually took over that role, fixing the last few operating system bugs and writing new programs: "The machine is very close to twice as fast as when Tandy sold it. . . ." Left to their own devices, the users have continued to shape and improve the physical artifact and also to develop new technologies which continue to enable the current use of the TRS-80.

In an interesting twist, and bringing the past directly into the present, some users have developed emulators. These are software programs that run on computers, other than the TRS-80, and which make that computer behave as the TRS-80 did. The emulators form yet another link between the old computers and current hardware. One user even runs a TRS-80 emulator on his Pentium computer! The original push behind the development of such emulators was to access data produced when using the TRS-80 that could not be loaded onto today's PCs. Some of today's emulators are very accurate and, according to one interviewee, "replicate the look and feel of those wonderful old machines." In keeping with the hacker culture's emphasis on mutual support and free software, the TRS-80 users worked together to adopt a standard format for the emulated floppy disks—"and that," said one user, "means that anyone can trade all their old floppies with everyone else."

Emulators are not the only link between the 20-year-old TRS-80 technology and modern personal computing. In a direct relationship between the early days and now, one of the popular current uses of the TRS-80 computer is for e-mail and to access the Internet, though only textual and graphical information can be obtained.

However, the linking of the TRS-80 to the Internet and to state-of-the-art PCs raises interesting questions about the relationships between the old and the new technologies here, and about any dependencies between them. One enthusiastic user of the TRS-80 stated:

The TRS-80 community is experiencing a rebirth and emulators are a large part of the reason why. The Internet being the other. . . . The Internet is almost single-handedly responsible for the rebirth of the TRS-80 community. It is how we found each other again, and how we trade information amongst ourselves. The real glue is e-mail. That's actually where most things are actually going on. We're pretty much in constant contact with each other, exchanging news, making more contacts, and working out plans for the future.

The TRS-80 has taken on a new lease of life on Internet and the World Wide Web. There are newsgroups in which information is exchanged and problems posted and solved; there are personal web pages, which include offers of software and operating system "patches" and fixes; there are web advertising emulators and software for sale, and advertisements about buying and selling these computers; as well as e-mail and mailing lists, information is disseminated through at least one online publication; there are computer history web pages and virtual museums online, in which people display pictures of their home computer collections.

The Internet plays a large, if not an essential, role in keeping the TRS-80 alive and well. This is paradoxical in that current computer technologies—the very technologies from which the TRS-80 users are careful to dissociate themselves—are an essential means for their continuation.

A few TRS-80 support groups still meet "in real life." The Nybblers meet monthly in Hayward, California. And there is an annual Vintage Computer Festival, also in California; its mission is "to promote the preservation of 'obsolete' computers by offering attendees a chance to experience the technologies, people and stories that embody the remarkable tale of the computer revolution" (Ismail 1998).

Time is therefore an important element in the complexity of the co-construction of users and technology. The subjective identities of the TRS-80 users are linked to both technologies and user representations simultaneously in the past and in the present, and the TRS-80 world continues to be a masculine culture. This story is somewhat paradoxical in that the users have constructed their identities as TRS-80 users in a large part by aligning themselves with the past and by differentiating themselves from current personal computer users. However, these TRS-80 users need these very same technologies for the communication and community that support the continued maintenance and future development of the TRS-80.

Conclusion

I learned three valuable lessons from my case study. The first is that user representations are dynamically constructed by different groups, including the users themselves, and by using different methods throughout the whole life history of the technology. The users I interviewed were, and still are, very active in the co-construction of their own identities and the TRS-80, and have taken on the many roles of developers, producers, retailers, advertisers, publishers, and technical support staff.

The second lesson is that there is a complex relationship between user representations and technologies of the past and those of the present. The relationship between the two time periods examined here is not linear and unidirectional, but exists in both directions and continues to be both interdependent and iterative. Studying the entire life history (or at least all of it so far), rather than just the design or use of a technology, enabled me to see these linkages across time.

The third and final lesson is that the disappearance of a technology from mainstream public view is not necessarily the end of that technology's life. Just because a technology is no longer being produced or sold does not mean that it is no longer being used or even, as in the case of the TRS-80, developed further. The disappearance of everyone except the users may, in fact, mean that the technology has been given a new lease of life through the skills, knowledge, interest, and expertise of those people previously considered and constructed to be just the recipients of the technology.

2

Resisting Consumer Technology in Rural America: The Telephone and Electrification
Ronald Kline

In this chapter, I ask how the resistance of users to new technology mattered in the creation of technological change and social transformation in the twentieth century. My material comes from a large study of how consumers, producers, and mediators of technology interacted during the introduction of the telephone, the automobile, radio, and electrification into rural life in the United States from about 1900 to 1960. I argue that what social scientists and reformers called "modernization" was a contested process. Farm men and women were not passive recipients of the transfer of technology from the city to the country; they were active consumers who resisted, modified, and selectively adopted these technologies on an individual basis. The telephone, the automobile, radio, and electrification did not determine social change in a prescribed manner. Instead, farmers contested efforts to urbanize rural life by resisting each technology, then weaving it into existing cultural patterns to create new forms of rural modernity.[1] My study contributes to the growing body of literature on users and technology (e.g., Cowan 1983; Douglas 1987; Marvin 1988; Woolgar 1991; Akrich 1992; Fischer 1992; Cockburn and Ormrod 1993; Kline and Pinch 1996) by focusing on resistance as a common aspect of the process of creating technological and social change.

In contrast, much of the literature on resistance to new technology deals only with social responses. In this vein, scholars have interpreted resistance in a variety of ways, ranging from irrational opposition to progress to a heroic act of defiance against an oppressor. Descriptions of resistance run the gamut from the Luddite revolts in Britain in the early nineteenth century to organized protests against nuclear power and biotechnology, and consumer resistance to information technology, in the late twentieth century (Bauer 1995b). Slaves (Dew 1994), artisans (Randall 1986), peasants (Scott 1985), and other marginal groups resisted being exploited by technology. In rural America, the subject of my research, Midwestern farm laborers broke machines during the economic crisis of the 1870s (Argersinger and Argersinger 1984). Farm men and women, both black and white, resisted the imposition of scientific-farming and domestic-science practices before World War II (Danbom 1979; Ferguson 1998; Walker 1996). The Old Order Amish in Pennsylvania still refuse to own telephones and automobiles (Kraybill 1989).

Other scholars have interpreted resistance to new technology as part of an interactive process of sociotechnical change. Martin Bauer, the editor of a recent book on the topic, gives a functionalist definition of resistance. Resistance "affects socio-technical activity like acute pain affects individual processes: it is a signal that something is going wrong; it reallocates attention and enhances self-awareness; it evaluates ongoing activity; and it alters this activity in various ways to secure a sustainable future" (Bauer 1995a: 3). Rather than seeing resistance as a market's error signal, Foucault (1979: 85; 1980: 162–165) includes it as a constitutive element of the distributive exercise of power in disciplinary techniques in schools, factories, and prisons (Dreyfus and Rabinow 1982: 146–147, 206–207, 211). Akrich and Latour (1992) discuss resistance, without using this term, in their semiotic approach to analyzing mutual relationships between users and artifacts. They include "subscriptions" based on "antiprograms" (resistance to a prescribed interpretation of an artifact) and "re-inscription" (a response to the antiprograms) in the vocabulary they use to de-scribe the "scripts" of artifacts, which result from the interaction between designers and users.

Although my approach is much closer to the latter group of authors than to those who focus on social responses, I take a somewhat different tack. I examine producers, mediators, and consumers symmetrically, rather than privileging the producers' interpretations of success over that of users, as Bauer tends to do. I agree with Foucault that resistance is an element of power relations, but my work shows that resistance is transformative, not ineffectual. I also pay much more attention to radical uses of a technology than does Akrich.[2]

In my view, resistance is a common means of negotiation among producers, mediators, and users that helps to create socio-technical change. I use the term to refer to my interpretation, as an analyst, of actions that contemporaries saw as resistance to technology. I look at who viewed what acts as resistance and why, the different interpretations of a technology that can help explain these views, the responses taken to actions labeled as resistance, and the accompanying changes in technology and social relations.

The extent of these actions in regard to the early history of the telephone, the automobile, radio, and electrification may surprise some readers of the larger study. But I do not claim that resistance dominated the relationships among promoters, mediators, and users. American farmers, after all, did adopt these technologies in large numbers throughout the twentieth century. My point is that paying attention to the many actions viewed as resistance helps explicate the contested aspects of a social transformation, usually termed "modernization," which is often seen as overpowering and foreordained. I call these actions "transformative resistance" because they helped to create technological and social change.

The actors in my story regarded the following sorts of actions as resistance to a new technology:

opposing the introduction of a technology into a community

not purchasing a technology (what is generally called "consumer resistance")

not using a technology in a prescribed manner.

My approach grew out of Kline and Pinch 1996, which extended the original SCOT (social construction of technology) framework by analyzing power relations among social groups and reciprocal relations between the use and design of artifacts.[3] Because the actions which the actors in my larger story viewed as resistance resulted from differing interpretations of a technology, they come under the category of "interpretative flexibility" in the SCOT framework. But I restrict my analysis to those actions in which measures were taken to overcome or respond to a perceived resistance. Resistance can thus be in the eye of the promoter, the mediator, or the user.[4]

I examine these forms of resistance and responses to them for two of the four technologies discussed in the larger study: the telephone and electrification. Promoters initially sold both technologies as urban luxury goods in the late nineteenth century, then marketed them more widely to urban and rural middle-class groups in the twentieth century (Fischer 1992; Nye 1990; Jellison 1993). Because their aim was to modernize and reform rural life by making it more urban, promoters initially imagined (Akrich 1992) and attempted to configure (Woolgar 1991) farm users as urban consumers.

Visiting on the Party Line

For the telephone, I focus on resistance to the use of the device pre-scribed by telephone companies because there seems to have been little rural resistance to acquiring telephone service at the turn of the century. From all accounts, middle-class farmers, especially those in the Midwest, welcomed the telephone and took steps to get the service. When the American Telephone and Telegraph company (AT&T) and independent companies were slow to serve the rural market, farmers built their own lines, bought their own equipment, and organized cooperatives to estab-lish a neighborhood communications system, which they often connected to town through a switchboard in a farm house. When commercial com-panies began to build more "farmer lines," they also built party lines, with from 10 to 20 telephones per line. These efforts were so successful that by 1920 the US Census Bureau reported that a slightly larger percentage of farm households (39 percent) owned a telephone than did non-farm households (34 percent) (Fischer 1987a,b).

Yet there were contested meanings of the new technology that con-temporaries viewed as resistance to the prescribed use of the telephone. Telephone companies, no matter who ran them, usually objected to two practices: playing music on the line and eavesdropping or "listening in." When patrons continued both practices in the face of rules instituted against them, telephone companies viewed the actions as resistance and took steps to overcome it. Although playing music and eavesdropping both tended to tie up the lines and wear down batteries supplied by the telephone companies (the main sources of complaints), I will focus on the more prevalent practice of eavesdropping.[5]

In 1907 a North Dakota newspaper described the party-line culture of eavesdropping, or "rubbering" as it was also called at the time, as follows: "Usually when a country subscriber rings anyone up several of his neigh-bors immediately butt in—not to talk—just listen. . . . Then there are a number of persons gossiping by the way of the telephone, and the busi-ness of T. Roosevelt, even, would have to wait, once they get started, till the matters of the entire community have been wafted over the wires. And occasionally a real talkfest occurs when there isn't much difference in the cyclone of conversation and the flow of soul of a sewing circle."[6]

Following the custom of publicly expressing gender stereotypes in this period, men viewed women as the main gossipers and eavesdroppers.[7] In 1902 an Indiana man wrote in a farm journal about a tenant who had sold "his last family cow to secure a telephone. . . . It robs his family of what they need much more, just because his wife probably delights in gossip and can't stand it to see others chatting over it and talking about their 'fun' without enjoying it herself." A writer in a Kansas newspaper complained in 1911 that "when two old windy sisters on a party line once get astraddle of the wire, nothing short of re-enforced lightning will ever shake 'em loose under an hour." In 1914 the *Literary Digest* published a photograph of a woman sitting at her sewing machine with a telephone receiver tied to her head by a piece of cloth so that she wouldn't miss a word while performing her daily tasks.[8]

Recent scholarship has criticized these gendered stereotypes. In her study of women and the telephone in a rural community in Illinois, the anthropologist Lana Rakow (1992: 33) argues that what looked like gossip to men, in the past and in the present, was work that held together kith and kin. It was "both gendered work—work delegated to women— and gender work—work that confirms the community's beliefs about what are women's natural tendencies and abilities."

There is some evidence that men enjoyed rural party lines as much as women. "All day long there is the chance of friendly gossip that is dear to the hearts of all women, and of many men as well if the truth were confessed," a journal admitted in 1905. "The phone company is going to take out the phone of a man who lives on a party line," a Kansas newspaper reported in 1911. "He is said to butt into all conversations on the line and to make things so unpleasant that if his phone isn't removed, the other patrons will have theirs taken out."[9]

Another reason telephone companies disliked extensive eavesdropping was that it tied up the lines and wore out batteries, which companies had to replace. Telephone companies tried all sorts of measures to stop eavesdropping on the farm telephone. These included passing rules against the practice, fining eavesdroppers, giving priority to business use, and limiting conversations to 5 minutes.[10] Several states (including Ohio and Indiana) passed laws making it a crime to repeat the contents of a telephone conversation. The journal *Telephony* printed poems, cartoons, and newspaper reports describing the neighborhood strife that could, and did, result when people "listened in."[11] In 1910, *Wallaces Farmer* reminded its readers: "There is an old maxim that eavesdroppers seldom hear anything good of themselves."[12]

But many farm people viewed eavesdropping in a favorable light, as a way to transplant the rural custom of "visiting" onto the new technology of the party line. When caught eavesdropping by her boarder in 1907, a farm woman replied: "We all listen. Why shouldn't you listen? I heard four receivers go up just when I stopped talking. There's lots of people [who] wanted to know about that [neighbor's] chimney [that caught fire]. It's all right of course." A telephone company man could not understand the custom, telling a convention in 1909: "It may be a strange code of etiquette that would actually defend eavesdropping, but they defended eavesdropping in defending their beloved telephone. This explains why so many farmers can be found who, at first thought, say they do not want a strictly private service." A technical writer observed in 1914 that farm women "meet and talk in company on the rural lines in a way which should be regarded as perfectly legitimate."[13]

Recent interviews of elderly farm people reinforce the view that listening in was socially acceptable in many communities before World War II. A few women interviewed in the 1980s thought it was not polite to listen, as did several women and men interviewed in New York state in the 1990s. In New York, Eva Watson hung up if she heard a click, George Woods could tell when one woman was on the line because he recognized the pattern of her breathing, and Lina Rossbach recalled that she and her husband would speak German when they heard someone on the line in the mid 1940s. This infuriated one listener, who blurted out "You cheat, speak English!"[14]

Yet many women thought listening in was a friendly habit. Helen Musselman of Indiana recalled: "It wasn't really nosiness; it was just neighborliness. I know I missed it when we had our new phones put in." Another Indiana woman, Pearl Snider, said: "We had a party telephone— probably thirteen or fourteen on one line. When the bell would ring, why if you wanted to listen, you could run in and hear the conversation of the neighborhood." Opal Cypert from Arkansas recalled: "When you'd get a ring, why everybody would take the receiver down and they'd listen. They was welcomed in on the conversation then, if they wanted to." (Arnold 1984: 150, 152) Edna Dagnen from Washington State remembered: "A lot of times when you were in a conversation, somebody would come on the line and say, 'Is that you Mabel? Do you know your cows are out?' Or, 'Are you going to be home?' or something like that. Pretty soon you'd have three parties on the line and sometimes four." (Arnold 1985: 188–189)

Redefining the Rural Telephone
Telephone companies responded to their customers' refusal to give up the habit of eavesdropping, which they considered to be resisting the prescribed use of the telephone, by going beyond rule making to create new techniques and hardware. In this case, producers responded to users'

resistance and adapted the new technology to fit the social patterns of rural life.

Some independent companies attacked the problem of eavesdropping by using the technique of battery testing. One telephone man suggested adding an extra charge to phone bills if batteries ran down after 10 months. Another told a convention in 1913 that he tested local batteries twice a year and informed customers that the company could calculate from the remaining battery life, and the number of calls, how much time they had spent eavesdropping. He suggested another means of "disciplining" customers by telling them that eavesdropping reduced the volume of a receiver. Afterwards, he overheard customers saying: "'We can't hear the Browns, they listen too much,' etc. Every subscriber is making a record for himself. Then when you have an occasion to speak to them about it, they don't hop up, get angry and say 'Why, we never listen.' They are aware you know what you're talking about."[15]

An Oregon newspaper reported in 1906 that the local telephone company was going to install a device that would let operators know whose receivers were off the hook when they were not being called. But the device did not stop eavesdropping. A Nebraska man announced a more elaborate invention in 1914. The device "sounds a warning when a third party breaks in on the wire, and also identifies the culprit to both the legitimate users of the telephone." The inventor supposedly got the idea from a farm woman who said it would make him a fortune. But a technical expert predicted that farm women would not adopt an invention that broke up their visiting on party lines.[16]

Many telephone companies wanted to do more than rely on customers to stop eavesdropping when told about it; they wanted a piece of hardware that would prevent eavesdropping. One way was to use selective signaling, a system than rang the desired telephone and not others on the party line. In 1896, an engineering manager of the Chicago (Bell) Telephone Company invented a four-party line with currents of different polarity. It became universal in the Bell System by 1905, especially when modified to work with common-battery switchboards (Miller 1905, chapter 24). The polarized ringing circuit replaced a more complex lockout design and a troublesome and little-used harmonic system, both of which were difficult to maintain (Fagen 1975: 121–123).

Telephone companies placed a few of these systems on rural lines before World War I. In 1903, Chicago Bell offered a selective signal plan (only two telephones would ring at the same time) on its eight-party farmer lines for the high price of $18 a year. Stromberg-Carlson sold a step-by-step mechanism for twenty-party service by 1905. In that year an Indiana newspaper reported that a local man "who invented a device to detect and cut off eavesdropping on a telephone line" had "the thanks of all legitimate users of the telephone." A large independent company installed a "full selective or secret lockout service" for 300 rural telephones in 1909. The system supposedly kept boys from monopolizing the line by playing music. Only one user preferred the old system, but the firm claimed that her eavesdropping "had been largely responsible for the adoption of the more modern secret service." A few lockout systems were installed before 1914.[17]

In a striking example of the social construction of technology, some Bell Company engineers decided early in the century not to fight party-line practices, but to modify farm telephones so they would work under these social conditions. In late 1903, P. L. Spalding of the Pennsylvania Bell Company wrote Joseph Davis, chief engineer of AT&T, that replacing the standard #13 induction coil with a #10 coil improved the volume on farmer lines, where "there is likely to be a great deal of listening in on the circuits by a third party." But since the #10 coil "was not designed to mount in the subscriber's set," was there a more suitable coil? Davis could "not recommend its use except in cases where it is found absolutely necessary to do so."[18] In 1905, C. E. Paxson of the Chesapeake Bell Company wrote AT&T's new chief engineer, Hammond Hayes, that "when we ring on a [farmer] line, a sufficient number of subscribers along the line take the receiver off the hook to prevent us from ringing again." Would it be feasible to solve this problem by placing a condenser in series with the receiver? Hayes noted that several Bell companies were doing this, but he did "not recommend the use of condensers for this purpose unless absolutely necessary" because of the added expense and the slight drop in transmission volume.[19] Even though AT&T resisted making these changes, the examples show how the local practices of farmers resisting the prescribed use of the telephone could lead engineers to redesign rural telephony in the early part of the century.

Rural Electrification

Resistance to rural electrification was more widespread than resistance to the telephone. Here I discuss opposing the introduction of electricity into a community and consumer resistance. Although there was a large demand for electricity in the countryside and not all farm people opposed electrification, enough did for promoters to take several measures to overcome resistance to it. I discuss these actions in the context in which most farm people decided whether or not to electrify the farm—through the appeals of the Rural Electrification Administration (REA), which was founded in 1935. One of the most successful government programs of the New Deal, the REA made low-interest loans to cooperatives, which bought electricity from private companies and distributed it to farms. Although the REA praised the co-ops as a public-power, grassroots movement, the agency designed the power system, organized the co-ops, micromanaged them, and pushed the widespread use of electricity so that the co-ops would pay off their loans in time (Brown 1980; Person 1950).

One of the first tasks of an REA cooperative, securing rights of way for its power lines, led to much resistance. The REA had to obtain these rights from landowners because its co-ops could not condemn property for this purpose, as could public utility companies. Farmers expected the government to pay them large sums for easements, lines crossed farms in ways that renewed family quarrels, and poles interfered with field work. When issues could not be resolved peacefully, some farm people took matters into their own hands, as an earlier generation had done against the "devil wagon" automobile at the turn of the century (Kline and Pinch 1996). An Illinois woman guarded a freshly dug hole with a shotgun to prevent the REA from completing the pole-setting job (Severson 1962: 143), Iowans tried forcibly to prevent REA crews from erecting poles along a highway (Severson 1972: 136), and a Minnesota man said he would chop down poles if they were put on his land (Tweton 1988: 143). A farm couple carried out the threat in Monroe County, Wisconsin. When the co-op put up poles on their land without obtaining an easement, the husband chopped them down, parked the family car on top of the (unelectrified) fallen wires, then brought food to his armed wife stationed inside the car to keep the REA at bay with a shotgun, while he finished the plowing (Richardson 1961: 46–47).

Most protests were more peaceful and consisted of simply not giving a right of way. In 1938, the Administrator of the REA complained: "Everybody says he wants electricity, but when it comes to locating the lines and locating the poles, many people either refuse to hand out essentials, thereby denying their neighbors electricity, or make it so difficult and so costly that certain lines cannot be built at all."[20]

Often husbands and wives disagreed about electrification based on a flexible interpretation of gender roles. An Indiana woman was heard to exclaim at an REA meeting: "Daddy had better sign up [for electricity] or things are going to get hot around home. . . . This is one time when he is going to make up his mind in a hurry."[21] In contrast, an Iowa woman got upset with her husband for granting an easement after holding out for 6 months, even though the poles would interfere with his plowing. An organizer recalled that the man "stamped his foot and shouted 'Now, mama, it's not you who is going to farm around those REA poles it's me'" (Severson 1964: 25). In a Texas household with a different, gendered division of labor, a male co-op director did not want poles placed in his fields because it would interfere with his wife's plowing (Chessher 1964: 27).

When REA canvassers went from farm to farm in the mid 1930s to sign up members for the co-op and get them to wire their houses and agree to buy home appliances, they encountered a good deal of "consumer resistance." Although the net income of farm operators had recovered from the depths of the depression by then, limited cash incomes and a reluctance to go into debt were major reasons why large numbers of farm people chose not to electrify the farmstead. Yet refusing to join the co-op was not limited to income. Middle class farmers in Illinois refused to sign when a co-op required that they install two outlets in the kitchen and one in each room. Putting electric lights in the bedroom seemed like a luxury when you could light the way to bed with a kerosene lamp. Many people feared losing their farms if the co-op folded, some had been burnt in disreputable co-op schemes in the 1920s, and most farmers refused to sign up for their tenants (Severson 1965: 71–71, 84, 188, 220, 248; Severson 1964: 17, 24, 26–27, 32). In 1941, some prosperous farmers in western Kansas were waiting until their home generating plants wore out before joining the co-op.[22]

There were concerns other than economics. An REA staff member reported: "Many projects do not serve 30 to 50 percent of the people along their lines merely because the people do not feel that electricity is worthwhile to them even though they can often well afford it."[23] Others (especially McKinley Republicans in the Midwest) distrusted a New Deal agency, thought the simple scheme of paying a co-op membership fee to get electricity was too easy, or resisted giving up older technologies. Some would not sign a membership application for religious reasons. More than a few farm men and women were afraid of a mysterious force they could not see (Severson 1965: 143; Severson 1972: 106, 114, 130; Severson 1962: 31; Sørenson 1944: 268; Chessher 1964: 25). Some thought power lines would attract lightning, others worried that lines would fall across barbed-wire fences and electrocute their cattle (Caro 1982: 525). The fears were not groundless. Accidental electrocutions of linemen, farmers, and cattle were common enough for the REA to embark on a major safety program in 1939.[24]

Mediators of technology also resisted electrification. The archives of the REA are full of field reports complaining about boards of directors and superintendents of co-ops who resisted implementing the agency's full-scale sales efforts, such as hiring home economists to push appliances as load builders. A field representative observed in 1937: "When board members are encountered so frequently who don't have their own premises wired or who think electric ranges are foolish luxuries, it seems that something should be done" to enforce bylaws that all members buy electricity from the co-op.[25] Many local appliance dealers resisted joining forces with the REA because they resented the co-op's aggressive sales efforts or thought most farmers could not afford electrical appliances.[26]

A report about an uncooperative board of directors illustrates this form of resistance. Reporting on a Pennsylvania co-op's annual meeting in early 1941, REA field representative Elva Bohannan recommended to the meeting that the co-op adopt sales and education programs, and elect some fresh faces to the board, including women. Then "Mr. Proctor, the chairman and President of the Board leaped to his feet—screaming 'These accusations made by Mrs. Bohannan call for an answer.' He grabbed [the] book *Rural America Lights Up* [and] screamed—'The only reason Washington wants to dictate to us about who shall stay on the Board of Directors is—This book was sent to us to sell (waved book frantically in the air) every member a copy of this thing. We refused!' . . . 'Washington cannot dictate to us because we are men of our own opinions and none can change us.' He claimed that Washington was not the friend of the members of the Cooperative because they now had on hand [enough money to make an advance payment, but Washington] 'wouldn't take it. They want your interest payments.' . . . He then screamed—'Do you know what Washington Demanded we do—? Why, stick a one-can milk cooler down your throats. But did we refuse—? I'll say we did.'"[27]

This report indicates some of the rhetorical resources used in the local battles to control the REA program. Bohannan employed the rhetoric of rural uplift to complain that the co-op's board was under the domination of a president who tried to thwart the REA's "progressive" measures of co-op education (e.g., by selling the book), democratic elections, and agricultural modernization (e.g., by selling the one-can milk cooler). President Proctor drew on gendered Populist rhetoric to resist the inroads of Washington and uphold an independent rural way of life.

Co-op members throughout the country further resisted the appeals of the REA by not purchasing a full ensemble of electrical appliances.[28] In this manner, farm men and women had something to say about what an electrical modernity would look like in the countryside. An indication of what appliances farmers bought is given by surveys of newly electrified farms on REA lines, which show a remarkable consistency in which appliances were first purchased in these years (table 2.1). Because the order is not based simply on purchase price or operating cost, the data give an idea of which appliances were valued on the middle-class farm.

Despite increased farm income in this period and the best efforts of REA agents, new co-op members bought mainly radios and irons. Only about one-half bought washing machines and refrigerators. All four items were familiar technologies that prosperous farm people had run on either batteries (for radios) or petroleum products (for irons, washing machines, and refrigerators) before they had electricity. The house was

Table 2.1
REA-serviced farms reporting ownership of selected appliances, 1938–1941 (percentages). The number of projects in each survey varied from 46 (1938) to 123 (1939), the average length of service from 8.4 months (1938) to 19.3 months (1941), the number of customers responding from 17,100 (1938) to 70,893 (1941), and the percentage responding from 63.5 percent (1938) to 68.8 percent (1939). Sources: Rural Electrification News, July 1938: 4–10; January 1940: 68; October 1940: 10–11; J. Stewart Wilson to Robert Craig et al., August 8, 1941, REAA, Entry 17, Box 4.

	1938	1939	1940	1941
Radio	86	82	88	90
Iron	81	84	84	85
Washing machine	47	59	55	55
Refrigerator	26	32	33	42
Toaster	24	31	29	32
Water pump	17	19	15	18
Vacuum cleaner	16	21	21	21
Hot plate	12	19	15	15
Motors of less than 1 horsepower	9	18	15	5
Coffee maker	6	6	8	9
Range	5	3	4	4
Cream separator	5	14	8	8
Milking machine	2	4	2	3
Chicken brooder	1	3	4	7
Water closet	—	6	6	6

electrified long before most agricultural operations were. Other items that were more popular in the city, such as vacuum cleaners and coffee makers, were often seen as "foolish luxuries."[29] The electric range found little favor on the farm or in the city.

Buying habits during the more prosperous years immediately following World War II show similar patterns, except that younger couples tended to buy more electrical appliances and liquefied petroleum gas became common in the Midwest.[30] Interviews of elderly farm women in Southern Illinois found that several women "added indoor bathrooms [in the 1960s] after their husbands died. Many men (but not, apparently women) resisted bringing toilets into the house; it violated their notions of sanitation." (Adams 1994: 210)

Responses to Resistance

Promoters responded in a variety of ways to what they viewed as resistance to electrification, primarily by creating new organizations and techniques. The REA hired home economists to demonstrate electrical appliances to farm women at kitchen parties, established a circus-like Farm Equipment Tour to entice farm men and women to buy electrical appliances, and produced the promotional movie *Power and the Land* (Kline 1997a,b).

The high installation and operating costs of the electric range seem to account for its unpopularity. But in the 1930s and well into the 1960s, most farm women were extremely reluctant to give up their coal or wood stoves because they served a variety of purposes. Women used them to cook and bake food, to heat their (usually large) kitchens in the winter, especially in the North, and to heat water for washing dishes, doing the laundry, and bathing. Even when women bought a kerosene or gasoline stove, they did so primarily for canning food in the summer and increased cooking during harvest time. They kept their coal or wood stove for the colder months (Rapp 1930; Brackett and Lewis 1934). Another disadvantage of the electric range was the unreliable delivery of electricity—outages were frequent.

An innovative response by appliance manufacturers was to sell a combination coal or wood stove and electric range. A 1930 study by Purdue University found that five of 41 farms with electricity had purchased a combination coal-electric range. The women told the researchers that they were satisfied with the hybrid device, with which they cooked mostly by electricity and heated the kitchen and hot water by coal (Rapp 1930: 12–13). In recent interviews, three Indiana women said they owned combination stoves (wood and gas, coal and gas, and wood and electric) (Arnold 1984: 18), as did two men and one woman in New York (two wood and electric, one wood, coal and electric). The New York woman, Thena Whitehead, recalled that she and her husband bought a combination wood-coal-electric stove when they got electricity in the mid 1940s for heating purposes, but also because she thought gas stoves were too dangerous.[31]

These combination technologies help us parse what modernity meant to farm people. There is nothing essentially "modern" about using wood, coal, gasoline, kerosene, or electricity as a fuel, because all of them were, at one time, considered to be the latest, up-to-date fuel, that is, they were socially inscribed as "modern." As the coal or wood stove began to seem old-fashioned in the 1920s and the 1930s, kerosene, gas, and electric stoves looked more "modern" to both urban and rural eyes because they were more compact, had knobs, seemed cleaner, and usually were made of enamel and steel. When looking at photographs of these devices today, it is difficult to distinguish the combination coal-wood-electric stove from these newer stoves, but not from the coal or wood stove. The combination stove looked similar to the newer ones because of their smooth-line cabinets and the fact that cooking and baking duties could be relegated to the smaller electrical part of the stove.

The hybrid stove is a good example of farm people weaving an urban-style technology into rural life. In this case, the material culture (and economic conditions) on the farm in this period encouraged rural folk to heat the kitchen and hot water with the same source used for cooking and baking. One way to bring electricity into this part of rural culture—without upsetting the entire farmstead—was to combine the new with the old technology (an almost literal weaving together of so-called urban and rural technologies).

The invention of the combination stove shows the interactive aspect of the contested urbanization of the family farm. In response to widespread resistance to the electric range, entrepreneurs developed and sold combination stoves. The result was a change in technology (new forms of stoves) and changes in rural life (new summer cooking patterns and the addition of a symbolically "modern" device in the kitchen). Yet the changes occurred within a familiar structure. Winter heating habits and the size and design of the farm house remained the same. Thus farm people's consumer resistance contributed to the shaping of a technological system and the creation of a new form of rural modernity—as had occurred with the telephone, the automobile, and radio.

Conclusion

In my account of the early history of the telephone and rural electrification, I have tried to consider what was being resisted, by whom, in what manner, and from whose perspective. I argued that the interactions among promoters, mediators, and farm people, involving what some groups interpreted as resistance, created new technologies and social relations. Some Bell Telephone Companies responded to farm people not using the telephone in a prescribed manner by altering the system to permit extended periods of listening in, which allowed farm people to transplant the rural custom of visiting onto the party line. The REA responded to farm people not signing up for electricity, and not purchasing a full complement of electrical appliances, as well as to cooperatives not aggressively selling appliances, by creating new social technologies. Manufacturers created new artifacts like the coal and electric combination range, which farm people wove into an altered fabric of rural life.

In general, I found less evidence of resistance to the telephone than to rural electrification. One reason is that REA's records of interactions between mediators and users are much more extensive than those of AT&T and other telephone companies. It also seems that farm people thought they had much more control over the telephone at the turn of the century than rural electrification in the 1930s. Because telephone companies initially took little interest in building farmer lines, farm families cooperated to build and operate their own telephone systems. Some prosperous farmers took a similar route with rural electrification before 1930 by hooking up to a nearby utility company line, forming an electric cooperative, or making their own electricity with a home generating plant (Lee 1989). But when these efforts failed to reach more than 10 percent of farm houses by 1930, New Dealers set up an aggressive REA program which farmers resisted in a variety of ways. REA staffers explained the poor attendance at annual meetings, for example, by saying farm people complained that the REA co-op was not a local organization, but a government program emanating from Washington.[32] We have seen how co-op leaders drew on this anti-Washington sentiment to defend their resistance to REA's programs.

These sorts of cultural resources help us understand the patterns of resistance to both the telephone and electrification. Throughout the twentieth century, middle-class farm people made decisions about whether to accept new technologies and how to use them in the context of such enduring, yet historically contingent, rural practices as making do with the materials at hand, sharing work, visiting, avoiding debt, distrusting urban culture, and defending property rights—all of which were constructed within a gendered system of social relations (Neth 1995). When these practices conflicted with values of an urban-based consumer culture, it is little wonder that promoters (as well as farm people) viewed many actions as a form of resistance. Farmers chopped down power lines to defend their version of property rights. Many farm people listened in and talked with their neighbors on telephone lines as a form of visiting. Farmers refused to go into debt to buy appliances until older technologies wore out. In this respect, middle-class farm men and women acted like workers in interwar Chicago, studied by Lizabeth Cohen, who interpreted technologies like the radio, phonograph, and movies in terms of their ethnic, neighborhood cultures, rather than in terms of mass consumer culture (Cohen 1990).

The agricultural prosperity and the depopulation of the countryside following World War II made the REA's job easier in many respects as it finished the job of electrifying the countryside by 1960. But most middle-class farm families did not become a father-knows-best suburban family depicted by the US Department of Agriculture on the cover of a 1965 yearbook titled *Consumers All* (US Department of Agriculture 1966). The outward signs of farm life were a pickup truck in the driveway, a long propane tank nestling against a clapboard farmhouse, and farm machinery parked everywhere. The material culture in the farm house resembled that in urban and suburban homes, but work patterns and leisure activities were still those of farm men and women. Although postwar farmers demanded electrical service, the REA still vigorously promoted appliances in order to overcome the consumer resistance of a supposedly consumerist era. Farmers bought television sets long before installing bathrooms, for example, and used automatic washing machines to free up time for farm women to help plow the fields (Jellison 1993, chapter 6).

Like their grandparents and parents before them, who resisted the telephone and electrification before World War II, middle-class farm couples in the supposedly consumerist 1950s resisted and used consumer technologies selectively. This transformative resistance helped to create sociotechnical change in the form of rural versions of modernity.

3

Non-Users Also Matter: The Construction of Users and Non-Users of the Internet

Sally Wyatt

On Not Driving a Car

I have never owned a car. I am, however, very well qualified, having passed two driving tests. The first took place during a snowstorm in Toronto a few months before my seventeenth birthday. I think I passed only because I successfully navigated the course without skidding. I needed two attempts when I was 25 to pass the British test, in a more complex machine with a manual transmission. I have never driven much. My parents sold their car soon after I passed my test in Toronto. I have driven only twice since I passed the British test in 1984.

I simultaneously inhabit the same world as car drivers and a different one. My life is affected by cars: as a pedestrian and a cyclist, I see them as a threat to my health and well-being, and as a user of public transport I find that they slow me down. The reach of public transportation constrains where I can live and where I can visit. My knowledge of London is very much based on public transport routes convenient to where I lived and worked. The London underground map is a better and more useful representation of my experience than a topographical one. In 1999, when I was driven from London to Amsterdam. I was fascinated, to the amusement of my driver friend, by the alien world of motorways, petrol stations, motorway services, and drive-on ferries.

Not driving is regarded by many as a deviant and bizarre choice. One of my friends thinks it reflects a failure to grow up on my part, that "real adults drive cars."[1] There are, of course, advantages to not driving—it saves money, time, and stress; it reduces one's chances of being killed or killing others; and, in these post-Rio, post-Montreal days of greater environmental awareness, it provides a tremendous feeling of self-righteousness. Occasionally, policy makers accept that non-car-users also have rights and interests which need to be taken into account in the formulation of transport, industrial, and land-use policies. Non-car-users are increasingly rejecting the land claims made by the producers and users of cars who want to turn ever more land over to roads and parking lots. Since 1995 and the growth of protests against cars and against road building around the world, militant non-drivers are becoming more active in asserting the desirability of car-free spaces (Reclaim the Streets! 2001).[2]

I have presented this autobiographical information in order to make two analytic points regarding non-use of technology.[3] First, the existence of individuals who choose not to own a car even though they could afford to do so, raises questions about the explanations put forward for non-use. Second, and related to the first point, voluntary rejection of a technology raises the question of whether non-use of technology always and necessarily involves inequality and deprivation. In other words, is the policy assumption that all non-users of a particular technology wish to become users appropriate?

The assumption that non-use or lack of access is a deficiency to be remedied underlies much policy discussion about the Internet. For example, the meeting of European Union heads of government held in Lisbon in March 2000 was dubbed the "dot-com summit," reflecting the realization by heads of state of the importance of information and communication technology generally and the Internet in particular for the economic well-being of Europe. The aim of the summit was to discuss how to increase employment by promoting enterprise, competition, and a dynamic, knowledge-based economy. To this end, the leaders agreed to reduce the cost of Internet access to US levels within 3 years, to connect all schools to the Internet by 2001, and to train teachers in its use (Tisdall 2000). Everyone is clearly understood as a potential user of the Internet. Access to the technology is seen as necessarily desirable, and increasing access is the policy challenge to be met in order to realize the economic potential of the technology. Concern about the social inequalities that may arise if the 'digital divide' is allowed to grow was expressed by some heads of state. While such concern about social exclusion is laudable, it is nonetheless based on the assumption that Internet "haves" will be in a better socio-economic position than Internet "have-nots." As Neice (2002, 67) argues, "it is simply presumed by those advocating the elimination of the "digital divide" that having Internet access is always better than lacking it." The reasons why private companies selling Internet-related products and services wish to promote the Internet as a universal medium are self-evident; the reasons why policy makers support them are less clear.

The March 2000 European summit is only one of many examples of politicians' and policy makers' assuming that access is the problem. From that perspective, making Internet access cheaper and providing more education and training are among the obvious solutions. It is assumed that once these barriers to use are overcome, people will embrace the technology wholeheartedly. Access to the Internet is seen as necessarily good, and more as necessarily better (though, as with champagne and chocolate, moderation is advised). Politicians hope that people will use their knowledge of and from the Internet to create wealth and employment and to become active citizens, but maybe they will use it to look at pornography or play games. The most popular online activities in 2001 were e-mail and finding information about hobbies, travel, music, books, films, news, and health (Pew Internet Project 2002). Perhaps some people will not use the Internet at all, and perhaps a lack of Internet access does not have to be a source of inequality and disadvantage.[4]

The contributors to this volume are, quite rightly, keen to emphasize the importance of users in technological development. As Oudshoorn and Pinch argue in the introduction, users are not simply passive recipients of technology; they are active and important actors in shaping and negotiating meanings of technology, which is significant both for understanding design processes and the relationship between the identities of technologies and their users. Users have been neglected for too long.[5] Including them helps to overcome the problems associated with approaches to science and technology studies and innovation studies that emphasize the roles of powerful actors such as scientists, engineers, politicians, and financiers in producing technologies. But restoring the dialectic between production and consumption by establishing the importance of use and users may introduce another problem: by focusing on use, we implicitly accept the promises of technology and the capitalist relations of its production. Users are increasingly introduced into technology studies to counterbalance the emphasis on producers found in much of the literature, but all categories involve exclusions. Therefore, users should be seen in relation to another, even less visible group, that of non-users.[6]

To what extent is not driving a car analogous to not using the Internet? "Superhighway" metaphors specifically and travel metaphors generally have played an important role in discussions of the Internet. This chapter is premised on the idea that there is something to be gained from exploring the limits of such metaphors through an examination of those who choose not to travel down particular technological roads. All metaphors and analogies have limits. Perhaps it would also be valuable to pursue the similarities between the telephone and the Internet, insofar as both are communication media. Nonetheless, I shall deploy the car analogy in this chapter because it is a better illustration of the use/non-use dichotomy. Also, both cars and the Internet have been loaded with cultural significance. The car was a paradigm case of a symbol of modernity in the twentieth century. To many people, cars reflect wealth, power, virility, and freedom. The Internet promises many of the same attributes on an even larger scale, with its possibility of global reach. The symbolic value of having Internet access is often presented as a sign of inclusion in a high-technology future.

Users of the Internet

The dramatic increase in the number of Internet hosts since the development of the World Wide Web tempts many commentators to conclude that this rate of growth will continue, or even accelerate.[7] It is assumed the Internet is following a path taken by many other successful technologies before it. Economists refer to this path as "trickle down," meaning the process whereby technologies which are initially expensive to use become cheaper, simultaneously providing more people with the benefits of the technology and enlarging the market. In the case of the Internet, the early users were a small number of academics who used computers paid for largely from university budgets or defense contracts. Academics are now in the minority as firms, governments, administrative bodies, political parties, voluntary groups, and individuals at home all use the Internet for a huge variety of applications and purposes.

According to the trickle-down view, there may be inequalities of access and use during the early stages of a technology, but these disappear, or are at least much reduced, as the technology becomes more widely diffused. Internet enthusiasts often claim that connection is a global process, albeit an uneven one. This is not unique to the Internet. Similar claims can be found in much literature and in policy statements about industrialization and modernization more generally. Individuals, regions, and nations will "catch up"; those not connected now will be or should be connected soon. This is the real annihilation of space by time: the assumption that the entire world shares a single time line of development, with some groups ahead of others but with everyone on the same path.

The evidence for the catching-up assumption, and ultimately a more even distribution of access, is furnished—at least within so-called advanced industrial societies—by time series of statistics relating to own-

ership of consumer goods such as motor vehicles, televisions, and refrigerators, all of which were once owned by very small percentages of households but are now much more widely diffused. Globally, however, the
catching-up effect is less obvious, and measurement of ownership of consumer goods per se says nothing about inequalities in the type and quality of goods possessed. Moreover, the economic prosperity enjoyed in
OECD countries during the 1960s and the early 1970s is not likely to be
replicated.

Collecting and interpreting data about Internet use is not straightforward. Defining a host, ascertaining its location, and identifying users and
their demographic characteristics are all fraught with difficulty. Jordan
(2001) demonstrates how estimates of the size and growth of the Internet
are often motivated by commercial needs and are not well informed by
reliable sampling methods. Although I am critical of the data, some are
presented below both because such data have had practical significance
in policy discussions and because, despite the limitations, some patterns
can be discerned.

Graphs with numbers of hosts, connections, or users along the vertical
axis and with time along the horizontal axis often illustrate news reports
and policy documents about the growth of the Internet which forecast
exponential growth, even though the rate of growth has been quite stable over a long period. Despite the growth, differences between countries
remain stark. In mid 1998, industrialized countries—with less than 15
percent of the world's population—accounted for more than 88 percent
of Internet users. The United States, with less than 5 percent of the
world's people, had more than one-fourth of the world's Internet users
(UNDP 1999). This picture has worsened slightly, according to 2002 data
available from Nua.[8] Even within the European Union there is a divide.
In the Scandinavian countries and the Netherlands more than one-half
of households have home Internet access, whereas in Spain and France
less than one-fourth of households do (Altevie 2001). Global and
regional digital divides remain.

The stereotypical user remains a young, white, university-educated
man. However, closer examination of the available data indicates some
weakening of this stereotype, at least in the United States. Gender differences have shown the most dramatic reduction since the development of
the World Wide Web. Georgia Technical University has been conducting
online surveys of Internet users approximately every 6 months since
January 1994.[9] In the first survey, only 5 percent of users were women. By
October 1998 (the last such survey placed in the public realm), women

represented just over one-third of users worldwide. The Pew Internet Project (2001) claims that indeed half of US Internet users are women. In Europe, over 40 percent of Internet users in Sweden and the United Kingdom are women, but in most other countries the percentage of women users is well under 40 (Nua 2002b). There are differences in patterns of use, men spending more time on line and logging on more frequently (CyberAtlas 2002a)

The size of the divide may vary between countries, reflecting national traditions of difference and exclusion, but social divisions in Internet access continue to exist. As the most advanced user of the Internet, the United States may offer some interesting lessons for the rest of the world. Manuel Castells (2001) uses US data as a paradigm case for the rest of the world. Castells examines the ways in which differences between social groups have changed and concludes that differences are narrowing, especially gender differences. Castells is optimistic about the disappearance of all digital divides, not only that based on gender but also those based on ethnicity, income, and education. But the sources he cites are less sanguine. The US Census Bureau conducted large-scale surveys (of approximately 48,000 households) on behalf of the National Telecommunications and Information Administration in 1995, 1998, 1999, and 2000. The analysis of these surveys highlights what the authors call a "persisting digital divide." They note substantial increases in Internet access, but then they state: "Nonetheless, a digital divide remains or has expanded slightly in some cases, even while Internet access and computer ownership are rising rapidly for almost all groups. For example, the August 2000 data show that noticeable divides still exist between those with different levels of income and education, different racial and ethnic groups, old and young, single and dual-parent families, and those with and without disabilities." (NTIA 2000, summary)

Non-Users of the Internet

Will the cyberworld come to dominate the physical world to anything like the same extent as cars and the associated socio-technical system? Is it possible to turn off the machine? Or will everyone's choices come to be shaped by the Internet, just as many people's transport choices are influenced by the automobile whether or not they own one? The shift of retail outlets from town centers to out-of-town shopping centers makes life more difficult for non-drivers. Similarly, will the disappearance of offline information sources limit people's ability to participate in public life?

The surveys referred to in the preceding section are all concerned to demonstrate growth, and of course growth has been impressive according to all available indicators, including the numbers of hosts, domain names, and users. Nearly all of the academic and policy literature focuses on how to increase the number of users, and takes the additional step of assuming that once a user an individual will always be a user. For example, Hoffman and Novak (1998: 9) write: "Ensure access and use will follow [and] access translates into usage." Moreover, Hoffman and Novak conclude that "programs that encourage home computer ownership . . . and the adoption of inexpensive devices that enable Internet access over the television should be aggressively pursued, especially for African Americans" (ibid.).

I shall leave aside for the moment the question of indirect use of the Internet (e.g., people making a query in a shop or agency where an employee uses the Internet to provide the information needed). Despite the huge global inequalities in Internet access referred to in the previous section, I shall focus here on the evidence for non-use in some highly industrialized countries where non-use could be a choice. There are some data available which suggest that providing access may not be the sure, simple solution it appears.[10] Cyber Dialogue, an Internet research consultancy based in the United States, has found evidence of a slowdown in Internet growth (Cyber Dialogue 2000). They claim that the rate of growth is slowing down overall and that there is evidence of an absolute decline in the number of users aged 18 to 29. In part, they attribute non-use to cost: some people cannot afford a computer and Internet access. They also claim that approximately one-third of all US adults simply do not believe they need the Internet and what it offers.[11] Even more significant is the growth in the number of adults who have tried the Internet and then stopped using it, only one-third of whom expected they might use it again at some point in the future. In early 1997, Cyber Dialogue estimated there were 9.4 million former users; by September 1999, they calculate that there are as many as 27.7 million former users. In 2001, the Pew Internet Project found that half of all the adults in the United States do not have Internet access and 57 percent of those non-users are not interested in getting it (Pew Internet Project 2001). A survey conducted in the United Kingdom in 2000 found that one-third of British adults has no intention of ever using the Internet (Ward 2000).

Based on two national random telephone surveys conducted in the United States, James Katz and Philip Aspden (1998) suggest there are patterns to Internet non-use. Their analysis of "Internet dropouts" was a side

effect of some research about barriers to Internet use in the United
States. They candidly admit they included the category of "former user"
in their surveys only for logical completeness. They were surprised to dis-
cover in October 1995 that former users and current users each
accounted for about 8 percent of the sample. They did another survey in
November 1996, by which time the proportion of current users had more
than doubled to 19 percent of the sample. The proportion of former
users had also increased, but only to 11 percent. People who stop using
the Internet are poorer and less well-educated. People who are introduced to the Internet by family and friends are more likely to "drop out"
than those who are self-taught or those who receive formal training at
work or school. Teenagers are more likely to give up than people over 20.
The reasons for "dropping out" vary by age. Older people are more likely
to complain about costs and difficulties of usage; younger people are
more likely to quit because of loss of access or lack of interest. Katz and
Aspden (1998) draw attention to this important category of non-users,
though their explanation for non-use remains very functionalist, focusing
on issues of cost and access. Their use of the term "drop out" is rather
pejorative, suggesting again that use is to be preferred to non-use. As
Ronald Kline argues in this volume, viewing resistance to technology
from a functionalist perspective reinforces the promoters' interpretations
of success. In his analysis of resistance to the telephone and electrification
in the United States at the beginning of the twentieth century, Kline
defines resistance in the terms of contemporaneous actors and demonstrates how resistance itself can induce socio-technical change.

The Cyber Dialogue data and the results of Katz and Aspden need to
be treated with caution as former users can, of course, become active
users again at a later date. Nonetheless, they are interesting because they
call into question the assumption of never-ending growth. They also suggest that public access provision, quality of information and training
remain important policy issues. If the results about teenagers are replicated elsewhere on a large scale, certain assumptions about the rate of
exponential growth have to be re-examined. Turkle (1995) draws on
Erikson's theories of adolescent identity development to explain some of
her observations of the behavior of young people in multi-user domains
(MUDs).[12] She suggests that MUDs provide a safe environment in which
adolescents and young adults can experiment with different forms of
interaction and relationships. Maybe the Internet is one of many things
with which teenagers experiment only to abandon or use in moderation
as they become older.

Other new technologies indicate patterns of use and non-use. Leung and Wei (1999) examine the use and non-use of mobile phones in Hong Kong. Mobile phones have a much longer history than the Internet as a consumer technology. Leung and Wei identify the factors that are important in determining the take-up of mobile telephony, although they do not distinguish between those people who have never used a mobile phone and those who have stopped using them. Age, income, gender, and education all work in expected ways. However, age dominates—if one is older (unspecified), having more money and more education does not make much difference. Income levels are declining in significance, thus providing some support for the effectiveness of "trickle down." Intensity of use of mass media is not significant, but belonging to social groups that use mobile phones is. Equally unsurprising is the finding that non-users perceive the technology to be unnecessary because they have an alternative or because they find mobile phones either complex to choose and use or intrusive. Leung and Wei's results confirm a growing gap between the communication rich and the communication poor, with users of mobile phones more likely to possess a range of alternative and complementary forms of telecommunication (pagers, answering machines, etc.), whereas non-users had only one reasonable alternative. Leung and Wei accept the premise that having multiple communication devices is intrinsically good, whereas having only one adequate communication device is a sign of deprivation.

Leung and Wei's results are not very surprising: people do not use mobile phones if they have alternatives, find them intrusive, and/or think them expensive. By extension, maybe some people decline to use the Internet because they have alternative sources of information and forms of communication that are appropriate to their needs, or because they think it is cumbersome and expensive.

Re-Constructing the Categories of Use and Non-Use

The question of "dropouts" may be a transient issue if all the non-users given that label eventually return to the Internet, perhaps when their income rises or when they regain access by means of a television or a mobile phone. In any event, in the United States alone there are millions of former users about whom very little is known. They may be a source of important information for subsequent developments. Even within the rhetoric of increasing access, it is important to know why such people leave and whether anything should be done to lure them back. Internet service and content providers as well as policy makers might have much to learn from this group.

There are different categories of non-use. As Bauer (1995: 14–15) points out, there is a difference between passive "avoidance behavior" and active resistance. Also, care should be taken to distinguish between non-use of a technological system (such as the Internet) as a whole and non-use of specific aspects of it (Miles and Thomas 1995: 256–257). In a preliminary taxonomy of non-use, we (Wyatt, Thomas, and Terranova 2002: 36) identify four types of non-users. The first group consists of "resisters" who have never used the Internet because they do not want to. The second group consists of "rejecters" who have stopped using the Internet voluntarily, perhaps because they find it boring or expensive or because they have perfectly adequate alternative sources of information and communication. The third group consists of people who have never used the Internet because they cannot get access for a variety of reasons; they can be considered socially and technically "excluded." The fourth group consists of people who have effectively been "expelled" from the Internet; they have stopped using it involuntarily, either because of cost or because they have lost institutional access.

The policy implications are different for the different groups. It might be appropriate to develop new services in order to attract the resisters and the rejecters. If Internet access is seen as inherently desirable, this might be accompanied by the provision of measures to facilitate access. Another possibility is to accept that some people will never use the Internet. This could lead either to a focus on existing users or (moving away from the perspective of the suppliers and promoters who see non-use only as a gap to be filled) to policies that would make alternatives to the Internet available to people who want or need them. The access issues identified at the European Union meeting in March 1999 related to cost, skill, and location are more relevant for the third and fourth groups (the excluded and the expelled who would like access). At the very least, it is important to distinguish between "have nots" (the excluded and the expelled) and "want nots" (the resisters and the rejecters).

Once one has made the step of including "former user," as well as "current user" and "never a user," it is not too much more of a leap to begin to take apart the notion of "user." What exactly does it mean to be a user? How is it defined? Is it possible to distinguish between non-users and non-owners? In a well-established UK survey conducted by the NOP Research Group (1999), estimates for total numbers of users are based on answers to the question "Have you personally used the Internet in the last twelve months?" This allows for an enormous range in frequency of use. The notion of Internet usage should be treated in a rather more nuanced way, distinguishing between those who spend almost every waking hour online and those who check their e-mail once a week. The CyberAtlas data now distinguishes between those with access to the Internet and those who actually use it at least once a month. In 2002, it was estimated that there were 149 million Internet users in the United States but just over two-thirds of them had used it within the past month (CyberAtlas 2002b). (I occasionally use taxis and accept lifts from friends; while this does leave me open to the accusation of hypocrisy, it does not make me a car user in any meaningful sense.) Frequency of use and the range of services used are both important to consider when conceptualizing Internet use. There remains a great deal of interpretative flexibility about what problems the Internet is solving and for which social groups.

The Internet "user" should be conceptualized along a continuum, with degrees and forms of participation that can change. Different modalities of use should be understood in terms of different types of users, but also in relation to different temporal and social trajectories. The latter include changes in lifestyle determined by processes such as aging, changing jobs, educational history, and geographical mobility. Internet use encompasses not only different types of use, but also the possibility of reversals and changes of direction in the individual and collective patterns of use. In addition to the usual demographic variables, details about the frequency and nature of use help to construct a fuller image of the multiplicity of uses and users of the Internet. Many authors have pointed to the ways in which producers and designers of technology draw on the "I-methodology," using themselves as the paradigm of a user (see the chapter by Lindsay in this volume), or the singular, undifferentiated user, or users in the plural as a homogeneous group. Including the variety of non-users also helps to open the way for subtler description and analysis of the multiplicity of users.

Incorporating Non-Use into Technology Studies

In this chapter, I have presented some of the data about use and non-use of the Internet, not in order to provide a definitive snapshot of its current level of diffusion but rather to begin to explore the category of non-use and what it means for science and technology studies. Much more research is needed to understand the variety of reasons people resist and reject technology. Analyzing users is important, but by focusing on users and producers we run the risk of accepting a worldview in which adoption of new technology is the norm.

Cars are not simply wheels, engines, and steel; they exist within a socio-technical infrastructure that includes test centers for drivers and vehicles, motorways, garages, the petrochemical industry, drive-in movies, and out-of-town shopping centers. The more people use cars, the greater the infrastructure to support them, and the lessening of car-free space. Similarly, the Internet is not just web content. It includes many other applications as well as computers, telecommunication links, routers, servers, educators, and cyber cafés. The more people use the Internet, the more pressure there is to develop user-friendly interfaces and to provide more access equipment, greater bandwidth, and faster switching and routing. But there is a paradox here: as the network expands and becomes more useful, it may also become more difficult to create well-working communities. It is thus important to analyze the Internet not only along a single dimension or characteristic but as a large technical system (Mayntz and Hughes 1988; Summerton 1994; Coutard 1999). In this chapter, I have argued that it is essential to consider the role of non-users in the development of large technical systems such as the Internet rather than focusing only on the changing relationships between system builders and users.

Acknowledging the existence of non-users accentuates certain methodological problems for analyzing socio-technical change. At the beginning of the chapter, I highlighted the importance of incorporating users into technology studies as a way of avoiding the traps associated with following only the powerful actors. Another way of avoiding such traps is to take non-users and former users seriously as relevant social groups, as actors who might influence the shape of the world. Elsewhere in this volume, Jessika van Kammen explores user involvement in design and Ronald Kline analyzes the ways in which people who resisted the use of the telephone and the spread of electrification affected subsequent design choices. Maybe it is also possible to explore the possibilities for including non-users in design processes. For example, would mobile phones make such irritating noises if non-users had been involved in their design? There are obvious methodological problems to be overcome here as non-users may be particularly difficult to locate. Non-users may not be a very cohesive group as people may have very different reasons for not using the Internet. This invisible group is another instance of the difficulties posed by an over-literal interpretation of the dictum to "follow the actors."

Kline (this volume), Kline and Pinch (1996), and Bijker (1995a) have vividly demonstrated the important roles played by, respectively, resisters of the telephone and electrification in the rural United States in the early twentieth century, anti-car farmers in the US in the same period, and anti-cyclists in Europe in the late nineteenth century. Kline and Pinch explore the significance of rural inhabitants of the US, who initially opposed the use of motorcars and who even after accepting its presence used the car for a variety of agricultural purposes. Kline and Pinch demonstrate the significance of this for subsequent designs of both cars and roads. Bijker examines the ways in which anti-cycling groups influenced the design of bicycles, contributing to a safer configuration around which bicycles eventually stabilized. These and other histories of technologies demonstrate how resistance and rejection play an important shaping role; however, it seems possible to allow this only with the safety of hindsight when technologies have been stabilized and normalized.

The use of information and communication technology (or any other technology) by individuals, organizations, and nations is taken as the norm, and non-use is perceived as a sign of a deficiency to be remedied or as a need to be fulfilled. The assumption is that access to technology is necessarily desirable, and the question to be addressed is how to increase access. Sometimes the answer involves investment in infrastructure, public education to overcome ignorance and fear, or training and standardization to improve ease of use. Informed, voluntary rejection of technology is not mentioned. This invisibility reflects the continued dominance of the acceptance of the virtues of technological progress, not only among policy makers but also within the STS community.

Acknowledgments

The work on which this is based was supported by the Virtual Society? Programme of the UK Economic and Social Research Council under grant no. L132251050. I am grateful to Tiziana Terranova and Graham Thomas, my colleagues on the project, for many stimulating discussions and for the fact that neither of them drives a car. I am also grateful to the following for comments on some of the ideas in this chapter: Brian Balmer, Flis Henwood, Helen Kennedy, Tim Jordan, Ian Miles, Lera Miles, Nod Miller, Dave O'Reilly, Hans Radder, Els Rommes, and Paul Rosen. In nearly all cases, comments were provided via the Internet.

4

Escape Vehicles? The Internet and the Automobile in a Local-Global Intersection
Anne Sofie Laegran

The automobile and the Internet are both technologies that enable communication across distances, symbolizing freedom and mobility. The automobile symbolizes and enables physical mobility and freedom of movement in material space, but with a limited reach. The Internet enables mental mobility and freedom of movement in virtual space; it is believed to make locality less pertinent, the freedom of movement being, in principle, unlimited. This chapter will analyze the co-construction of these technologies and youth cultures: how these technologies are reinterpreted and given different symbolic and utility values among youth cultures, and how the use and non-use of the technologies figure in the shaping of the users' identities. It is based on observations, interviews, and informal chats with young people in and around an Internet café in a village in central Norway during August and September of 1999.

Traditionally, users have been considered important actors in the diffusion and acceptance of new technologies; however, they have been viewed mostly as passive recipients of the technology. More recently, scholars pursuing constructivist studies of technology have begun to look upon users as creative agents of technological change. Even more recently, non-users have been made visible in studies of technology. Similarly to the contributions of Ronald Kline and Sally Wyatt in this volume, I intend to show how non-use and use of technologies pertain to the co-construction of technology and users.

In this chapter, I take a perspective within the thinking about users and non-users that perceives the use of technologies as a domestication process. Etymologically, domestication is related to the domestic. Silverstone, Hirsch, and Morley, who introduced this concept related to technology appropriation, used it in relation to how technology is integrated into the "moral economy of the household." They did, however, highlight the possibility that the concept might be used analytically to understand the use and the appropriation of technology in other settings (Silverstone et al. 1992).[1] In the broader understanding, domestication has to do with how individual users, as well as collectives, negotiate the values and symbols of the technology while integrating it into the cultural setting. This process may be thought of metaphorically as "taming" (Lie and Sørensen 1996). Through domestication, technology changes as well as the user and, in the next step, the culture. More than within other constructivist theories on technology and users, such as script theory (Akrich 1991) and the SCOT model (Pinch and Bijker 1987; Bijker 1987; Kline and Pinch 1996), the domestication perspective enables a thorough analysis of the users without relating directly to the design and manufacturing of the technology. It allows for redefinitions of practice and meanings even after the construction of the technology is closed from the producers' and the designers' points of view and even if the shape and the intended use of the technology have been stable for a long time.

Silverstone et al. identified domestication as a process of four stages. Sørensen, Aune, and Hatling (2000), however, refuse to talk of stages implying a defined succession, but define four dimensions of the process of domestication. Their dimensions are also more applicable in analyses that go beyond the household and its moral economy as the unit of study. In this model, domestication is a multi-dynamic process in which the artifact must be acquired (that is, bought or made accessible in some other way), placed (that is, put in physical space as well as in mental space), interpreted (in the sense that it is given meaning within the household or the local context, and given symbolic value to the outside world), and integrated into social practices of action.

Strategies of domestication have a practical, a symbolic, and a cognitive dimension (ibid.). The practical dimension is focused on action and on how the technology is used and integrated into social practice. The symbolic dimension is focused on how the technology is interpreted and given various meanings, which the user may identify with or reject. The cognitive dimension is focused on the learning aspect of the technology use: what kind of competences are needed and created in the appropriation process.

Domestication is a contingent process, depending on local resources as well as on structural or global intersections. It is sensitive to local conflicts, friction, and resistance (ibid.). In this chapter I will show how these aspects are relevant in understanding how young people use or reject technologies and how they integrate them as symbols as well as practical instruments into their youth cultures.

Technology and youth are highly influenced by, as well as being driving forces of, cultural globalization. Transportation technology makes long-distance mobility faster, cheaper, and easier, and cultural products in form of signs and artifacts are spread to any part of the (Western) world at tremendous speed—a development often referred to as time-space compression (Harvey 1990). At the same time, modernity is marked by time-space distantiation (Giddens 1990), whereby social relations are stretched out in space. In pre-modern society, Giddens argues, "space and place largely coincided, since the spatial dimensions of social life were, for most of the population, dominated by 'presence'—by localized activity" (ibid.: 18). Modernity, however, includes, to an increasing degree, relations between what Giddens calls "absent" others, locationally distant from any given situation of face-to face interaction" (ibid.). Transportation and new communication technologies become more important as people try to cope with these increasingly complex time-space relations, .

Youth is defined here as a liminal stage in the transition from a child to an adult. In practical terms, the youth period implies that an individual is undergoing education or job training and has freedom and independence to try and do various things before settling into adulthood. This period corresponds to what Ziehe (1989) terms a culturally defined category of youth. This is a modern phenomenon, resulting from what Ziehe conceptualizes as cultural unleashment—an erosion of "traditional" social structures that increases opportunities for choice but at the same time puts considerable stress on a young person.

Youth is a period of reflection on the questions "Who am I?" and "Who do I want to become?" The construction of identity is a collective as well as an individual process, balancing between individuality and identification with a group. Cliques and subcultures tend to flourish (Epstein 1998). Groups may develop in opposition to the hegemonic and/or the parent culture. Hegemonic culture is defined as the culture created by the most powerful groups in society, parent culture as the more class-specific responses to the hegemonic culture (ibid.). Hebdidge (1979) shows how artifacts may play an important role for identity and identification with groups, where the acquisition of meaning of these artifacts involves a dual process of appropriation from the parent culture and transformation within a subcultural context. The Internet and the automobile—technical artifacts appropriated from the hegemonic culture as well as the parent culture—are transformed in different ways in the youth culture as symbols as well as means of practice, acquiring different meanings and implications for identity.

My particular focus in this chapter is on the spatial aspects of identity. Identity is not only about who one is and who one wants to become; it is also about where one is and where one wants to be. Spatial metaphors are often used in conceptualizing identity (Pile and Thrift 1995). Metaphors such as "roots" are often used to emphasize that identity is a steady and stable nucleus. One is "rooted" in a specific place, such as where one was born. Complex spatial relations, however, makes us relate to other places. Not only does this give us the choice to reach out of the local; in addition, we have to relate to more complex spatial relations, whether we want to or not.

There are large differences among young people in how they cope with spatial complexity. This is the case, first, in regard to their relationship to particular places with which they are affiliated and to which they create personal attachments. For many, anchoring to particular places is important for identity, whether to the place they were raised or to other places of meaning. Second, there is the consideration of relation to places that are out of reach in everyday life. For many young people, the journey itself, or the venturing out, is as important as particular places in their identity work. Thus, a more fruitful way to conceptualize identity in the modern and globalized society is to talk of identity as routes (Hall 1995). Young people take different routes in their construction of identity, some remaining close to one place for most of their life and others exploring over greater distances (perhaps later returning to the old place and seeing it in a new light).

Cultural unleashment and spatial complexity may be particularly relevant for identity construction among rural young people who are exposed to a global youth culture and a society full of opportunities but who are able to realize few of these opportunities in the local context. Research on youth has tended to focus mostly on urban subcultures (Thornton 1997; Skelton and Valentine 1998). However, in Scandinavia quite a few studies have been done on young people outside cities, and these have revealed tension between identifying with the local community and a desire to reach out and "see the world" (Jørgensen 1994; Heggen 1996; Waara 1996; Fosso 1997).

This article will focus on two particular youth cultures in a village: the friks and the råners. "Frik" is derived from and pronounced like the English word "freak." Its roots are in the American and British hippie subculture of the 1960s. Today in Norway this notion is most prevalent in urban areas, where young people express countercultural artistic and political preferences to varying degrees. "Råner" (pronounced "roaner") actually means *boar*, but in youth slang it has come to mean *cruiser*—that is, one who drives up and down a street. This masculine youth culture has been evident in the countryside and the small towns of Norway and Sweden for years (Bjurström 1990; Garvey 2001; Lamvik 1996; Rosengren 1994). New Zealand has a similar culture of "bogans" (Nairn et al. 2000). These cultures may be related to the American car culture of the 1950s and the 1960s (Lewis and Goldstein 1983; O'Dell 2001). The råner culture shows less of a connection to the United States, even though some råners use the American flag and other American symbols. Despite this and the fact that the preferred cars are Swedish, the råner culture must be understood as a local, rural, and Norwegian phenomenon.

In the rest of this chapter, I will analyze how the Internet and the automobile are constructed in different ways in the frik and råner cultures, and how the spatial aspects of identity figure in this construction.

Communication Technology in the Norwegian Context

In the Norwegian political context, regional policy is of the utmost importance. There is a strong emphasis on protecting the countryside from population decrease, and on the right of people to choose to settle in peripheral areas. The state has played an active part in the development of industries and that of the communication infrastructure. The automobile and information and communication technologies (ICTs) have been seen as devices to help people living in rural areas remain there or to make it attractive to move there from urban areas.

The automobile and the roads and bridges that accommodate it have been important in maintaining a dispersed pattern of settlement. The automobile has made it possible to live a rather "urban" life in rural areas, and also to live a rather "rural" life in small towns and suburbs built on the "garden cities" model (Sørensen and Sørgård 1994). However, as people become more mobile and as access to consumer-oriented mass media becomes widespread, the gap between the availability of services and commodities and that which people expect increases, particularly among the young and the more highly educated (Dale 1995). In line with this, bridges and roads built to connect remote places or to improve communication between places also serve as routes of escape, and access to the center increases. Despite the strong focus on the countryside in regional policies, centralization and urbanization have increased greatly in recent decades, though not as greatly as in some other European countries.

In Norway the automobile has been integrated into everyday life as a necessity. It is without real competition when it comes to everyday transportation. In addition to this, the automobile has been given symbolic values not necessarily related to transport; it is often a transportation medium for meanings as much as for persons and goods (Østby 1995; Lamvik 1996). According to Lamvik, the first subcultures in Norway were car-oriented cultures inspired by those in the United States. This coincided with the deregulation of automobile sales in 1960.² Boys and young men with particular interests in automobiles are still found in rural areas and small towns, often in association with traditional cultures.

As Wyatt notes in this volume, for many people cars reflect wealth, power, virility, and freedom. The Internet, Wyatt states, promises many of the same things on an even larger scale. The Internet may be perceived as overcoming spatial barriers to the extent that what is available somewhere becomes available everywhere. Where Internet access is available, it makes the global accessible to the local. The November 1999 monthly survey on access to and use of the Internet conducted by Gallup Intertrack showed that access had become available to 2 million people in Norway either at work, at school, or at home (Norsk Gallup Institutt 1999)—nearly half of the population. Because access is increasing more rapidly in towns and cities than in the countryside, the gap in Internet use between rural and urban areas seems to be increasing (Hetland 1999). The gap is smaller among young and more highly educated people, however. Thus, the rural/urban difference may be attributable to differences in age distribution and labor markets (ibid.).

The concept of the Internet as a space-annihilating medium is implemented in Norwegian policies on ICT as well as in regional development, in the sense that the policies emphasize the importance of using the ICTs to diminish regional imbalances.³

In the 1980s, optimism on behalf of the countryside was manifested in telecottages—multi-purpose telecenters providing commercial services based on telecommunication and some educational activities for the communities. Surprisingly from a technologically optimistic and technologically deterministic view, the telecottages closed one by one (Hetland et al. 1989, 1996). In regard to Internet technology, too, it seems that the potential of the rural districts was exaggerated. The increase in ICT-based industries, as well as in ICT use among common people, has come in urban areas, mainly near Oslo (the capital city). Regional and ICT policy papers state the importance of trying to develop a counter-policy. In contrast with the 1980s, however, it is now recognized that this depends on how the technology is put to use, not on the technology itself. Among the aims are the creation of competence-intensive jobs for the more highly educated, the promotion of telecommuting, and the establishment of small and medium-size enterprises.

In addition to their economic and job-creation aspects, the policy papers focus on how the technology may provide telemedicine, shopping, the arts, and culture to the countryside. In this way the government tries to encourage optimism in regard to how communication technologies can improve the quality of life for people in remote areas. However, the possible cultural urbanization and increased access to the "world out there" which this provides may create pessimism in the countryside. It may tempt young people to orient themselves away from the local even more. "When I look at how my daughter uses the Internet," the father of a 14-year-old girl living in the region in question told me, "she seems to become more global and urban. And if this continues, then this region will be no alternative for her." This quotation describes a fear of some of the inhabitants of rural districts. Despite more facilities and opportunities in the local region, the plans and expectations of young people increase accordingly, and the gap between what young people intend to achieve in terms of career and spare time activities and the possibilities available within the local region increases.

In recent years, volunteer-run Internet cafés meant to attract young people have popped up in rural areas and small towns. These can be seen as technical and cultural intermediaries in the process of innovation and diffusion of the Internet from the global market to the local community, where the new technology is integrated in an existing institution: the café (Stewart 1999). In Great Britain, several of these cafes have been established in urban areas in recent years, whereas in rural areas telecottages still provide Internet access in a social environment (Liff 1999). Wakeford (1999) shows how in Internet cafés the computer becomes an element of social relations involving the staff, the guests, and the café's atmosphere, décor, location, food, and drink. Stewart emphasizes that most of the activity at a Scottish Internet café—using a computer as well as drinking coffee and chatting with people—can be done as well at home, and that the café becomes a home away from home. It is interesting to see how in Norway this is constructed as an urban phenomenon in the cultural sense. "Cafés" were first established in the cities of Norway during the "yuppie" urban trend of the 1980s, and have up until now been found only in urban areas. The construction of "urban Internet cafés" in rural districts is, however, rather paradoxical, since Internet cafés rarely exist in the larger towns and cities.

Places to Meet: The Internet Café and the Automobile

The village in question is situated on an idyllic peninsula in a fjord of mid Norway, about 600 kilometers north of Oslo, 400 kilometers south of the Arctic Circle. The village is the main center of an extended municipality with 6,000 inhabitants. Although considered a pleasant place of residence, the migration balance has shown a slight deficit for several years, which is a problem not only for this place but also for the region as a whole. Agriculture and the food industry are the main activities. It also serves as a dormitory village for commuters to neighboring towns. Entering the village on an ordinary day, the calm, almost dull atmosphere is striking. The only street, despite a couple of shops, a police station, and a bank, is rather empty even during what would usually be rush hours. A few people are sitting in the co-op cafeteria, mostly retirees or mothers with small kids. In the afternoons you find some school children and other young people hanging around or rushing on their way to organized activities. Later in the evenings it is quiet. A restaurant attracts some people, especially on weekends, as they serve alcohol, and the age limit is 18. The only interruption of the calmness is some cars playing loud music, driving up and down the street once in a while—sometimes rather often. These cars are owned and driven by the råners—young men between 18 and 22 years of age—sometimes with girls sitting in the back seat.

My interest in the youth culture of this village developed when I heard about the establishment of an Internet café, the first in the region. Originally coming from this same area, I would not have expected to find an Internet café here. However, this village differs in several aspects from most other villages and towns in the countryside. In addition to its beautiful location in terms of natural landscape, the municipality has received prizes for the preservation of its architectural style and settlement structure. There are several rather large farms, many modernized in an alternative way cultivating rare species and processing their own products on the farm. The community has a strong cultural profile based on what we may call "high culture." A modern comprehensive cultural center, a professional jazz ensemble, a fine art gallery, and a secondary school with a special branch for the performing arts all contribute to a rather impressive cultural life compared to even many urban areas in Norway. One of the people interviewed (a 19-year-old male) remarked: "We even had a drag artist working out of here—that's something hardly any city in Norway can come up with!"

In my interpretation, the notion of the drag artist here symbolizes that this is a community open to including not only unusual forms of entertainment in their cultural profile but also alternative gender relations as compared to many other rural communities.

The municipal administration actively uses their cultural profile to attract tourists and new inhabitants. The slogan "The Golden Detour" signals that it is worthwhile turning off from the main north-south highway to visit the village. Although the majority probably supports the cultural profile, there is tension in the community between supporters of "high culture" and those with more popular orientations. This can be formulated in terminology borrowed from Bourdieu (1995), as tension between those who are rich in cultural capital and those who are not. As we shall see, these tensions are also found among young people.

The village offers several organized athletic and musical activities for young people; however, there has been a lack of informal meeting places in the public space. The cafeteria at the co-operative store does not really meet the demand, nor does the cultural center as the school is also located there and the place is thus associated with duty rather than leisure and pleasure. The car-interested boys, the råners, have solved this by creating their own meeting place outside the gasoline station, the only kiosk open at night. As a place to meet and hang out, this is the territory of the råners and some girls who socialize with them. The girls rarely have their own cars and do not show nearly the same interest in cars. The gas station (known as the "auto") is the meeting point, but most of the interaction takes place in cars. Cars line up side by side, and people do not even have to go out of the car to interact—they just roll down the window and start chatting. And although the car is a means of transport that gets them to parties in the area on weekends, the journey often becomes the party in itself. The car means a great deal; however, the råners claim that if they had an alternative meeting place they would prefer to go there to socialize.

The activities around the gas station and the apparently unnecessary driving are widely discussed in the village community. Much of the talk concerns the distribution and drinking of homemade liquor and the violence of some of the råners.

In 1998, a group of young people at the secondary school and some supportive adults started a café in a former shop in the center of the village. Young people wanted a place to meet, and the adults saw a potential to create something for them as an alternative to the gas station and the cruising. The initiative was taken as a purely voluntary matter with support from local business affairs and from the municipality. Some of the adults involved have children in the targeted age group; others got involved because of interest in computers or in community activities. According to its guidelines, the café aims to create a sound environment free of drugs, drinking, and prejudice. The name of the café, "e@," is hardly used except by its initiators. The place is generally known as "the Internet café."

The café is a non-profit establishment run by volunteers and one conscientious objector. According to its initiators, places like this village lack a touch of "urban" culture, which young people naturally long for. The café is an attempt to satisfy this demand. As I interpret the guidelines, the Internet is one way of attracting people. In addition to being useful, it is a symbol of openness, knowledge, and curiosity about the world. The Internet café can thus be characterized as an attempt to give the young people something urban in the rural, as well as access to the global in the local environment.

Intended as a meeting place for all young people and to break down some of the barriers that exist between different groups, the Internet café soon acquired a reputation as a hangout for friks. Many of them participated in the restoration of the building and developed a sort of ownership of the place, and they also take turns as volunteer attendants. However, according to many of those hanging around the café, just by doing so they are constructed as friks although they do not necessary identify with that style themselves. Anyhow, the råners feel that this place is not for them. One of them (a 19-year-old male) said: "The café is just for the friks; they got it, whereas we don't get anything."

The conscientious objector, who knows the råners as well as the friks has encouraged the råners to come. Nevertheless, the råners feel that the friks dominate the place. They do not feel welcome. The fact that the café opened at the same time as the local newspaper published a series of articles and debates pertaining to problems and clashes among young people probably did not work in favor of unity of the two groups.

Co-Construction of Youth Cultures: Råners and Friks

Before I continue, I think it is necessary to give some clarification of the friks and the råner, the constructed users of the technology in this story. The co-construction notion used in this context implies that, at least in this particular village, these cultures are not constructed just in relation to similar cultures elsewhere, but very much in relation to each other.

The friks are mainly students—male and female, between 16 and 19 years old—in the secondary school's performing arts branch. They share a strong interest in culture and in performance, and several of them aspire to be professional artists. Musical instruments and audio equipment are important technologies for marking identity in this culture. However, the friks' new meeting place is centered on the Internet. Many of them come from neighboring towns and rent rooms in homes, as this is the only secondary school in the area with a program for the performing arts. Others have grown up in the village or settlements within the municipality, and thus live with their parents. Once in a while there are rumors about the smoking of cannabis among some of the friks. These rumors, however, concern only a few individuals and do not create a negative image of the group as a whole. With dyed hair and second-hand clothes, they are rather visible, and their visibility seems to provoke the råners.

Whereas the mix of girls and boys among the friks is close to fifty-fifty, the råner culture is mainly male. Girls are more or less invisible among them, except on weekends. According to the boys, the girls hanging around with them either are, have been, or want to be dating one of them. The råners claim there used to be a few girls hanging around more often who they would actually define as råners, but they have all moved away from the village.

Almost all of the råners have been trained in mechanics or similar subjects. At the time this study was conducted, some had apprenticeships at garages and some were unemployed. The råners have lived in the municipality more or less all their lives and still live with their parents. They have a strong attachment to the municipality, they have been raised here, their family and friends lives here, and they hope to settle here. They seem to feel a kind of responsibility for protecting the place from too much input from outside. Several of those I interviewed came with such statements as "The school brings so many strange people here."

"Strange people" refers to the friks, who, the råners claim, dress like tramps and have long and untidy hair. The story goes that this is a culture brought in from outside. However, a lot of local young people at the school have adopted the frik style. This is a bit provoking to many of the råners:

They have changed—they did not use to be like this before, it seems like they just have to become like that when they go to that school. (male, 20)

The råners claim that they themselves dress like everybody else:

Jogging suit and Levis—nothing special about that. (male, 19)

The friks identify the jogging suit and the Levis as the uniform of the råners. Like the råners, the friks deny that they have a particular style:

I just don't want to use a lot of money on fashion clothes. (female, 18)

Although the village is small and rather dull, most of the friks express positive feelings for it. For them the positive attitude lies in the cultural activities and social environment as much as in belonging and attachment. This is not contradictory to a rather urban and outward orientation in terms of interests as well as in their future plans, which implies taking part in discourses and cultural trends that takes place in cities in Norway as well as abroad. They travel when they have time and money to do it, and would like to do it more often. They can hardly wait to finish school and move to one of the cities in Norway or abroad for further education. In fact, all those I have talked to really underline that within 10 years they hope to have been able to live abroad for a while.

The råners, on the other hand, hardly travel out of the local area, except for weekend trips to Sweden once in a while. They go to places in southern Sweden with similar car cultures, to attend car fairs or to buy spare parts. Oslo, and to a large extent Trondheim (just 100 kilometers away), are however, unknown territories in terms of socializing. There is no car culture in the cities, they say. Most of the råners dream of a job, a house, and a family in the village, and hope to avoid moving away. Some of them work as far as a 3-hour drive from the village during the week but make sure to "come home" on the weekends. The local attachment is, however, in defiance of, rather than caused by, the cultural profile of the village, as this statement from one råner (an 18-year-old male) shows:

The municipality only promotes activities for a few, and these are the people who will move from this place anyway. As many of them are going to be performing artists, they will have to move to Oslo where things happen. It is us that are going to stay and work here who are overlooked and scapegoated.

The råners are not organized as a club. However, some of them have started to work on getting a garage and a place to meet. Contacts have been made with the municipality on this matter, but so far without result.

In many ways, the friks and the råners may be labeled subcultures. However, most of the friks and the råners deny that they are distinctive groups, different from the mainstream in the village. They claim they do not differ much from "ordinary" people. When you walk around the village, though, it is not hard to distinguish the two groups. Neither is it hard to find that those affiliated have common interests, values, and meanings, in addition to their clothing. As I see it, however, the internal identification and the opposition to the hegemonic and parent culture are too loose for these groups to be considered subcultures. I would classify them as youth cultures or style.

The Automobile—Ambivalent Necessity, or Affective Toy?

Kline and Pinch have shown how users in rural America reinterpreted the automobile, used it for purposes not intended by the producers, and gave it various symbolic values. From their point of view, the core technology of the car stabilized around 1945 (Kline and Pinch 1996). Here we shall see that, although the car may be stable as seen from the producers' point of view, the symbolic as well as the practical aspects around the car are still open, and the domestication of the car differs in various ways.

The car is a necessity for physical mobility in this village, as buses do not run frequently. Hence, most of the young people get their driving license when they are 18, and establish some kind of relations to driving and to the car as an artifact. However, as we shall see, the practical, symbolic, and cognitive dimensions of domestication of the car vary a great deal, even among people of the same age in a local area.

The friks rarely acquire their own cars. They borrow them from their parents, saying that they don't want to spend money on cars themselves. They show a lot of resistance to the car in both practical and symbolic terms. Several of the friks express moral doubts attached to driving, and they make sure to drive only when they must. They never "just go out driving." The moral doubt has to do not only with environmental consciousness but also with money. This is an argument often used toward what they call "unnecessary" driving done by the råners:

It pollutes a lot, and besides—where do they get the money? (male, 19)

The most prevalent argument against the driving of the råners is that they do not understand the meaning of it:

Don't they have better things to do than just drive meaninglessly up and down? (female, 18)

The friks use (borrowed) cars just once in a while for trips to neighboring towns, or to visit friends living in areas not accessible by bus. As I have noted, the friks primarily orient themselves toward cities in Norway and abroad, which also implies traveling. The means of transportation then are trains and planes.

With reference to the cognitive aspect, auto repair is not really within the field of competence of the friks. If a car breaks down, they do not know what to do. This is admittedly a bit embarrassing, but they do little to learn how to remedy it. The ability to fix a car is not a kind of competence that brings status or is seen as important among the friks.

For the råners, however, the car is important, and it is domesticated as more than a means of transportation. Many of the råners acquire their first car by the time they are 14 or 15 years old. This is normally an old car that needs quite a lot of refurbishing. They work on it until they become 18 and get their license.

A typical råner car is a large Volvo, and not a stock one. ("Original equipment is boring.") The engine and the gearbox are modified in order to get more horsepower. This has a practical aspect in that it makes the car run faster. However, the symbolic is also important. A real råner puts a lot of affection into the car, making it as beautiful as possible. The car is an aesthetic object. It is lowered slightly and has new wheel rims, dark-shaded windows, white lights, spoilers, and various pieces of extra equipment produced specifically for car enthusiasts. The råners do most of the mechanical work and rebuilding themselves. The competence, learned from older brothers and friends, is developed from a very early age. Being good at this brings them status within the group.

Although they do most of the work themselves, the affective relationship to the automobile demands a good share of their income. One of the råners interviewed informed me that, out of his 6,000 NOK ($650) a month earned as an apprentice, 2,000 NOK goes toward paying off the car loan and 3,000 NOK is used for petrol. 1,000 NOK a month for everything else does not stretch that far, so it is advantageous to live with their parents. All the same, it is frustrating not to have one's own place to meet friends. Hence the car becomes a home away from home and a space of freedom. After a weekend, the odometer shows several hundred kilometers traveled by each car. Nevertheless, the radius of mobility is limited to up and down the main street of the village, and to other nearby towns or villages.

As previously said, the råners find the frik culture threatening to the local culture. There is actually a fight for control of public space and territory going on. From the Internet café you hear the engine and the sound system quite a while before the car actually comes by. The råners admit that they drive up and down the street not only because they are bored but also in order to annoy people. And people do get annoyed— especially the friks, who find the driving to be a source of pollution, noisy, and unnecessary. On the other hand, one of the råners (male, 20) asked: "What is more disturbing, some cars driving around or people walking around wearing old curtains?"

So it is to a large extent an aesthetic struggle that goes on. The car is a weapon in terms of being visible as well as audible. There have also been incidences in which a car has been used more as a weapon in a real sense. Some of the friks tell of having had close calls with råner cars when walking along the road. The råners claim that this is over-exaggerated. They admit though, that some may make a small detour toward frik groups once in a while, but in order to tease rather than to threaten.

The Internet—Stranger, or Ambivalent and Cool Time Killer?

According to the råners, the friks are also "computer freaks," as the café is based around the Internet. I thought so too before I visited the community. As one of the main purposes of the café is to make accessible computers that are linked to the Internet, I expected to find cultures that would be similar to what is written about in literature on hackers (Turkle 1984; Nissen 1993; Håpnes 1996) and heavy Internet users (Turkle 1995). Surprisingly I found no traces of such cultures at the café. The friks hanging out at the café are actually not that interested in computers. However, the Internet is important as a symbol. One of the regulars (male, 18) said:

It's really cool that we have Internet at the café. Internet is in a way something that goes along with this kind of café—like those in the cities. But I go there just to meet friends—to do nothing with somebody.

Internet access is not the main attraction and reason for people to go there. It is considered a "cool" thing to have at a café, creating an image of being up to date. So when friks go to the café, it is mostly to "do nothing"—that is, to meet friends and play table games. The café becomes their home away from home. The computers are occupied by a group of very young kids, mostly boys but there are also some girls who chat or play games.

The råners do not have access to the Internet at work or at home. Most of them have never tried it. Others have been using it a bit at school. Indifferent to further experience, most of them refuse to use this technology. They have an interest in mechanics, but are not interested in the technology of the computer. One of them (male, 20) said: "You can't mend a computer! Then you really have to know what you are doing." Repairing a car, on the other hand, is rather logical, they tell me. This is to a large extent tacit knowledge, developed through practice or informally transferred from older brothers and friends from when they were very young. For these boys, using computers is more associated with knowledge acquired at school and in formal settings.

Since the Internet café is so central in the symbolic battle over public space, the Internet for the råners is symbolically attached to the urban culture coming from outside. The fact that their only possibility of access to the Internet is (from their standpoint) the territory of the enemy does not, of course, make the Internet more attractive.[4] All the same, they have been told that there may be things on the Internet that are of practical use to them. In fact, several of Scandinavia's dealers in used auto parts are on the Internet, which makes it easier to get spare parts for their cars. Thus far, they say that they prefer to use text-TV and magazines for this purpose.

The friks hanging out at the café do not actually use the Internet much, and it seems to me that they may even under-report their Internet use. It seems that the Internet, like other mass media, is used, but not too much, and it is not the kind of activity of which friks are proud. As with the car, there are moral doubts attached to it. Friks are eager to emphasize that they are not interested in computers and the Internet, and claim they have no problems imagining spare time without the Internet. Several of them express concern about youth culture's becoming more of a consumer culture than actually initiating and creating cultures. This is interesting, as the image of the café is very much affiliated with the Internet, an image that is supported also among the friks.

Nevertheless, friks are on the Internet once in a while, to find sources for assignments at school or to kill time. Among those who use it most frequently, e-mail is important. For the others, looking up sites of interest on the web is the most common activity. None of the friks, however, seem interested in creating home pages, and none of them have even heard of multi-user domains, multi-user object-oriented domains, or the possibility of creating environments and characters on the Internet. They use the Internet as an information service and a communication tool.

It is interesting to see how Internet use becomes a means to reach out from the local context, be it socializing in chat rooms or surfing the Internet. When friks are online for fun or to kill time, e-mail seems to be the most important for those who use it regularly. Some spend time in chat rooms. Some e-mail "net friends" whom they have met previously in chat rooms. They do not use chat rooms or e-mail to communicate among themselves in the local community. The local communication is face-to-face or is mediated by stationary and cellular telephones. They surf the web for sites related to music and films, which are also popular topics to chat about on the Internet. Most of the friks have yet to travel a lot abroad, but they dream of a great journey to more exotic places, and many have started the mental journey by looking up information about foreign places and cultures. They also use the Internet to help realize some of their plans of moving out in real terms by looking up schools abroad, or getting information about places they would like to visit or stay at.

The material also shows that after some time young people stop using the Internet or particular Internet services; that is, it is more interesting when it is new (see Wyatt, this volume). One of them (male, 18) said: "I use e-mail; that is useful. But I'm not on the Internet surfing." When asked "Did you do that before?" he replied "Yes, you know, that's kind of something you do in the beginning, but you soon get tired of it." A girl of 18 said much the same about chatting: "That is something you do when the technology is new and exciting." Thus, after the initial excitement, the Internet is constructed as a trivial but nevertheless an ambivalent technology. "Real" socializing is considered better than socializing online.

Among younger people, the generation we know as having "grown up digital" (Tapscott 1998), the pattern of use and the symbolic value of the Internet are somewhat different. I observed this among the younger boys at the café, as well as through interviews with young people who do not attend the café. It was interesting to find young boys between 13 and 16 dreaming of modified Volvos and at the same time having a life on the screen.[5] Contrary to the friks however, the young "perhaps to become råners" do not use the Internet as a means to reach out. Similar to the older råners, these young boys have a strong local attachment, and the Internet is used to reinforce this. When looking up news on the Internet, they read the local newspapers. They go into Norwegian chat rooms, or even regional ones where they can use their own dialect. The purpose is to meet friends from school and people from neighboring villages and towns. They play games with friends and link the computers together. And they surf the Internet, mainly for sites about cars and cellular telephones.

It seems that the integration of the Internet into society at large influences the way it is domesticated in the local youth culture. Among the younger generation the Internet does not have a label of something strange coming from outside. The question is whether this will apply also for the older people as the technology becomes more available and trivialized in the media and popular discourses.

Escape Vehicles?

Elsewhere in this volume, Sally Wyatt shows how, to a large extent, the automobile and the Internet connote the same values. I have shown how these technologies may also connote different and contradictory values, and how their use and non-use may be linked to local conflicts and symbolic battles.

For the råners, the Internet is interpreted as coming from outside and interrupting the local, particularly through the way it is connoted to the urban-like Internet café. The friks perceive the Internet as a medium that enables communication in a global context, having a touch of urban culture. Thus the Internet is given similar symbolical meanings among the friks as it is among the råners. But, whereas the råners totally reject this, the friks like to be affiliated with it. Looking at the Internet as practice, however, we see that even the friks do not use it much. There is a practical aspect of time and having better things to do, but it also seems as if the Internet has a double edge: it is a cool and urban thing, and at the same time it is associated with a consumer culture they are reluctant to support. The Internet is nevertheless more in line with the image of being outwardly oriented and up to date on global discourses and competencies. It is largely a mental escape from the local and a tool to enable traveling out at a later stage. The Internet thus becomes an "escape vehicle to the global." The råners, in contrast, do not have the interest of "going global" and do not see the point to using the Internet for communication.

Now let us look at the symbolic meanings of the automobile in these two cultures. The friks express moral doubts and dissociate themselves from it. Pollution is the first thing that comes to mind when they think of the automobile, and the "unnecessary" driving of the råners annoys them a lot. For the råners, the automobile is the prime icon of expression and identification, and it is a second home. The automobile is used by both groups. The friks drive only when they really have to. The råners drive for fun, getting a feeling of freedom. It is also important to them to be visible and audible—to show off the car, which they have remade and decorated. For the råners, the practice of driving is limited to the local. It has a strong symbolic aspect and becomes a vehicle of local identity as well as of power in the local space. However, much of this local activity may also be understood as escape. It gives freedom to know that you may leave whenever you want—be it from their parents, friends, parties, etc.

The technology is used to reinforce identity oriented toward stronger affiliation to the local (as is the case with the råners) or to identify with the world outside (as with the friks). It is, however, not given that the technologies are interpreted that way. The younger råners-to-be interpret the Internet as just being a means of communication without the connotations of new and strange. They use it as a communication tool in the local context. One could argue, with evidence from diffusion of other communication technologies, that it is just a matter of time before the råners too start to use the Internet, probably as a local means of communication. However, as long as the Internet as technology is co-constructed with the Internet café and the style of the friks, it is not seen as an appropriate technology for the råners.

Youth and subcultures are often involved in struggles over place and space (Thornton 1997; Skelton and Valentine 1998). As I see the case of this village, the visibility of the råners in terms of driving and the friks in terms of style is a way of marking territory to express "here we are." These are expressions in a struggle between different youth and parent cultures against the hegemonic culture dominating the village. Although the style of the friks has a slight touch of expressing something alternative, their behavior and interests are to a large extent in line with the hegemonic culture. The slight oppositional aspects of the frik culture go against consumer society in general rather than against the locally prevailing norms and ideas. The friks get a lot of attention for their cultural activities, for instance in the local newspapers. Their identification with global discourses and orientation out of the village, including the use of Internet, is not in conflict with the hegemonic culture. This is also observed by the råners, several of whom expressed frustration at being overlooked. The public debate and the news reports about the råners are limited to discussions of driving, violence, and drinking, creating a rather negative image. As I see it, much of their behavior may be interpreted as positioning in a symbolic struggle—as marking their territory to show that they are part of, and want to belong to, the village.

Whereas the automobile and especially manual mechanics to a large extent is technologically affiliated with the industrial society and the "past," the Internet is said to be the technology for the "future." In this respect the råner culture may be understood as a culture not adapting to but rather falling behind the development of the modern society. The competence gained within the field of mechanics, as well as the symbolic capital developed and achieved within that context, is not necessarily compatible with other contexts out of the village or region, or for the future. The friks, on the other hand, take part to a larger degree in global discourses and achieve a symbolic capital easily convertible to other contexts. This is important, as their future plans are oriented out of the village, to cities in Norway and abroad. As I said in the introduction, the spatial complexity, relation to "absent others" is influencing everybody; we cannot escape from that. It is, however, still possible to chose to live an everyday life mainly in the local. As we saw, the råners are not interested in moving to other places; the village is their home. In that sense, not using the Internet is a rational choice as much as a sign of not adapting to society; the råners do not want to affiliate with an "urban" and "global" culture, and do not see the need for the Internet in their everyday life.

Acknowledgments

I thank the following for constructive comments at different stages of the writing of this chapter: Anders Löfgren, Knut Sørensen, Hanne Helgessen, Barbara Rogers, Kari Arnesen, Els Rommes, and the editors. Special thanks to Christina Lindsay for enhancing the English.

II

Multiple Spokespersons: States and Social Movements as Representatives of Users

5

Citizens as Users of Technology: An Exploratory Study of Vaccines and Vaccination
Dale Rose and Stuart Blume

It is not difficult to see that the development, design, and production of a material object entails what has come to be known as the "configuration" of its (intended) users (Woolgar 1991). This manifests itself in countless ways across a range of different objects. For example, any person who has felt severely constricted in a tourist- or economy-class airline seat comes to realize that such seats have been designed with someone other than him or her in mind. Individuals trying to buy clothes from the stores of chic Parisian or Italian designers may very well come to the same conclusion, even before looking at the price tag. Airline seats and couture clothes are designed with certain characteristics of their intended users in mind—in these cases, norms that bear some relation to the distribution of bodily shapes and sizes within populations. Such objects, of course, are traded or exchanged in markets, and a person of sufficient means can choose to fly first class and can have his or her clothes made to order.

Now consider another example. Von Hippel (1976) and Blume (1992) have shown that potential or intended users play a major role in the development and design of sophisticated technological artifacts such as instruments. What is vital here is less the bodily characteristics of users (though this is certainly not to be forgotten, and in the case of diagnostic imaging technologies, for example, it may be of central importance) than their preferences, or needs, or interests. For what purposes and in what ways might scientists or clinicians be inclined to use these new devices, and what implications does this have for their design? Alternatives and their implications are typically explored in collaborations between manufacturers and scientists or clinicians. In these cases, scientists and clinicians who participate in the development of particular technologies represent potential users not only passively (by virtue of their presumed characteristics) but also actively (through the skills, knowledge, and norms which they share with professional colleagues and which they bring to bear). If the device seems promising enough, an attempt may be made to bring it into commercial production. The field of instruments, scientific and medical, is replete with examples of these sorts of processes. And although these devices too are traded in markets, users, in order to express their interests or needs effectively, require not so much a deep financial reservoir (although this is important) as an ability to wield specialized cognitive and technical resources.

A third example, different again, is provided by assistive technologies (such as mobility aids), artificial organs (including implants and prostheses), and biological materials (such as genetic material). As with other "high-tech" products, much of the work of configuring the user takes place in the course of research and development. What kinds of patients is the technology to help, and how? In many cases, specification must not be too precise, and clinicians, innovators, and manufacturers recognize that a certain degree of "tailoring" may be required. For example, organs, tissues, and cells for transplant from either human or non-human donors need to be (rendered) immunologically compatible with the recipient (see, e.g., Hogle 1999; Cooper and Lanza 2000); artificial limbs need to be adapted to the anatomical specifics of their intended users (e.g., Kyberd, Evans, and Winkel 1998; Iannotti and Williams 1998); and when cochlear implants were to be offered to young children, the design had to be adapted to allow for growth of the skull. In other words, these technologies are designed so as to allow for human variability, a notion that is at the heart of fast-developing fields such as pharmacogenomics and tissue engineering. Of course there is room for discussion regarding the range or types of variability to be allowed for. Many critics have suggested that in the design of rehabilitation technologies the "person" and his "needs" have been conceived far too narrowly. Critics of cochlear implantation, for example, have argued that, in having recourse to the device, the deaf person is treated as no more than a deficient set of ears (see, e.g., Blume 1999). And of course variability as conceived in pharmacogenomics is variability in a passive sense; it is different than allowing for the fact that members of different racial or ethnic groups suffering from the same chronic illness may have different preferences regarding desired therapy.

Clearly the development of all of these kinds of technologies, and in particular those provided by our third example, entails different sorts of configuration work, and that work does not begin and end within the confines of the laboratory. Often issues surrounding a technology's provision in a marketplace tie in directly to the configuration of its (intended) users. How are these kinds of technologies made available, and to whom? What consequences are attached to their use? In light of the fact that some technologies serve a collective as well as an individual need, what is the role of the state in all of this? All these questions lead us to conceptualize technological innovation and use along a development-provision continuum or trajectory in which users are configured differently—by different actors in varied contexts—at each step along the way. Of course, contexts in which configuration occurs vary from country to country (and from culture to culture), and many of those differences can be captured only through detailed comparative (perhaps ethnographic) study. Here we will argue that, nevertheless, a rough distinction between essentially collective (i.e., state) and essentially market-like forms of provision helps us make some sense of the kinds of user configuration that occur. In most industrialized (welfare) states, the technologies we have in mind are provided through a mix of both of these forms of provision, with each itself in evolution, and the balance between the two a matter of continuing debate. Yet, as we shall see, both forms of provision often presume users to be (configured as) active consumers of technologies—regardless of the specific market mechanisms that states have put in place.

A related set of questions revolve around the extent to which users fit their configurations, as well as the activities they may undertake which act to solidify, modify, or reject those configurations. How can expression be given to values, needs, or interests that run counter to how users are configured? How are these divergent interests or needs expressed in market and non-market mechanisms? The market in couture clothes allows the rich consumer to avoid many of the more obvious constraints of standardization: given the resources, neither stature nor socially perceived deviant taste need be a barrier to self-expression. Yet self-expression or dissatisfaction may also take other forms. For example, individuals may resist using particular technologies for any number of reasons, regardless of the type of market in which they are exchanged. And collectively, groups may organize to protest either the limitations of what is available or under development, or the unavailability of what is not (see, e.g., Epstein 1996).

What difference does it make when we focus on technologies that are in large measure developed, and their use facilitated by, the state and its institutions? More specifically, how can we begin to discuss technologies that are developed to serve both individual and collective needs, and/or which act to fulfill state policies and goals?[1] Here, resistance or protest to the use of a technology may disrupt not only configurations of users as active consumers in a market, but may also highlight significant tensions between individuals as (intended or actual) users of technologies and the state of which they are members and citizens. Some of the most complex instances of this emanate from policies revolving around women's health and reproductive choice. For example, a number of tensions are contained in the debate regarding prenatal genetic testing. Consider the following:

> Policies that would in any way penalize those who continue pregnancies in spite of knowing that their child will live with a disabling trait must be avoided. Those prospective parents who either forgo prenatal testing or decide that they want to continue a pregnancy despite the detection of a disabling trait should not have to contend with losing medical benefits for their child, nor feel obliged to justify their decision. (Parens and Asch 1999)

There is obviously something to be said for making such tests widely available to those who want them. Part of that "something," it is often said, is to give individuals (in this case, women) the possibility of choice in their reproductive decision making. Choice thus becomes choice among products, or technologies, or even use itself. Yet there is a concomitant risk associated with the possibilities of choice: that some choices become less acceptable than others (Bauman 1996). To resist the configuring, and disciplining, effects of this and other technologies may be to protest the social policies implicated by the technologies themselves.

Users of Technologies

Early understandings of users were developed in economics-oriented innovation studies, which concluded that users—collectively—play an important role in technology development (see, e.g., Von Hippel 1976; Lundvall 1985, 1988). Users were seen by these and other researchers as highly active and agentic in processes of innovation; they are able both to influence market demand (by the expression of product preferences and by their purchasing power) and to articulate needs either about new products they desired (in order, perhaps, to fulfill functional needs) or about improvements they wished to incorporate into already existing technologies. More recently, traditional notions of users have come under empirical scrutiny and subsequently have undergone theoretical refinements (e.g., Cowan 1987; Akrich 1992; Akrich 1995; Kline and

Pinch 1996; van Kammen 1999; van Kammen 2000a; Oudshoorn 2000). However, the STS literature on users, rich though it is, pays little attention to the role of the state in processes of their configuration, either in the laboratory or once a technology has reached the market. This had led us to look at users slightly differently: as consumers in markets, but also as citizens in states.

Decades ago, to have considered the relationship of the state to technological change meant one of two things. It was to have referred to innovation policies, and the relative merits of direct as against indirect (e.g., fiscal) modes of support for industrial innovation. Or, insofar as government policies seemed to affect the form or content of the technology, it was to have referred (almost inevitably) to technologies associated with Cold War politics. In other words, it was the era of Big Science writ very large. But where military aircraft, weapons systems, space vehicles and (at least in Europe) nuclear power plants were the "state technologies" of the 1960s and the 1970s, matters are now very different.[2] Compare the (putative) "user" of a sophisticated weapons system with the putative user of a genetic test! Today, the state may try to influence the conditions under which the technology is used (for example by providing incentives for use, or by regulatory mechanisms or reimbursement mechanisms in the health area), but it is no longer the principle end user of the technologies that it helps to develop. It does, however, continue to play a role in configuring the end users of these newer technologies. Where the state is the principle innovator of a technology, it may act to configure users in ways similar to the kinds of user configuration that occur in the private sector.[3] However, the state may also enact policies that, while not configuring the user, per se, nevertheless create or maintain an environment which helps to shape how users are configured—either in the laboratory, or as we have already discussed, as a consequence of different modes of market and non-market provision. For example, the state may try to determine who is entitled to access to an organ transplant, or to a promising AIDS therapy, or to gene therapy.

Recent social science scholarship has taken issue with and critiqued the presumption that users can be categorized as a singular group, and has questioned the extent to which and how both potential and actual users of technologies actually participate in processes of innovation (e.g., van Kammen 2000a; Oudshoorn 2000). The complexities that define human users across various axes, for example, across gender, ability status, race/ethnicity, sexual orientation and class, often remain invisible to those who (often unintentionally) act to lump users into a single, unified, conceptually whole group—often without their knowing. This does a disservice to individuals who constitute these categories of people, and has the effect of homogenizing members of those groups by uncritically ascribing certain characteristics to them. One consequence of this is that neither heterogeneity among individuals nor diversity within a group, or population, are given due consideration in either processes of innovation or policy decision making.

Madeleine Akrich (1992) observed that assumptions about users are "inscribed" into technologies—in the form of a "script"—by the innovators that develop them. Thus "scripted," the resultant artifact embodies these assumptions, which in part consist of the values, beliefs, attitudes and norms that potential users are presumed to have—or should have. According to Akrich, "designers thus define actors with specific tastes, competences, motives, aspirations, political prejudices, and the rest, and they assume that morality, technology, science, and economy will evolve in particular ways" (ibid.: 208). Presumably users will (or should) fit themselves and their prescribed roles into this broader worldview: they should follow the script. While Akrich acknowledges that scripts can change, Oudshoorn (1998) argues the point more emphatically, noting that "in principle the possibility exists that a user [will] interpret differently, modify, diverge from, or totally reject" the scripts that have been written for them (p. 12, our translation). Users can resist their configurations, and in ways that go beyond the decision not to purchase a product on a market. They may indeed be unwilling (or even unwitting) users, a thought which should make us stop and think about our existing conceptions of users of technologies. Are users not presumed to want (or at least, not to not want) to use most technologies for which they are configured? (Is this not connoted in the very term, user?) How does all of this connect to the configuring work of the state? What of users who resist or protest? How, and under what circumstances are users, as citizens, actually coerced to use a technology?

All of what we have said so far leads us to our final theoretical concern: how to forge conceptual links between notions of individuals as users of technologies, consumers of commodities, and citizens of (for our purposes, welfare) states. Recent work in this area is revealing. Brown (2001: 63), for example, rejects the somewhat arbitrary consumer/citizen distinction that posits that "as consumers . . . individuals make decisions according to self-regarding preferences, aiming to maximize their private welfare [whereas as] citizens, people base their decisions on shared values, aiming to promote the public good." For him, the distinction is faulty because of the rather implausible presupposition that individuals occupying, say, the private sphere can "radically transform themselves" when shifting to the public sphere. We agree, yet for us the distinction is faulty for other reasons as well. If one accepts that consumers and citizens act according to the prescriptions noted above, then what we find implausible is the ability to make such an instantaneous shift from one to another sphere in the first place; our study supports the probably not so contentious argument that individuals are always and necessarily implicated in both spheres simultaneously.

Our more fundamental critique, however, relates to the behaviors prescribed by theory for citizens. In particular, we wish to interrogate the long-standing correlation between citizens and their desire for shared values to achieve a common public good.[4] For us, the issue is less whether this should be the case, and more the conditions under which this actually is or is not the case. It is here that the study of (the use of) technologies can provide such fascinating insights. In light of what we have said so far, we hypothesize that citizens, as users of certain types of technologies, are configured in such ways (and the technologies scripted) as to be presumed to share certain values to achieve a certain type of common, public good. From this, we can further hypothesize that when citizens use certain technologies—or use them in certain, specified ways—they fit with their configurations and follow the technologies' scripts. In other words, they actualize their potential as "good" citizens.[5] Of course, moral and social sanctions can follow for those who choose not to use certain technologies—or use them in ways other than prescribed. In these cases, it may very well be that individuals not only become inappropriate users of technologies, they also fail in their civic responsibilities to use them, or to use them appropriately; that is, they become "bad" citizens. What seems to be at stake, then, is no less than what it means (and what it takes) to be a good citizen of the state itself.

These rather lengthy expository remarks now set the stage for our discussion of a single class of artifacts, or technologies, with a long history of use and also of innovation, and with certain rather special features. Our work focuses on vaccines against human infectious diseases. What sort of work goes into "configuring the user" of a vaccine? In what ways and to what extent is the (welfare) state involved in the development of vaccines and in processes of their provision and use? How are we to conceptualize markets for vaccines as institutional relationships in the vaccines field have shifted? Where do the tensions lie between users, states and markets? Related to these questions are a parallel set of questions surrounding the challenges that consumers and citizens mount as con-figured users of these technologies. Indeed, it is often only when indi-viduals (singly or collectively) resist the development of these technologies, or when they actively choose not to use those already devel-oped and deployed by the state, that it is possible to highlight the ten-sions that seem to exist not just between users and developers, but also between citizens and the state itself.

Vaccine Development

Vaccination is generally considered one of the great success stories of public health. Indeed, vaccines against infectious disease are widely viewed as the most effective and most cost-effective medical technology ever developed, as measured by deaths prevented (Plotkin and Mortimer 1994). Thanks to a determined (some would say ruthlessly efficient) vac-cination program, smallpox has nearly disappeared (Greenough 1995b). Poliomyelitis has officially disappeared from most of the world, and the goal of eradicating it globally has nearly been achieved (Sutter et al. 2001). International organizations, including the World Health Organiza-tion and UNICEF, support the governments of poor countries in vacci-nating their populations and routinely encourage and monitor efforts to extend vaccination coverage to reduce the incidence of infectious disease (Greenough 1995a). Cooperation by national authorities is expected, and more often than not it is granted.[6] At the same time, the armamen-tarium of the world's public health establishments is gradually being extended. Despite the insistence of politicians, the hopes, and the claim, there is as yet no proven vaccine against AIDS, although a number of can-didate vaccines are in clinical trials (Veljkovic et al. 2001). There is no proven vaccine against malaria or any other parasitic disease to which populations of developing countries succumb in their multitudes. Never-theless, new vaccines for other diseases are regularly under development, primarily in industrialized nations. Consider the following example.

In the United States, work has been under way for a number of years on a vaccine to prevent group B streptococcus (GBS) infection, primar-ily in newborns. The bacterium is particularly harmful in humans, and until only very recently it has been the leading cause of morbidity and mortality due to a bacterial infection in newborn infants.[7] Moreover, GBS infection affects a disproportionately high number of black women (and their babies) relative to the general population. After extensive research on the pathogen and the host immune response to it, scientists theorized that the most effective way to prevent this disease in newborns is to immunize the expectant mother while pregnant.[8] Early trials of a candidate GBS vaccine were only moderately promising; it was reported that "the vaccine was safe but not highly immunogenic" (NIAID 2000: 46). The trial was conducted in a small group of third-trimester pregnant women because, theoretically, it is during this period that vaccinated women should "induce sufficient antibodies and [subsequently] passively protect their newborns" (ibid.). The development of newer techniques, coupled with a more refined understanding of GBS itself has led to the development of newer vaccines. Currently, a candidate GBS vaccine is in phase III clinical trials, and should the data from these trials be favorable, it is possible the product will be licensed.

In the last 10 years, GBS disease incidence has waned because of the introduction of prophylactic antibiotic use during the latter stages of pregnancy in "high-risk" women, which includes women in labor who have fever, prolonged rupture of the membranes or pre-term delivery, as well as women who have been screened vaginally and rectally at 35–37 weeks gestation period for GBS colonization.[9] According to the Division of Bacterial and Mycotic Diseases branch of the Centers for Disease Control (CDC), "between 1993 and 1998 the incidence of group B streptococcal disease during the first week of life declined by 65 percent to an incidence of 0.6 cases/1000 live births. Additionally, the excess incidence of newborn disease among black infants as compared with white infants decreased by 75 percent. We estimate that in 1998, 3,900 neonatal GBS infections and 200 neonatal deaths were prevented."[10] This has occurred without the use of a GBS vaccine, yet its development still continues for two reasons. First, there are significant risks associated with increased antibiotic use to combat GBS, which leads to antibiotic-resistant strains of bacteria. Second, GBS still persists as "a leading cause of neonatal sepsis, resulting in approximately 2,200 infections each year among children aged <7 days in the United States."[11]

As one might imagine, the idea of administering a vaccine to pregnant women has presented some problems. By 1997 no commercial developer had stepped forward to carry the vaccine through product development. Industry officials have cited the potential for liability as a reason not to proceed with development: officials in both the public and private sectors understand that, given the route of administration (to the fetus via the expectant mother), this potential is enormous, despite the fact that legal and administrative mechanisms are in place to protect the vaccine industry from most lawsuits.

Who are the users (or potential users) of this vaccine, and what are their needs? How are they configured, and who is doing the configuring? If we were to think in terms of end users, the most intuitive answer would likely be pregnant mothers. Of course, in no insignificant way, the fetus also "receives" the vaccination and so, as with other antenatal technologies, could also be considered a "user" (Saetnan 2000: 16). To our knowledge the prospective end users, pregnant women (and certainly fetuses) have not articulated any need for a GBS vaccine, and the extent to which women (and fetuses?) would want to use it should it become available is by no means clear. The case of the GBS vaccine provides a particularly potent example of a technology in which users are configured largely by the state. After all, industry has steered clear of significant involvement in its development, and work has proceeded largely in state-run or state-funded laboratories.[12] Users are seen as willing to open both their checkbooks and their wombs to protect their newborns—actions that correspond with being active consumers and good citizens. User need is articulated as scientifically identified necessity and technological feasibility [through risk discourses; see Gabe 1995]: the CDC considers GBS to be a considerable threat to newborns, and NIAID considers a vaccine (administered to the fetus via the mother) to be the most safe and efficacious manner in which to counter this threat. Yet in all the data we have collected, there is no explicit mention of users, or women users, or women as members of racial minority groups; and there is certainly no mention of a (potential user) group or groups who have organized to advocate for this technology's development (as in the case of AIDS vaccines).

State involvement in the development of vaccines historically has been fundamental to their success. Government activities have ranged from basic research, through product development, the facilitation and the conduct of clinical trials, and bulk production. Today, as we discuss below, the role of the public sector has become more focused—and limited. The GBS example shows the state at work in the area where it retains an important presence: that of research. The example shows the state engaged in its work of configuring the potential user of its vaccine technology. It also shows the presumption of beneficence: of general approval and universal uptake underpinning that work. End users are being configured, as it were, as both active consumers and "good," passive citizens.

The role of the state in bringing new vaccines into use has been substantially reduced. Despite continued, even increased commitments to basic vaccine research by NIAID and other national laboratories, the role of public-sector institutions in vaccine product development has declined vastly in the course of the past few decades. The roots of this trend go back much further. Nearly a century ago research on, and the production of diphtheria antitoxin the United States shifted primarily to private pharmaceutical firms. Prior to this, these activities fell under the auspices of public-sector entities, namely state and local boards of health (Rose 1999). In this case, as the private sector gradually increased its production capacity, it was able to place its product on the market for purchase by those very same public-sector entities that initially developed it. More recently, the commodification of vaccines has gone further and deeper, with important consequences for inter-institutional research, paradigms, and research programs within* vaccinology, as well as for providers and users.

In a number of countries prior to the 1980s, institutional relationships were rooted in commitments to public health. Hans Cohen, who was for many years Director General of the Netherlands State Institute (RIVM) responsible for producing and supplying the country's vaccine needs, tells of his earlier relationships with industry, specifically with Pasteur Mérieux: "[Mérieux] got all our know-how, and we weren't always happy about that, but on the other hand we got a great deal of know-how back in return. For example, I got a rabies vaccine. We exchanged. It took three minutes. A matter of 'what do you want from me?' then the boss says 'I'll have some polio, and what do you want?' And I'd say 'Give me a measles strain, and some of that and some of that. . . .' It was good. Really a free exchange."[13] Similarly, vaccine researchers were a relatively homogeneous and small group, consisting largely of microbiologists and virologists. Knowledge was freely available and freely exchanged irrespective of one's place of work. This, of course, has changed, and corresponds with what Gibbons et al. (1994) have termed a shift to "Mode 2" knowledge production, which characterizes a shift away from strictly disciplinary research conducted within relatively rigidly defined public sector / private sector boundaries, to that which is carried out in multi-disciplinary settings across both public and private spheres (Blume and Geesink 2000). Vaccine-related research is now pursued by molecular biologists, geneticists, immunologists and organic chemists, among other relatively "newer" (sub)specialties and (sub)disciplines, as well as by microbiologists and virologists, all of whom work in competing networks and organizations that jealously guard their findings. The knowledge generated in these newer "vaccinological" networks is no longer freely available, and is increasingly protected by patents. For example, by 1983 a government survey found that only two patents for 27 vaccine products existed; a decade later, SmithKline Beecham had to assemble 14 patents to produce and market its recombinant hepatitis-B vaccine (Mowery and Mitchell 1995). The work of developing new vaccines, even in the public sector, has changed fundamentally. Take, for example, the following quote from a leading vaccinologist, commenting on the ways in which communicating information and disseminating knowledge has changed:

> In terms of the way in which the whole vaccine community talks to each other, my experience in going to meetings in the last 2 or 3 years is that in the vaccines field the number of commercial companies involved is really quite large. In the old days, you'd go to a conference and it would be mainly your colleagues, people from universities throughout the world. Now you see a lot of representation from companies, who are certainly willing to talk about their data, often talking about their data far more freely than academics would. [And] probably knowing that their basic technologies, or basic ideas, have been covered by patenting anyway. I'm sure that that's a key issue in the whole thing.[14]

The changes and trends we have just described have led a number of respected commentators to view vaccine development and production as having been "privatized" (Freeman and Robbins 1991), which corresponds to our assertion that a shift toward commodification in the vaccines arena has occurred. The state, that is, the "welfare" state, is retreating from traditional responsibilities in the area. Some countries, like Australia and Sweden, have sold their state vaccine institutes to the private sector. And, as we have just pointed out, private industry is coming increasingly to dominate the development as well as the supply of vaccines (Institute of Medicine 1985). What values, what commitments, guide the search for new vaccines, and how does this impact potential users? The historian William Muraskin (1998: 117) quotes a British public health official as saying that "the manufacturers were developing new [vaccines] without any regard for public health priorities, and by ignoring the problem of need, left public-sector officials open to being pressured into switching to new vaccines that had been designed to meet commercial, not public health needs."

State Immunization Practices and Vaccine Use

New vaccines can be expensive, as firms look in the first instance to recoup development costs and in the second instance to make a profit. Adding them to already-overburdened vaccination schedules can further strain the resources of health ministries beyond their breaking points.

Unsurprisingly, countries do not react in the same way to the availability of a new vaccine, any more than they do to the emergence of any other new medical technology. The nature and extent of state involvement in the provision and finance of health care greatly affects the introduction and the spread of vaccinations as it does of other health care interventions (Hollingsworth, Hage, and Hanneman 1990). How might this work in the specific case of vaccines? What might lie behind national differences? Linda Bryder (1999) has compared responses to the development of a vaccine (BCG) against tuberculosis, first developed by two French scientists (Calmette and Guérin) at the Pasteur Institute, roughly around the time of the First World War. She questions why the Scandinavian countries adopted BCG vaccination so quickly and enthusiastically, while the United States adopted it hardly at all (with Britain somewhere in between). Her answer has to do with distinctive conceptions of social welfare and the responsibilities that states see themselves shouldering: "Above all, the different policies appear to mirror the respective social welfare traditions and systems. Scandinavia adopted the socialistic policy of treating everyone in the same way, and chose to focus on preventive measures. In Sweden, for example, public health programmes have been described as 'constructed within a strongly egalitarian context.' . . . In the USA emphasis was placed on tuberculin as a diagnostic tool and the detection of early cases of treatment. . . ." (Bryder 1999: 1165) Thus the earlier practices in the Scandinavian countries, but also, for example, the Netherlands, had been to regard vaccination practices—and the development and manufacture of vaccines—as expressions of the state's responsibility for the health of its citizens. This is in partial contradistinction to the United States, which, as we noted, soon began to shift responsibility for the production of vaccines to the private sector.

In all countries, including the United States, state institutions (as well as individual health care professionals), mediate many of the relationships between the vaccine technology and the end user. The nature of this mediation is quite complex, and occurs along a number of different dimensions. Consider state involvement in the provision of vaccines, which can occur under the auspices of public health establishments (in a highly regulated market with, for example, price controls) or in the private marketplace (with relatively few price controls). In Europe there exists wide variation in the state provision of vaccines. For example, in Spain, 95 percent of vaccines are provided via public health authorities, whereas in the United Kingdom that number drops to 70 percent, and in Greece only 35 percent of vaccines are made available in this manner. In the United States, matters are more complicated as there exist a variety different federal and state programs for the provision of vaccines for persons who may or may not require assistance in their purchase. For the most part, though, the federal government facilitates vaccine provision by funding the purchase of vaccines in bulk at discounted prices. The CDC has been given primary responsibility for this (through the Vaccines for Children program), and presently federal and state purchases of vaccines amount to around 70 percent of total vaccine purchases.[15] Just as an ideal-typical end user is seen to impact processes of innovation through the representation of his or her needs, so too does the CDC, as an organization, act as a user in this regard. The CDC's role as a mass purchaser of vaccines suggests that a powerful state organization might influence vaccine development and subsequent use in two distinctive ways: both in advance (through its enunciation of needs based on its technical—for example, epidemiological—expertise) and through its purchasing power. This of course occurs within the context of specific vaccine marketplaces, and manufacturers may be less than enthusiastic about the market power of so large a customer (Rose 1998: 158–159).

States' immunization-related activities are not limited to the purchase and delivery of vaccines. Health ministries and other state-sponsored organizations have established standardized protocols and recommendations for use for most vaccines currently available on the market. Emphasis has been given to the establishment of schedules for routine pediatric vaccines (which vary by country), and these are now firmly rooted in most public health institutions across national contexts.[16] In the Netherlands, for example, a child is vaccinated for the first time at the age of 2 months with five different antigens (a combination vaccine against diphtheria, pertussis, tetanus, and polio, and a vaccine against hemophilus influenza-B). In the first 10 years of life, the child will have been given seven inoculations, including vaccines against eight different diseases. In some countries, including the Netherlands, vaccination is officially "voluntary," while in others (e.g., France and Italy) it is obligatory. However, the boundaries between what is obligatory and what is voluntary have become increasingly blurred as states have enacted policies which have tied together the provision of crucial social services and benefits with vaccination. Again, the United States provides an example here.

At the urging of the Department of Health and Human Services (DHHS), which under the Clinton administration greatly emphasized (and funded) the development of programs designed to improve vaccination rates and coverage, individual states began to establish statewide immunization registries and databases in order to track children's immunizations. The CDC (a branch of DHHS) provided grants totaling millions of dollars to the states to set up these registries, and to link the information contained in them to the records of federal and state aid and services programs. Recent developments in the immunization infrastructure of the various states have focused on tying certain necessary social services to immunizations. This entails, for example, linking supplementary nutrition vouchers administered under the Women, Infants, and Children program to immunizations, so that the former cannot be fully acquired without the latter. The Head Start program for at-risk children is similarly tied to immunization. Except under certain proscribed circumstances, parents must have their children fully immunized on schedule in order to receive this aid and these services (see e.g., NVAC 1991). At present, only 14 of the 50 states require that individuals consent to be tracked via the registry; almost all others presume implied consent.[17] Of this remainder, 12 states have no provision to opt out or limit access by other organizations or agencies.

In many countries, recommendations issued at the highest levels of government are designed to be implemented at lower (regional, provincial, state, county) levels. How, precisely, depends on the ways in which authority to carry out health-related or vaccination activities is vested in various departments or ministries of government. The following excerpt from a 1998 interview with two Swedish health officials in 1998 (conducted as part of a research project examining shifts in vaccine development and production from the public to the private sector) highlights some of the complexities of how authority is delegated at various levels of government.

DR: Let's talk about the regulatory powers that this agency does have. (pause) You may start anywhere!

AB: Now we're again back to recommendations.

DR: So that is regulation in a sense. . . .

AB: We say, "One ought to offer . . . " but we actually [mean], "You shall . . . "

JK: We are not allowed to say "You [shall . . .]."

DR: But for the most part, this agency carries the history, if nothing else, to have its recommendations followed through?

AB: Yeah, that's the basic premise, yes. If you compare it to [the United States], I think most of these regulations are made in the [executive] departments. And this is done so in most European countries, but not in Scandinavia. And as I said, this is a 300 year old way of balancing power. I mean, the Ministry would very much like to tell us what to do once in awhile, but to do that in Sweden it will be put up in a Constitutional Court. There is no one who could tell us what we should say. They can tell us what to do, but not in what way.

JK: Likewise, you could not call the counties and say, "You should [start] the vaccinations."

AB: It's very different with different legislation. I mean, with the Communicable Disease Act, there are actually no other limitations, which is actually a little bit peculiar, but we can actually say whatever we want. When it comes to the health legislation, it is more precise what we can do and not do. I think there are about 30 pieces of legislation which we are supervising. Not all of those follow here. . . But it depends on which area you're going into, which powers are given to the National Board.

Despite AB's declarations to the contrary, the situation in the United States is similar—at least insofar as state authority to carry out immunization activities rests at lower levels of government. Individual states are charged constitutionally with the health and well-being of their respective citizens, so while federal policies can be designed to improve the health of the nation's citizens, it is the states that actually implement those policies—often in uniquely different ways. With regard to vaccinations, the Advisory Committee on Immunization Practices (ACIP) of the CDC sets a recommended schedule and timetable which individual states have the option to follow or not. As authorized by various federal and state laws and mandates, state- and county-run public health clinics provide immunization services to much of the US population (primarily to poorer households and families), while private health-care provider organizations serve the remainder. Whether they receive their shots from public or private providers, most individual states now require that children receive 34 doses of ten different vaccines before their fifth birthday.[18]

What does it then mean to say, as many in the area of public health would be inclined to do, that pediatric vaccinations are typically a matter of parental choice: that use is a voluntaristic act? Consider the following (Streefland, Chowdury, and Ramos-Jimenez 1999: 1712):

People have their children vaccinated because everybody does so and it seems the normal thing to do. There are not necessarily deep reflections behind mothers taking their infants to the child health clinic. They do so because everyone else does, and because it is what good mothers seem to do. . . . And unless adverse effects or rumours about "bad" vaccines intervene, each collective visit to a vaccination session will reinforce the notion of normality. In this sense, all vaccination users are interdependent, as they support and are supported by each other's decisions.[19]

Some scholars have attempted to deconstruct traditional understandings of vaccination acceptance by pointing to differences between a community's active demand for vaccinations, and its passive acceptance of them. Nichter (1995: 617) delineates the difference as follows:

> Active demand entails adherence to vaccination programs by an informed public which perceives the benefits of and need for specific vaccinations. Passive acceptance . . . denotes compliance: passive acceptance of vaccinations by a public which yields to the recommendations and social pressure, if not prodding, of health workers and community leaders.

In either case the onus to establish and maintain (culturally "appropriate") vaccination programs ultimately lies with the state—whether or not there is an "actual" (active) public demand in either a formal market sense, or as an individual's expression of a socio-cultural need to be healthy or to protect one's child. Yet whereas most experts advocate more culturally sensitive strategies to increase vaccination acceptance and coverage levels (as a way to induce Nichter's notion of "active demand"), we should not forget that ultimately, the power to coerce is retained by the state. Greenough (1995: 633), for example, notes that "public health measures derive their authority from the police powers of the state." Thus, while vaccination acceptance may be perceived ideally as a consequence of active demand by an informed public, it often remains the case that the public is creating that demand partly in response to coercive measures.[20] In other words, the state resorts to compelling or coercing immunization, despite the fact that in many countries the decision to use vaccines is ostensibly a matter of choice.[21]

Kevin Dew (1995), in discussing proposed policies in New Zealand that would also tie together school records with immunizations, has commented on an important qualifier to the notion of choice in that country, noting that what parents really face is not so much "choice" as "mandatory choice." The contradiction inherent in the terms is intentional, and is meant to illustrate the fundamental tension that exists between the "choice" that citizens have to be vaccinated (or not), and the demands of the state that virtually require that parents make the choice to have their children vaccinated. The effect of all this is that "mandatory choice is as near to compulsory vaccination as one could get" precisely because, in the context of this example from New Zealand, if any breakout of a vaccinatable disease occurred, unvaccinated children would be required to stay at home ostensibly to protect them from getting sick. "In other words," Dew notes, "they will be discriminated against in the educational system" (ibid.: 23).[22]

Just as our example of GBS vaccine illustrated how users are configured in laboratories along market and citizenship dimensions, so too can we begin to see state policies and institutions act to configure users of technologies in similar ways. In the context of welfare states that have established markets in which vaccines are exchanged (regulated or not), individuals as end users are configured as demand-creating consumers of these products precisely because the state has the authority to compel the decision to choose for vaccination. And it is precisely this, in turn, that results in the configuring of users as good, passive citizens.

Individual and Collective Protection

There is a long-standing debate regarding the appropriateness of the market as a mechanism for the supply, and perhaps "harvesting," of human body parts and other biological materials (e.g., Titmuss 1971). In the case of blood, for example, controversy has involved appeals to efficiency, to equity, to safety and to innovativeness (see, e.g., Starr 1998). In regard to vaccines, just as to other biological products, the relative merits of market and non-market modes of supply and distribution have been debated. Yet there is an additional issue here that is peculiar to vaccines. Vaccines can confer both individual- and community-level protection against infectious diseases. At the individual level, vaccines typically impart partial to full immunity against particular diseases by way of stimulating the production of various types of antibodies. At the population level, protection is achieved through a mechanism known as "herd immunity," which operates as a sort of societal-level barrier against the introduction and spread of disease-causing antigens.[23]

The dual nature of the protection afforded by many vaccines is also a source of tension between the view of vaccination as a voluntary, individual act, and the view of it in terms of the health of the wider public. Occasionally vaccines are developed that ultimately work "too well": the incidence of disease in a community may drop, yet a small number of individuals may either become infected by the disease itself or suffer side effects attributable to the vaccine.[24] At the state level, this can present serious complications in efforts to effect meaningful public health policies. Where do states draw the line between a vaccine's benefit to the community and the risks it may impose on individuals? States as policy makers, but also as users of vaccines themselves (to effect those policies), are required to make very difficult decisions—and these decisions, in turn, can have awkward political consequences.

We can understand some of the implications of all this by reference to a long-standing (more than 30 years) debate surrounding the relative merits of the two polio vaccines: inactive polio vaccine (IPV, initially developed by Jonas Salk, which contains killed virus) and oral polio vaccine (OPV, developed initially by Albert Sabin, which contains weakened, or attenuated, virus). At the beginning of the 1960s, as the OPV became available, debate began. A major argument used by some health officials and medical authorities in favor of the OPV was the expectation that it would protect the community. Take, for example, this comment by the editor of the *British Medical Journal*, writing during the formative years of the debate (*BMJ* 1964):

Immunization against poliomyelitis should aim not only at protecting the individual but at establishing a community immunity which should lead to the complete eradication of the disease as well as the causative viruses. Many experts believe that the Sabin oral vaccine, when used on a wide enough scale, can bring about this state of affairs. . . . It is able to produce not only a substantial humoral immunity, as shown by the presence of circulating antibodies, but also a resistance of the intestinal tract to infection. This latter type of immunity, when vaccination is on a wide scale, is highly important in breaking the chain of transmission of infection and in leading to elimination of the poliomyelitis viruses from the population. . . . Although Salk vaccine is effective in protecting the individual from the paralytic disease the intestinal tract of the immunized person remains relatively susceptible to infection, unless the vaccine is of high potency.

The point here was that the OPV was thought to provide a more "natural" resistance to re-infection, but would also interfere with the spread of polio through fecal matter and sewage. Attenuated live virus, excreted, would help protect those who had not been vaccinated. In the United States, Britain, and in most of the world, OPV gradually replaced the inactivated vaccine. But not in the Netherlands or Sweden. How could this be? How could these arguments for the greater social effectiveness of the OPV fail to have swayed Dutch and Swedish public health officials?

It has partly to do with how effectively polio had already been beaten back and—perhaps still more importantly—with what still remained to be accomplished. In the Netherlands, as in Sweden, polio had virtually disappeared. "Virtually" because the Netherlands was still faced with outbreaks among groups with a religious objection to vaccination. Health officials looked at these figures, and they looked at what other countries had achieved. Perhaps there was no universally best solution. Thus the editor of the *British Medical Journal* wrote:

The best type of vaccine to be used for primary immunization against poliomyelitis in each situation in different countries and climates has yet to be clearly defined, though the tendency in most countries in the past year or two has been to give the oral vaccine to children in a large-scale programme because of the ease of its administration. (*BMJ* 1964)

And while such programs have almost always been extremely successful,

In Sweden, . . . exactly the same result has been obtained by the use of a potent inactivated virus vaccine, and live virus vaccine has not been used. (ibid.)

It is not difficult to imagine that the potential advantage of herd immunity conferred by the OPV—by protecting the unvaccinated through indirect means—would be more persuasive the greater the distance still to go. In that sense we can imagine that the Dutch and Swedish health authorities were not persuaded of the need to change to the OPV. However, matters were different in the United States and in Britain, where hundreds of cases still cropped up annually, and where the promise of herd immunity could be of considerable importance. But in the United States controversy would not die down. According to a summary published in Nightingale 1977:

The decline in the number of adequately vaccinated persons, new data bearing on the controversy about safety and effectiveness of live, attenuated as compared with killed virus vaccine, increased interest by consumers in information about and protection from adverse reactions, and pressure from manufacturers seeking adequate protection from liability were factors that led the Department of Health Education and Welfare to request a re-examination by the Institute of Medicine of poliomyelitis vaccine programs.

In March 1977 the Institute of Medicine, an organization chartered by the National Academy of Sciences, delivered that report.[25] A major issue was the small but politically significant risk associated with use of the attenuated vaccine. Between 1969 and 1976, 132 cases of paralytic polio had been reported in the United States, 44 of which were classified as resulting from the vaccine itself. Related to the numbers of doses of vaccine used, or the numbers of people vaccinated, the risk is estimated at one in anything between 4 million and 23 million, depending on the way risk is calculated. "Such a risk would be acceptable," Dr. Nightingale (the project's study director) wrote, "except that countries using only IPV report no serious complications." Would it thus make sense for the

United States to abandon the attenuated (Sabin) vaccine in favor of the inactivated (Salk) vaccine? European countries using only IPV had managed to protect their populations without this risk of vaccine-attributable disease . . . but countries like Sweden and the Netherlands had vaccinated more than 80 percent of their populations. This was not the case in the United States, where the percentage hovered around 60 percent.

Could we say, then, that in light of its "success," end users of the OPV were again configured both as active consumers of the product, but more importantly as individuals who could accept the risks associated with the live vaccine? In other words, did the state configure the intended users of the OPV as passive citizens who would (or had to) accept the risks associated with the use of the technology, and bear any subsequent burdens? How do these issues tie into the technology itself, which by its very nature provides two kinds of protection (individual and collective), and yet which somewhat ironically puts users in a situation in which the one kind of protection could come at the cost of the other? This point, to which we will return in the conclusions, has to do with a possible "affinity" between the characteristics of a given vaccine and those of a society in which it is to be used.

Resistance

Much of what we have discussed so far makes clear that processes of user configuration do not begin and end in the laboratory, but are articulated and modified in the rules and conventions that govern the use of vaccines (or other medical technologies) in practice. In addition to explicit legislation and recommended schedules, these rules and conventions are expressed in a host of different ways: in the inserts that accompany drug packets, in the labeling of drugs, in the criteria for reimbursement, and in the standard practices of health care providers. All of the things that constitute the assumptions regarding vaccine use which are embodied in practice clearly differ from one country to another. But we can also see similarities and patterns in the ways in which (potential) users, individually or collectively, reject these assumptions and try to redefine their relations both to the vaccine technology and to the state which encourages or obliges its use.

One way to engage this issue is to examine the context in which expressions of dissent and acts of non-compliance take place. In both cases, users, or potential users, are key actors, and their configuration as such becomes all the more important to understand. The tendency has been to assume that users "are configured"—a term that connotes that someone or something other than individuals themselves are doing, or are even able to do, the configuring. Yet it is easy to overlook the fact that individuals also configure themselves through their actions as users or purposive action as non-users. In the literature, much has been written on the former point (some of which we have already reviewed) without the countering perspective of the latter.

Second, and related, potential users of technologies are heterogeneous (as we have noted), both as a collective group (i.e., "polio vaccine users") and as individuals. Clarke and Montini (1993) have argued this already, as has Akrich (1992: 177). Yet Akrich's prescription for this—to "align" the relevant networks (ibid.: 177) in order to reach some sort of compromise, some stabilization of user configurations ("representations," in her terms)—seems to us too simple, and does little to address the sorts of fundamental tensions that arise precisely because of that very heterogeneity. Users as consumers should be able to express dissatisfaction with the products they buy—whether on an open or highly regulated market—which acts not only to modify a technology's script, but which as a consequence takes the technology itself either off the market or back to the drawing board. Users as citizens should be able to draw on the very discursive, political, or material resources that define and constitute them as such to effect changes in their relationships to the state and the public health establishments that "demand that they surrender their immune systems as a public duty" (Greenough 1995: 606).

Users and non-users alike have occasionally demonstrated a remarkable ability to define themselves on their own terms. This has manifested itself in a number of ways, but invariably intended users have shown the capacity to modify the scripts ascribed to them (i.e., through the technologies they use) by others (Akrich 1992; Oudshoorn 1998; van Kammen 2000a). Opposition to vaccination has as long a history as routine vaccination itself, although Greenough (1995: 633) points out that "resistance in the sense of overt acts of refusal appears less common in the present than in the past, when vaccination campaigns triggered both street riots and sustained struggles to overturn compulsory vaccination laws in nineteenth-century American and Europe." Nevertheless, while acts of belligerent disobedience may be on the downswing, organized civil opposition has sprung up virtually everywhere. Coalitions have been formed, and organizations set up either to resist policies that have the effect of compelling vaccinations in certain populations, or to oppose routine pediatric vaccinations altogether. In the context of states that in essence mandate vaccinations, these groups essentially represent conscientious objectors (Veenman and Jansma 1980: 24), that is, individuals configured as users but who are or become, in fact, non-users. Their acts of resistance are typically grounded in philosophical, religious or medical claims, yet in all cases these individuals and groups situate themselves in opposition to the claims of the state, which configures these individuals by drawing upon its monopolized authority over its subjects to compel or coerce immunization (directly or indirectly), or its evocation of indisputable medical knowledge as the basis for its policies.

In the United States, groups like the National Vaccine Information Center (NVIC) have been established to disseminate information on and advocate for increased vaccine safety, as well as policies that defend "the human right of all people to make informed, voluntary decisions about medical interventions which can cause injury or death, including vaccination."[26] The claims of the NVIC correspond with some of the tensions that we discussed earlier, and highlight some of the intricate links and perceived asymmetrical relations between the state, its citizens and the market in which vaccines are exchanged. Barbara Loe Fisher, co-founder of the NVIC, has argued publicly that the laws, regulations and policies that govern vaccine use run counter to the same policies that govern the rights of citizens to be aware of, and make decisions based on, their informed consent for medical interventions and treatments. "Parents," she has noted, "are tired of being forced, without their informed consent, to use every vaccine the drug companies produce and public health officials decide to mandate."[27]

In testimony before the US Congress, Fisher argued in essence that potential users of vaccines are not passive recipients of these technologies at all, and must not be treated as such. Parents, she would claim, are more than just obliging citizens beholden to the mandates of the state's public health establishment and willing consumers of new medical technologies: they are intelligent, agentic and ultimately capable of making appropriate decisions for their families with regard to the purchase (if necessary) and use of vaccines. Should parents come to the conclusion that immunization is not in their child's best interest, she would argue, then they should have the right to refuse it: "We're asking for the right to exercise conscious belief exemption if we believe our children are at great risk of having a reaction. . . . Parents have got to have the right to have the information and then make informed decisions for their children. . . . We've got to give parents more credit." (NVIC 1999)

Resistance to vaccination has also come from those organized to oppose the use of specific vaccines. For example, in France, a strong movement comprising various organizations as well as private individuals has arisen in opposition to the use of the hepatitis-B vaccine. The movement represents an assortment of prospective users: one set of users includes health care workers and students, who are required to be vaccinated against the disease; another set, parents, are strongly recommended the vaccination for their children. Both groups feel that the health authorities and the "vaccine producers" have strongly exaggerated the dangers of acquiring hepatitis-B (French National League for Liberty in Vaccination 1998). They have argued that a needless push for immunization has resulted, one consequence of which is a suspected (causal) association between the use of the vaccine and a higher observed incidence of multiple sclerosis and other neurological disorders. Their action, which led to a class-action lawsuit, further led to a revocation of the mandate for school children to use the vaccine.

In the United States, numerous hearings have been held to discuss the required use of the same recombinant hepatitis-B vaccine. Importantly, it is more than just parents who are voicing their concern over the policies surrounding it. In a statement before the House Committee on Government Reform and Oversight, the Association of American Physicians and Surgeons, which represents health care providers "devoted to defending the sanctity of the patient-physician relationship," emphasized the threat that universal mandatory vaccination programs present to that relationship. Their spokesperson, Dr. Jane Orient (1999), stated the following:

Once a vaccine is mandated for children, the manufacturer and the physician administering the vaccine are substantially relieved of liability for adverse effects. The relationship of patient and physician is dramatically altered: in administering the vaccine, the physician is serving as the agent of the state. To the extent that the physician simply complies without making an independent evaluation of the appropriateness of the vaccine for each patient, he is abdicating his responsibility under the Oath of Hippocrates to "prescribe regimen for the good of my patients according to my ability and my judgment and never do harm to anyone." Should a physician advise against a mandated vaccine, he faces increased legal liability should the patient acquire the disease. Moreover, he may risk his very livelihood if he is dependent upon income from "health plans" that use vaccine compliance as a measure of "quality."

Resistance to vaccination, growing in many countries, is a major source of concern to public health authorities and vaccine producers alike. Should we view it in terms of consumer activism . . . or of militant political individualism? Or is the answer "somewhere in between"?

Conclusions

Recent attention in STS on the users of technologies has focused on their various "configurations." Yet there has been little or no attention to the structures (markets or non-markets) within which technologies are developed and made available (Blume 1992), let alone to the influence of the state on these structures. It is thus no surprise that the state's role in the direct or indirect configuration of a technology's intended users has escaped attention. While this may be a late ripple of a trend which has occurred throughout the social sciences, we believe that it is time, here too, to "bring the state back in."

In the 1960s and the 1970s, when political attention was focused largely on military, space and nuclear technologies, the importance of the state and its agencies as users of technology was apparent. It makes sense, even today, to explore the redesign of a space station or a weapons system in terms of changing "user" (i.e., political) priorities. But today political preoccupations are addressed to other technologies, and the involvement of the state in their genesis, introduction and use is both less apparent and a less fashionable direction of social scientific inquiry. In this chapter we have chosen to look at a technology which embodies many of the characteristics of today's priority technologies. Vaccines are manufactured on a vast scale. They are global, portable, and biological. They are made (or at least some of them are) by the most modern tools of biological science. States and public health establishments have proclaimed them our major hope in confronting the emergent diseases (and other social "problems," such as fertility) that threaten us. In the context of this technology, not unrepresentative of today's biology-based technologies, we try to understand what sorts of configuration work goes on; how and in what ways the state itself helps to configure—directly or indirectly—the user of the technology.

The configuration of the user of a technology often begins in the laboratory, but it does not end there. The mechanisms, rules and conventions governing the technology's use in practice extend and perhaps modify the work of configuring started in the laboratory. How does this apply to vaccines? Vaccines are developed in complex configurations of public- and private-sector institutions and their provision facilitated through (mutually interdependent) mixes of market and collective mechanisms which differ from country to country. In this chapter, we have discussed shifts in each of these areas: in the development of new vaccines, as well as a shift toward the exchange of these particular products on the marketplace.[28] In these regards, commodification in the vaccines field is said to be taking place. Despite the fact that vaccines are considered among the mightiest weapons in the arsenals of the world's public health establishments (both at national and international levels), the competence to go forward with their development and manufacture has been effectively monopolized by the private sector. A tension seems now to exist between the notion of vaccines as commodities, on the one hand, and the notion that they are collective goods, on the other.[29]

Though the role of the state in vaccine development and manufacture may have declined from what it was three or four decades ago, it remains important. States not only play a major role in underwriting and per-forming basic research aimed at the vaccines of the future, they also deter-mine the provision of vaccines. This is crucially important to our argument that states help to configure users in certain kinds of ways. Across national contexts, the state permits the existence of markets in vac-cines and it regulates them (by and large) as it does other drugs and bio-logical products. It may also make vaccines freely available, while at the same time encouraging or even demanding their use. Though in public health discourses it is conventional to distinguish between obligatory and voluntary systems of vaccination, the distinction in practice is less clear cut. Through a variety of mechanisms, parents (at least) are typically encour-aged or coerced into ensuring that their children are vaccinated against a number of infectious diseases according to an officially determined sched-ule. In most Western industrialized nations, the state thus configures two types of vaccinee: the consumer of a commodity, and a more passively ori-ented public citizen, one whose actions as a user of these technologies defines that person (or those persons, in the case of two parent families) as fulfilling a civic responsibility—as being a good citizen.

We have also noted that vaccines confer protection against disease both on individuals and on communities. The dual nature of the market in which they are exchanged is in a curious sense mirrored by, and in some ways coupled to, the dual nature of vaccines' protective working. How does this duality affect the practice of vaccination, and what sorts of consequences does this have for individuals as configured users of these products? Some individuals choose to forego immunization because they perceive most other individuals to have been vaccinated already: the ben-efits, not to mention the time and possibly expense needed to secure a health provider appointment, do not seem worth the effort. This is an example of the economist's classic free-rider problem. We saw how some-thing like this may have played a role in the debate regarding the relative merits of the two vaccines against polio. Yet there is much more at work here, and we will return to the question of resistance to vaccination shortly.

The evidence we presented suggests that some states rely on herd immunity more than others to meet their collective health needs. In the United States, we saw how for decades the community and the state were able continually to reap the benefits which the OPV afforded through the mechanisms of herd immunity (via passive immunization) despite relatively low vaccination coverage. We could therefore see herd immunity serving a number of purposes: through its "technical" (biological/immunological) workings, it serves to protect populations; it acts as a counterweight to the market when state provision of the vaccine (and subsequent coverage) does not reach optimal levels; and it serves as a vehicle that allows states to require that its citizens do their part to ensure that the community is protected.[30]

All vaccines, like all drugs, present a potential risk to individuals. Some vaccines may present more risks than others—yet they may also confer more benefits. When states carry the (sole) authority to articulate the risks of vaccines (either directly or indirectly, e.g., via approved drug labeling), and further articulate the risks of diseases for which those very vaccines are being developed, they are in a very strong position to configure the users of those products. As states come to understand and accept that individuals will face risks associated with vaccines, and as they further institute policies that function to compel or coerce immunization, states act ultimately to configure users as "good" and passive citizens. In other words, citizens are configured as individuals who would submit to the mandates or coercive practices of the state's public health establishment, even if this means some risk or danger to themselves. For vaccines that induce herd immunity via passive immunization, this type of configuration becomes all the more solidified. However, this technical requirement is not a necessary factor to configure users as good, passive citizens. Our GBS vaccine example provided an example of this type of configuration work: women, and in particular pregnant women in their third trimester, have been configured as willing recipients of the vaccine. They, too, are passive citizens—and presumably active consumers. Who, after all, would want to put their expected child at risk of infection when a perfectly good and safe vaccine may very well become available—particularly when the state says so?

Over the course of this chapter we have made explicit mention of "the" state, and occasionally the "welfare" state. We of course acknowledge that the states we have considered vary quite a bit in their actual policies and practices. But in certain fundamental, if limited ways, virtually all Western industrialized nations may be characterized as welfare states. This is a consequence of (historical) shifts within these countries in which the state acts to intervene "in civil society to alter social and market forces," ostensibly to guarantee some type of equality of social provision, benefits and well-being (O'Connor et al. 1999: 12). The empirical question of actual progress toward equality in various aspects of social life is beyond the scope of this chapter. We can, however, begin to see differences among welfare states in terms of the roles they assume, and the responsibilities they take on or delegate in providing for the health of their citizens with respect to vaccination. In light of this, we can construct ideal-types that reflect these differences. On one end of the spectrum, the (socialized) welfare state will institute policies that guarantee basic rights to health, and access to health care, including vaccinations. The state assumes the responsibility for its citizens to achieve healthy living, including universal access to vaccinations, ideally at no cost to them. At the other end of the spectrum, the (liberal) welfare state is governed by a different set of principles. Most notably, while social rights like the right to health may be explicated by the state, its achievement becomes the responsibility of its citizens.

Part of our argument is therefore that citizenship is becoming a very important dimension with which to talk about users of technologies that are in large measure developed, and/or their provision made possible, by the state. In the context of our study, this is so precisely because the growth of vaccine development, provision and use in Western, industrialized countries has coincided to a large extent with the growth of the welfare state itself, and because our discussion of the welfare state presupposes that the individuals who constitute it are rights- and obligations-bearing citizens who in some manner realize those rights and obligations through their interactions with, and conformity or non-conformity with policies and practices meant to maintain social order and well-being. This is a fancy way of saying that vaccine users (or non-users), by the very decisions they make, contribute to their own configuration not only as users (or non-users), but as citizens of the state itself. Nowhere is this more apparent than when individuals act in ways that run counter to how they have been configured.

As we have already discussed at some length, intended users are typically configured as both passive (and "good") citizens and active consumers. Sometimes, and indeed increasingly, resistance to these configurations emerges. Parents demanding the right to purchase a two-component vaccine because one component of the obligatory three component vaccine is felt to be dangerous, is an example (Gangarosa et al. 1998). Examples of anti-vaccination movements can be found in many countries, and we have illustrated a few examples above (and there are, of course, countless more). Taken together, all of these examples lead us to conclude that traditional configurations of vaccine users are faulty. Naturally, this depends on context (not the least of which would take into account both national and cultural differences). Nevertheless, we have demonstrated not only that tensions exist when users are configured in these more traditional ways, but that individuals possess the kind of agency which could, for instance, configure them as non-consumers or non-users. We would like to emphasize, however, that individuals can also configure themselves as active citizens, as opposed to passive ones. Seizing upon the very "resources" which define them as citizens, namely as rights- and obligations-bearing individuals in certain relations to the state, they have been able to claim the right to be healthy and achieve health in opposition to the state's claims and policies (which often entails compelled or coerced immunization). In other words, by acts of resistance and opposition to the use of vaccine technologies, these individuals help to highlight the contours of citizenship itself. In light of this, what it means to be a "good" citizen is open—again—to reevaluation.

Acknowledgments

We thank Nelly Oudshoorn and Trevor Pinch for their efforts in organizing special sessions at the 1999 Society for the Social Studies of Science conference in San Diego, and subsequently for putting this volume together. We gratefully acknowledge the participation of our interviewees in the United States, the Netherlands, the United Kingdom, and Sweden. Dale Rose thanks Stuart Blume and Adele Clarke for being wonderful mentors; Renée Beard, Jennifer Fishman, Jennifer Fosket, Chris Ganchoff, and Sara Shostak for their intellectual prowess and unwavering personal and professional support; and the Graduate Division and the Department of Social and Behavioral Sciences at the University of California, San Francisco for its financial support and generosity.

6

Knowledge Is Power: Genetic Testing for Breast Cancer and Patient Activism in the United States and Britain

Shobita Parthasarathy

When discoveries of the BRCA genes, two genes linked to inherited susceptibility for breast cancer, were announced in the mid 1990s, attention turned almost immediately to the development of related diagnostics and therapeutics (Davies and White 1995). In the United States and in Britain, groups began to develop technologies to test for mutations in the BRCA genes that predicted an inherited susceptibility to breast and/or ovarian cancer. Because genetic testing for breast cancer was the first genetic testing technology for a common disease, however, a variety of political actors, including scientists, activists, clinicians, biotechnology companies, and government officials, struggled to influence its development. This chapter examines the politics of developing genetic testing for breast cancer in the United States and Britain and specifically, the role of patient advocacy groups, in order to understand how national contexts frame how users matter in the development of a medical testing technology. How did patient advocates try to influence genetic testing for breast cancer? Did their efforts differ in the United States and Britain? How did national context figure in the activism of patient groups regarding the new genetic testing technology?

Recent scholarship in the emerging field of science and technology studies has shifted from a focus on the construction of technologies themselves to the relationships between technologies and their users. Feminist scholars have demonstrated how particular understandings of the user, such as gendered assumptions and constructions of the female body, are embedded in the very design of technologies such as household goods and even synthetic hormones (Wajcman 1991; Akrich 1995; Oudshoorn 1994). Other scholars have focused on how activist communities influenced the directions of scientific research and technological development (Epstein 1996; Kaufert 1998). Steve Epstein has shown how American AIDS activists were able to pressure the Food and Drug

Administration (FDA) to speed up drug approvals and also were able to enter the previously closed doors of peer-review committees and influence research funding decisions.

This chapter adds a comparative dimension to the study of users focusing on the influence of patient advocacy communities in the development of genetic testing for breast cancer in the United States and Britain. The United States and Britain are particularly good sites for this comparative analysis for a variety of reasons. Both the United States and Britain are English-speaking and affluent Western capitalist democracies with very close ties to one another as well as many shared political traditions. There are also a number of aspects specific to the case of genetic testing for breast cancer that make the United States and Britain rich and compelling sites for this comparative analysis. First, scientists in both countries were heavily involved in the effort to find the breast cancer genes. Researchers at US genomics company Myriad Genetics were credited with finding the BRCA1 gene, while British geneticist Mike Stratton from the Institute for Cancer Research was credited with discovery of the BRCA2 gene. Second, incidences of breast cancer (and breast cancer gene mutations) in the populations of the two countries are considered to be equally high.[1] Both the United States and Britain have exhibited significant commitments to genetics and biotechnology research, as exemplified by their leadership in the Human Genome Project, the effort to map and sequence the entire human genome.

Despite these similarities, however, we might easily imagine that national approaches to health care could structure the politics of developing genetic testing for breast cancer in very different ways in the two countries. Britain has a government-run National Health Service (NHS) that guarantees health care to all its citizens while the United States relies on a private health insurance market for the provision of health care (Starr 1982; Klein 2001). Furthermore, many scholars have argued that the British NHS is based on principles of public health and equal access, while America's private insurance system is based on competition and consumer choice (Blume 1992; Ashmore, Mulkay, and Pinch 1989; Skocpol 1996).

This chapter begins with a brief description of the development of genetic testing for breast cancer in the United States and Britain and then explores how patient advocates responded differently to the new technology in both countries. It closes with some discussion about the relationship between national context, technology, and users and the utility of this comparative analysis for the study of users of technology. In addition, like

Jessika van Kammen's article in this volume, the chapter problematizes the relationship between activists and individuals for whom they speak.

Developing Systems of Genetic Testing for Breast Cancer

On September 13, 1994, Tom Brokaw opened his nightly newscast on NBC as follows: "There's an important breakthrough in breast cancer research. . . . A rogue gene could show the way to treatment and prevention. Scientists think the gene is responsible for one in twenty breast cancers." (Brokaw 1994) NBC News deemed the discovery of the BRCA1 gene, which was led by a team of researchers from Myriad Genetics, a genomics company based in Salt Lake City, so important to the American public's immediate welfare (and NBC's ratings) that it broke a press embargo imposed by *Science* magazine (Angier 1994b; Saltus 1994; Brown 1994). Its announcement emphasized the value of the discovery for both prevention and treatment for all breast cancers. NBC's excitement was not unique, however. Soon, news of the discovery had spread throughout the media and newspapers across the world announced the discovery on their front pages. The discovery of the second gene linked to breast cancer, BRCA2, by Mike Stratton at the Institute for Cancer Research in England in December 1995 only intensified the excitement. An article in the newspaper *The Scotsman* (Christie 1996) noted: "The discovery may result in the development of a screening test to identify those at risk of contracting breast cancer. They could then be monitored more closely, enabling the identification at the earliest treatable stage."

In the United States, four very different providers began to develop BRCA testing services on a large scale immediately after the gene discoveries in the mid 1990s. While their services varied considerably, each used an approach that had roots in the American biomedical context. As a research laboratory at an academic medical center, the University of Pennsylvania's Genetic Diagnostic Laboratory offered individuals who visited a genetics clinic at an academic medical center access to an experimental method of analyzing the BRCA genes. Oncormed, a start-up biotechnology company with close ties to the medical genetics community, offered high-risk individuals access to its laboratory analysis services through research protocols organized by itself or other investigators. The Genetics and IVF Institute, a reproductive services clinic, offered both clinical care and laboratory analysis of the BRCA gene mutations common among the Ashkenazi Jewish population under one roof. Finally, Myriad Genetics, another start-up biotechnology company who had been credited with finding the first BRCA gene, offered BRCA testing like any other medical test: individuals could use its DNA analysis service through any physician. Of these four, Myriad Genetics developed the largest-scale testing service. By allowing access to any individual who received a referral from any physician, Myriad ensured that the potential market for its service was quite large—it was available to anyone who could afford it.

In contrast to the variety of providers that developed BRCA testing services in the United States, BRCA testing services in Britain were provided on a regional basis through the state-run National Health Service. Its shape was reminiscent of other specialist services in the NHS, involving both risk assessment and triage. Individuals interested in testing would first provide their family history of breast and ovarian cancer to a primary or secondary care physician in their region. Then, using a standard that had been developed in consultation with geneticists across the country, these physicians would classify individuals into low-risk, moderate-risk, and high-risk categories and offer services accordingly. Only individuals classified as at high risk would be allowed to visit a regional genetics clinic and access both counseling and laboratory analysis. BRCA testing services were actively being developed in both countries.

Patient Advocacy in the United States

Breast cancer advocacy groups, who had been particularly powerful in the United States since the early 1990s, immediately got involved in influencing the development of BRCA testing in the United States. After securing a six-fold increase in federal research money devoted to breast cancer in 1991, the National Breast Cancer Coalition (NBCC), a Washington-based advocacy group, had ushered in an era of breast cancer activism where advocates were frequently invited to join government advisory committees, speak at congressional hearings, and sit on peer-review committees where decisions about research funding were made (Stabiner 1997; Love 1995). With regard to NBCC's presence in biomedical policy making, one member noted: "I think for the Coalition, I just think that we have a much more reasoned, analytic way of looking at problems. And I think we have, I know we have the respect of many people on the Hill, when they have a breast cancer issue, they call the Coalition to see what we have to say." (interview with National Breast Cancer Coalition representative 1999). Breast cancer activists were now called upon by both the media and government to comment on breast cancer advances, health-care controversies, and funding debates alike.

The discoveries of the BRCA genes and development of genetic testing for breast cancer were no exception.

Despite the excitement expressed in the media regarding the promise of BRCA testing, most breast cancer advocacy groups took a particularly cautious approach toward the new technology. They first spoke out vociferously in the media, arguing that the new information would not create the panacea that the media predicted and test providers promised. Nancy Evans, a representative of BCA was quoted in an article published immediately after the first gene discovery, saying, "It's a very mixed blessing to have this knowledge. . . . It's the first step in a long journey, and the journey is probably across a minefield." (Angier 1994b) Fran Visco of NBCC expressed concern as well, noting that "women will have to be very careful. . . . You're talking about giving them a test telling them they have an 85 percent chance of getting a disease that we don't know how to prevent, and for which there is no known cure." (Angier 1994a) Both groups argued that providing genomic information without effective therapeutics was potentially dangerous as it made women anxious while providing them with no avenues to relieve their worry. They felt that the new technology had unknown implications and required additional care by test providers, health-care professionals, and patients alike. Overall, they constructed BRCA testing as potentially dangerous and requiring caution in its integration into health care, while arguing that women needed protection and did not have the right to simply demand testing.

Many of these advocacy groups did not stop at comments in the media and issued more methodically argued and organized position papers, press releases, and articles through their newsletters, criticizing the rapid provision of BRCA testing as a commercial service by organizations such as GIVF and Myriad Genetics. "Because much more needs to be researched about the sensitivity, specificity, and reliability of the genetic tests," NBCC recommended (1996b), "and because not enough is known about the effectiveness of genetic education and counseling, genetic testing should only be available within peer-reviewed research protocols." Though BCA stopped short of recommending access to the test only through research, it too suggested that "no one should be tested without access to education and counseling concerning all benefits and risks of genetic susceptibility testing" (Breast Cancer Action 1996).The National Action Plan on Breast Cancer, a public-private partnership funded by the NIH but including activists and scientists, also recommended that BRCA testing only be made available in the context of research (National Action Plan on Breast Cancer 1996). The perspective of this public-private group was not unexpected, however, because an investigative testing system would surely benefit researchers funded by the NIH by increasing both funding and freedom to conduct research in this area while satisfying advocacy groups by ensuring a cautious approach to testing.

Advocacy groups defined BRCA testing as a new technology to which women had a right to controlled access as well as a right to be protected from bad medical choices. They suggested that individuals should only be given choices among good medical options, as opposed to options simply defined by the market, such as Myriad's BRACAnalysis service. Instead, they argued, citizens and patients deserved protection by health-care professionals and the state against potentially dangerous technologies.

As they enlisted the help of government and the biomedical community, activists also presented themselves as appropriate authorities that could distinguish between good and bad medical options. As BCA methodically detailed reasons why testing was dangerous, for example, it displayed its own ability to distinguish between good and bad science: "It is equally clear that the BRCA1 test for genetic susceptibility is not the early detection tool we need . . . a positive result from the BRCA1 test does not mean that the person tested will develop breast cancer. (Nor does a negative test mean she is not at risk.) And, even if a positive test meant a woman would certainly develop the disease, there is currently no known effective method of preventing breast cancer. . . ." (Brenner 1996) Like the AIDS activists Steve Epstein describes in his research exploring how the AIDS community gained power in biomedical policy making, these activists tried to develop their credibility by emphasizing their scientific expertise.

Even more explicitly, NBCC specifically identified itself as an expert in the definition of good science: "Together we can make certain we get the data we need. Too many medical recommendations in breast cancer—on how to treat women, what tests to give them—are made without a basis in good science. We must not add genetic testing and its followup to this category." (National Breast Cancer Coalition 1996a) Not only was commercial BRCA testing not in the category of good science, activists argued, but individuals should not be in the position of deciding what types of health care were best for them. Instead, they argued, individuals should be advised and protected by the state, physicians, and even knowledgeable activists about the appropriateness of particular health-care options.

As they advocated limited choice to BRCA testing, breast cancer activists distanced themselves from the individuals they represented.[2] They claimed a combination of scientific training and expertise in the patient experience that authorized them to distinguish between health care based on good science and a technology that was potentially dangerous. Meanwhile, they argued that their constituencies did not have a mastery of scientific knowledge and needed to be protected by them as well as physicians, test providers, and the government.

Both scientific and professional organizations and breast cancer activists suggested alternatives to the testing service marketed by Myriad. They called attention to the uncertain nature of BRCA gene information and argued that the test should be cautiously integrated into health care. Moreover, they offered a new definition of empowerment, arguing that women had a right to be protected from bad knowledge and that genomic information had the potential to be disempowering without effective therapeutics.

Empowering the Individual

The efforts of breast cancer advocacy groups to protect women from the new BRCA testing technology might sound surprising considering not only their history of encouraging women to take charge of their medical care but also a broader context of patient advocacy which always seemed to lobby for greater access to innovative medical care. However, the breast cancer advocates' attempts to protect women from BRCA testing technology was by no means unprecedented. In fact, this episode highlights the complicated definitions of empowerment, choice, and protection among patient advocacy movements that had emerged in the United States since the 1960s.

Although the American women's health and disease-based social movements that had developed in the late twentieth century advocated increased access to knowledge, medical care, and new technologies in order to empower the patient through additional knowledge about their bodies, they also fought against the use of knowledge and technology that they considered dangerous. Both of these efforts were considered empowering. A 1973 edition of *Our Bodies, Ourselves*, the book that launched a generation of women's health activism and popularized the phrase "Knowledge is power," emphasized the importance a woman's control over her body through knowledge, particularly in the face of what the authors perceived to be a paternalistic medical establishment: "Finding out about our bodies and our bodies' needs, starting to take control over that area of our lives, has released for us an energy that has overflowed into our work, our friendships, our relationships with men and women, and for some of us, our marriages and parenthood." (Boston Women's Health Book Collective 1973) As activists in the 1970s emphasized power through knowledge, then, they also lobbied for additional government regulation over drugs and medical devices. During this period, activists blamed the government, and specifically the FDA, after learning that diethylstilbestrol (DES) and the birth-control pill had caused serious side effects even after approval by the FDA (Ruzek 1978). Although they argued that women were capable of making decisions about their own bodies and lives, they also sought protection against what they perceived as dangerous medical interventions.

In this manner, these activists described their empowerment objectives in two very different ways. First, they were empowering patients by advocating increased access to new knowledge and technology, arguing that only they had the right to make choices about how their bodies should be treated and medical care provided. Second, they tried to limit their empowerment objectives by arguing that women should only have access to "good" science and medical care, as "bad" knowledge or technology could be potentially disempowering. Here, activists considered themselves expert to make these distinctions between good knowledge that could be empowering and bad knowledge that was potentially disempowering.

When breast cancer advocates recommended limited access to genetic testing for the disease, they were also defining the individual's agency by arguing that individuals would be hurt by access to useless genomic information and should only be provided with access to "good" medical options as defined by the state, the biomedical community, and activists. They argued that the average woman was not capable of making the distinction between "good" and "bad" choices and needed to be protected. While these activists focused on "empowering people to deal with the issues raised by a breast cancer diagnosis," NBCC advocated access only in research, and BCA stated that "we should be a long way from offering a test to anyone who wants it" (Breast Cancer Action 1997; Brenner 1996). Attempts by activist groups to limit access to testing not only highlighted this complication of empowerment but also demonstrated the distance between advocacy groups and their constituencies. As they limited and altered their definitions of empowerment, activists emphasized their own position as authorities determining the welfare of the uneducated public.

The conflicting empowerment objectives of the women's health movement (and subsequently the breast cancer movement) were also particularly clear as some women's health advocates explicitly disagreed with some of the breast cancer advocacy groups with regard to the provision of

BRCA testing. In 1996, the NIH's Advisory Committee on Research on Women's Health (ACRWH), which was made up of doctors, scientists, lawyers, social scientists, and public health officials primarily concerned with women's health issues, reviewed the availability of pre-symptomatic genetic testing for breast and ovarian cancer. Although it initially resolved to restrict testing to the research context and bar unlimited availability, some committee members criticized what they considered to be a paternalistic approach to medical care. Marjorie Schultz, a law professor from the University of California at Berkeley and a member of the committee, asked: "Can you imagine yourself saying to a woman who comes to a center to do testing, 'No you can't unless you're a research subject'?" (Pinn and Jackson 1996) The committee eventually recommended that testing be conducted in the context of counseling, rather than recommending that breast cancer testing be restricted within research protocols. Empowerment had a multiplicity of meanings, even among the advocacy community, and each of these definitions had different implications for both understanding the appropriate use of the technology and the representation of the patient community in the politics of biomedicine.

Responding to US Patient Advocacy Groups

How did test providers respond to such criticism from patient advocacy groups? Myriad Genetics, the largest testing provider, tried to gather support for its testing service by playing on the importance of empowerment rhetoric not only among patient advocates but also the public.

Myriad started to publicize its vision of a BRCA testing system immediately after the discoveries were announced. It defined itself as a commercial diagnostic laboratory providing women with the opportunity to inform themselves about their genetic status. Its 1996 annual report characterized testing as a life-saving technology that was important "for thousands of women who will soon gain access to genetic testing that can enhance and extend their lives" (Myriad Genetics 1996a). In an article published immediately after the discovery of BRCA1, Mark Skolnick, who led the winning University of Utah–Myriad Genetics research team, asserted that the diagnostic test would provide "knowledge that can allow [women] to make an appropriate choice about cancer detection and treatment" (Volland 1994). Genomic knowledge as provided by Myriad, the company argued, could help women make choices about their own health care. Mark Skolnick reiterated these sentiments at a conference on breast cancer genetics attended by both activists and scientists in 1996, stating that

BRCA testing should be as commonly available as a Pap smear (Brenner 1996). The company argued that it was providing an important technology that could help women make their own decisions, and invited both scientists and activists to join in this process of empowerment.

Using what Bruno Latour (1986) has called an "I want what you want" strategy, the company argued that women had the right to determine their own destiny. Myriad told women that it was providing them with power through information, an oft-cited demand of the women's health movement. Articulated by a number of women's health activists and championed in *Our Bodies, Ourselves,* "Knowledge is power" had become synonymous with women's empowerment. Now Myriad was using this rhetoric that was once solely in the domain of women's movements to develop a market for its new genetic testing technology. By convincing women that BRCA testing would provide information to help them take power in making decisions about their own health care, Myriad might be able to encourage them to use its service.

Myriad also tried to initiate a dialogue with the patient advocacy community about their concerns regarding the provision of a commercial testing service. The company reached out to advocacy groups, recognizing that these groups had tremendous control not only over their constituencies, but also among Washington players and media outlets. The company organized meetings with activists and consumer groups to "get good relationships with advocacy groups, to make sure we heard what they were thinking, get the tone of what the advocacy groups are thinking" (interview with Myriad Genetic Counselor 2000) and to explain its own position.

Many advocacy groups (including NBCC and BCA) refused to attend these meetings or even speak to Myriad representatives. A member of the National Alliance of Breast Cancer Organizations (NABCO), however, eventually agreed to sit on the company's clinical advisory board in an ad hoc capacity while helping the company develop educational materials (interview with National Alliance of Breast Cancer Organizations 2000). While the NABCO representative's participation allowed Myriad to include a visible patient advocate in their discussions, most advocacy groups continued to refuse contact with the company.

As the futility of efforts to convince activist groups, groups of scientists, and groups of health-care professionals became clear, Myriad largely gave up and began to market its testing service directly to individuals potentially interested in testing and physicians. By distinguishing between individuals potentially interested in testing and physicians and their representatives, the company hoped to stabilize its testing service.

Myriad began to market its testing service to the public by advertising in the mainstream media as well as sending targeted brochures and videos to physicians and individuals who requested them directly. Advertisements for Myriad's testing service appeared in such diverse locations as the *New York Times Magazine*, the USAirways in-flight magazine, and a Broadway playbill. This strategy suggests that Myriad defined its market as the entire population of American women. In order to develop this market, Myriad relied on the "Knowledge is power" strategy. Through advertisements in newspapers and magazines and promotional videos, the company told women that their BRACAnalysis services would empower them by providing information about their bodies. An advertisement in the *New York Times Magazine*, for example, showed a woman boldly staring straight at the camera and declaring "I did something today to guard against cancer." By taking the accurate and informative genetic test, Myriad argued, this woman would be empowered to take charge in the delivery of her own health care.

Educational materials distributed through physicians or sent to individuals who contacted the company expressed similar sentiments. The cover of one brochure sent to women curious about their BRCA risk states "Given a choice, would you rather deal with the known or the unknown?" The back of the brochure offers "answers" (Myriad Genetics 2000). The company promised women both the information and the opportunity to deal with the unknown risks of breast cancer. An educational video put out by the company (Myriad Genetics 1999) made the message clear: a woman who had undergone testing stated "Knowledge is power."

Myriad also tried to garner public support by developing a reimbursement structure for its expensive technology. Indeed, the only way for it to expand the market for testing was to develop a procedure for insurance reimbursement. While Myriad emphasized the woman's right to choice, this choice was severely constrained by economics. Myriad's testing services cost anywhere from $500 to $4,000, with the most common full sequence analysis of both genes costing about $3,000. The company recognized that the costs might be prohibitive if promoted as an unnecessary service, and worked with insurance companies through the Myriad Reimbursement Approval Program (MRAP) to encourage insurance reimbursement.

The company publicized this effort by reiterating its commitment to improving health-care choices. Announcing an agreement with a leading health insurance company, Myriad stated: "We are pleased that Aetna US

Healthcare is taking this step to provide women at risk of developing cancer with access to a test that provides information that might save their lives." (Myriad Genetics 1998) Although many women were still reluctant to ask their insurance companies to reimburse BRCA testing for fear of discrimination, Myriad used publicity from these insurance services to emphasize not only providing a state-of-the-art clinical service but also empowering women through genomic information (interview with US Geneticist 1999). Indeed, as the company developed its testing system, it defined a pivotal role for insurance companies who became responsible for maintaining the individual's right to demand BRCA testing.

The company's strategy clearly demonstrated an effort to entice interest in the new technology by contextualizing the testing system within the efforts of the women's health movements of the late twentieth century which had likened empowerment to increased access to knowledge. Realizing that it would be unlikely to gain activist support, it used empowerment rhetoric to convince average women to use its testing system. While Myriad tried to divide activists and their constituencies by capitalizing on a particular definition of empowerment, activists continued to work against the company by encouraging women to be careful about their BRCA testing decisions and to use testing only in the context of clinical care. Press releases, position papers, and statements on BCA and NBCC's web sites suggested an ongoing effort to strengthen an alternative approach to testing.

Patient Advocacy in Britain

Although Britain did not have a long tradition of patient activism like the United States, patient advocates in Britain had just begun to gain strength and visibility around the time of the BRCA gene discoveries. This was particularly evident as a number of groups tried to get involved in the development of BRCA testing. Unlike their American counterparts, however, most British activists supported the widespread availability of BRCA testing and wanted to get involved in ensuring its integration into health care. Wendy Watson, a middle-aged woman from Derbyshire who had a prophylactic double mastectomy (preventive removal of both breasts) in 1991 after learning of her extensive family history of breast and ovarian cancer, was particularly vocal in support of widespread development of genetic testing for breast cancer after the discoveries of the BRCA genes in the mid 1990s. Immediately after the discovery of the BRCA1 gene in September 1994, she was relieved by the promise of new technologies that she thought would be able to identify inherited susceptibility: "If there had been a test for me I would definitely have had it. But I thought at least now it might be available for people like my daughter." (Rogers 1994) Watson argued that the development of testing was particularly important because it would help women like herself who would otherwise be sick with worry to take action to deal with the inherited risk. When describing her decision to have a prophylactic mastectomy to the *Times* of London, she said: "Once I thought that [about the possibility of surgery], no one could have shaken me because my overriding worry was of dying of cancer. I went into hospital, had the operation, woke up and thought: 'Thank goodness for that, it's done, that's the gamble off.' I felt absolutely fine." (Laurance 1996)

By May 1996, Watson had begun the Hereditary Breast Cancer Helpline to provide information to individuals concerned about their family history of breast cancer. Funded initially by the UK Department of Health, Watson received thousands of calls a year from individuals anxious about their risk of breast or ovarian cancer, curious about how to access genetics services, and uninformed about the options after learning they tested positive for a BRCA mutation (interview with Wendy Watson, 1998). Bolstered by these interactions with individuals worried about their BRCA risk, Watson began to lobby for increased availability of genetic services. In an article in *The Scotsman* (Christie 1996), she noted: "Every woman has the right to discuss her future with informed and sympathetic professionals. . . . I think its ludicrous to say we cannot afford to fund these genetics clinics." Watson was a strong proponent of BRCA testing services across the United Kingdom. She argued that it had played such an important role in her health and happiness that other women should have the right to have access to the same services. This strong support of BRCA testing, of course, was in stark contrast to the position of American activists who characterized the new BRCA technology as useless and genomic information as potentially dangerous. Moreover, while Watson argued that women had the right to access genetics services, American patient activists limited this right to medical advancements that they classified as good science.

Like Watson, the Genetics Interest Group also sought to become involved in the development of BRCA testing services in Britain. Much like NBCC in the United States, GIG was founded in 1989 by a group of voluntary organizations and disease support groups who wanted to "coordinate action" on the issues they had in common. By the mid 1990s, GIG was issuing position papers and reports on a variety of issues related to genetic disease. Its approach to genetic testing for breast cancer was very similar to Watson's, supporting the development and wide availability of these services across the country. A GIG representative summarized its position as follows: "Now our view on testing is, everything is subject to the informed consent of participants, if somebody has reason to suspect that they may be at risk from a genetic disorder, then they should have access to services. . . . That testing should be done in a context where informa-tion is provided, where counseling is provided before, and after the test-ing process. . . ." (interview with Genetic Interest Group representative 1998) While American activists saw testing as a dangerous and uncertain technology from which the government should protect citizens, GIG lauded the potential benefits of the proposed British system. Supporting the risk-assessment and triage system helped GIG potentially to carve out a new and powerful role for the disease advocacy community in Britain.

   In fact, GIG specifically supported the idea that care by provided according to an individual's family history: "Because it is possible to say by drawing up certain protocols whether you are as an individual, at high, medium, or low risk. And it's inappropriate to waste health-care resources, testing people for whom there are no prior indications. As it is inappropriate to avoid using resources to ensure that people who fulfill the criteria do actually get that help and support. But it's a rational thing." (interview with Genetic Interest Group representative 1998) Not only did GIG advocate this risk-assessment-based testing system, but it argued that such a system provided a perfect opportunity to assure equal access within the NHS. It was inappropriate to waste scarce resources on individuals at low risk, they agreed, when testing high-risk individuals could confer significant benefits.

British Patient Advocates Making a Difference

What difference did British patient activists make in the development of BRCA testing? Advocacy group representatives sat on advisory commit-tees and attended meetings where the appropriate provision of genetic testing was discussed, and lobbied NHS purchasers to buy genetic testing services for their regions. The Genetic Interest Group (GIG) had a rep-resentative sit on the committees that developed national standard of risk assessment and triage for BRCA testing. A representative from the Genetic Interest Group that represented all individuals with genetic dis-orders commented: "I think people are realizing that the benefit will come by virtue of treatments for rare disorders being piggybacked onto the technology that cracks common disorders." (interview with Genetic Interest Group representative 1998) Indeed, GIG often calmed members of its constituency who had rare genetic disorders by arguing that its initiatives specifically with regard to common diseases such as breast cancer would improve their efforts to influence genetics policy writ large. Discussion about the appropriate provision of genetic testing for breast cancer was taking place at the highest levels of the Department of Health and the NHS, and could provide advocates with the opportunity to demonstrate their importance in the biomedical policy-making process. The Genetic Interest Group and Wendy Watson were the most vigorously involved in discussions about BRCA testing. Both strongly advocated access to testing, but they also supported the national standard involving risk assessment and triage. Rather than challenging the triage system and its limitations, they lobbied NHS purchasers to accept and provide BRCA services within this framework.

GIG was not alone in its involvement in trying to influence policies for genetic testing for breast cancer. Wendy Watson also supported the proposed BRCA testing system. Like GIG, Watson lobbied for increased access to testing. She sat on a variety of advisory committees and even gave seminars to NHS purchasers across the country. Watson involved herself directly in the decisions of NHS purchasers, arguing that BRCA testing offered not only life-saving benefits for patients but cost-saving opportunities for the NHS: "I . . . explain to them about the advantages of purchasing genetic services, how much money it saves them. In my family, genetic testing saved the NHS 68,000 pounds. Simply because four of us had the genetic test before we had breast cancer or anything like that. Three of us got faulty genes, and all of us had preventive surgery at a cost of between 2,000 and 4,000 pounds. My sister hasn't had a preventive mastectomy, but she didn't have to. So she didn't have a 3,000-pound operation that would have been unnecessary." (interview with Wendy Watson, 1998)

Watson took a more extreme position than GIG, however, arguing that every woman should have a choice of the test as well as subsequent medical options: "Everyone should have the right to have a genetic test and take whatever action is necessary to save their lives. . . . So that's been very important, to be able to empower people, give them the information, and then they do what they want, whether it is nothing, screening, preventive surgery, even radical preventive surgery. Whatever they choose, it should be their option and they should be fully supported." (ibid.) This advocacy of unlimited choice did not appear consistent with the restricted risk-assessment model. Instead, Watson adopted a position somewhat similar to Myriad: advocating limited choice for individuals who desired access to a potentially life-saving test. As she advocated this strategy, which appeared extreme in the British context, she proactively lobbied on the part of regional cancer genetics clinics in support of the national standard. For her, the more access to BRCA testing, even within the NHS's risk-assessment and triage model, was better than no access at all. As mentioned earlier, however, she was advocating access to a technology that was very different than Myriad's.

Watson and GIG's involvement in discussions about breast cancer risk-assessment services helped to encourage regional purchasers to fund cancer genetics clinics. Watson reported, for example, that her perspective has been received very well. Regional health authority officials were interested in hearing what she had to say, and she often inspired them to begin funding BRCA testing in their region. She recalled, "At the end of my speech [at a regional health authority] last week, I was inundated with people who wanted to chat to me and then someone from the health authorities said that they were mortified that they haven't had 'somebody pleading the case, because people are worrying. I find it mortifying that I haven't already purchased the service, and I should be doing it with a matter of urgency.' So I'm not greeted as being, not knowing what I'm talking about." (interview with Wendy Watson, 1998) Her activism on behalf of BRCA testing certainly influenced awareness and purchasing decisions among NHS officials.

While proponents of the NHS standard had initially excluded potential test users and their representatives from discussions about BRCA testing, Watson and GIG forced themselves to be recognized as legitimate participants in stabilizing the testing system. They could be important contributors to the policy-making processes, they argued, lobbying with health-care professionals on behalf of the public for more funds. Even though proponents of the NHS standard largely ignored them, these activist groups became quite relevant in the development of BRCA testing in Britain.

Why did these British patient activists take such a different position to the provision of genetic testing for breast cancer than their American counterparts? These are a few possible reasons. First, patient activists in each country were reacting to a very different provider of the new technology. While British activists were responding to the provision of BRCA testing by the trusted, state-run NHS, American activists were worried that commercial providers such as Myriad and GIVF would stifle research and offer testing as any other consumer product rather than a new and uncertain technology. In addition, the testing systems themselves differed in the two countries, Breast cancer advocates in the United States were reacting largely to the prospects of commercial services such as Myriad's that required no specialized counseling. In Britain, on the other hand, counseling was an important part of the BRCA testing services that the NHS offered. Second, American and British activists were operating in two very different health-care systems. American activists worried that within a market-driven system that encouraged the rapid availability of new technologies and the growing biotechnology industry, there were few mechanisms that would regulate or even monitor development of the new technology. In Britain, on the other hand, BRCA testing was going to be a provided by a trusted state-run system that was extremely popular among the citizenry.[3] Finally, the histories of patient activism were quite distinct in the two countries. In the United States, patient activists had been steadily gaining power since the women's health movements of the 1970s. Their opposition to Myriad's genetic testing system would be unlikely to seriously jeopardize the power and credibility that breast cancer activists had gained since the early 1990s. British activists, on the other hand, had not yet become major figures in biomedical politics. As a result, they would be much less likely to oppose NHS practices of powerful clinicians. Also, patient activists in the two countries had traditionally been oriented toward slightly different goals. While AIDS and breast cancer activists, for example, had lobbied for increased research money, patient advocates in Britain usually worked for better access to services within the NHS. Indeed, while BRCA testing was not offered in the context of research in Britain, only US activists raised this as a major issue in their lobbying efforts.

Conclusion

This comparative analysis has demonstrated how national specificities played a very important role in the responses of patient advocacy groups to the development of genetic testing for breast cancer in the United States and Britain. While the user of health care might seem at first glance to have uniform interests worldwide such as a desire for high quality care and access to services for which there is a demonstrated clinical need, this chapter has shown that national contexts shape the response of the user to the development of a new medical technology. In the United States, patient advocates' distrust of commercial providers and power in biomedical politics led them to take contrary positions toward the new genetic testing technology. Their positions might even be considered risky, as they reinterpreted women's empowerment goals in terms of protection. In Britain, on the other hand, where the health-care system was more trusted by the public and there was virtually no history of patient activism, patient advocates supported development of BRCA testing.

Comparative analysis provides us with important insight into the role that national specificities play in the relationship between medical technologies and users. Of course, the case studied here is particularly stark because BRCA testing was provided by a private company in the United States and offered by the NHS in Britain. Both the US and Britain offer other specialized health-care services in contexts which might lead to more similar experiences for the user in the two national contexts—consider the provision of cancer care services paid for by private insurers in Britain or by health maintenance organizations in the United States. Both local and national contexts frame the interests and experiences of the user of health-care services.

It is also important to keep in mind that this is a comparative analysis of patient advocacy groups, rather than individuals who actually had their blood analyzed for mutations in the BRCA genes. In fact, this analysis has emphasized the distinction between these two groupings. Patient advocacy groups in the United States, for example, adopted a protective stance toward the individuals they represented and also emphasized a knowledge gap between advocacy groups and individuals interested in testing. An NBCC representative made this distinction even clearer when describing the advice of NBCC staffers when their constituent grassroots advocates visit Washington for "Advocacy Day" once a year:

> . . . we try mostly to look at issues, and it's hard to do, but kind of take a step back and take the individual out of the issue, and look at it in more of a global public policy way. So when we have our advocacy day and we have advocates from around the country and a lot of them are people who haven't ever been to Washington, they haven't been to Capitol Hill, and they certainly haven't been in to meet with their member of Congress, and we really try to drill into them that when you go up there, this is not about your breast cancer, and your specific treatment, and your specific disease and your family, this is about breast cancer. (interview with National Breast Cancer Coalition representative 1999)

In addition, test providers in both the United States and Britain took advantage of the differences between patient advocacy groups and individuals interested in testing. This analysis forces us to question how we should understand advocacy groups in the context of user studies and perhaps more broadly, consider what makes a "user."

7

Who Represents the Users? Critical Encounters between Women's Health Advocates and Scientists in Contraceptive R&D

Jessika van Kammen

In recent years there has been a growing consensus among researchers, policy makers, funding agencies, and women's health advocates that the future users of contraceptive methods should be involved in the developmental process. The 1990s witnessed a major shift in the field of contraceptive research and development, from the paradigm of population control to this new paradigm. This culminated in the Program of Action adopted at the 1994 United Nations International Conference on Population and Development in Cairo. For the first time people's reproductive health and rights were placed center stage. During the decade, major organizations in the field of contraceptive development, such as the World Health Organization and the Population Council, took a number of initiatives in order to follow up the ICPD recommendation that "users, in particular women's, perspectives and women's organizations [*sic*] should be incorporated into all stages of the research and development process." Meetings between contraceptive developers and women's health advocates were organized, policy was formulated, and social scientific research into the needs and preferences of users was intensified in order to develop what was called "the integration of users' perspectives" (ICPD 1994).[1] At the heart of this new strategy lay the conviction that users had not previously been taken sufficiently into account by contraceptive developers (Bruce 1987; WHO/HRP 1992; Population Council 1990; Cottingham and Benangiano 1997). But while there was wide agreement about the need to involve users in the research and development process at an early stage, no specific strategies for achieving this had been developed.

In this chapter, I analyze how the integration of users' perspectives took place in the case of immunocontraceptives. These new birth-control methods are an especially interesting case, because the coordination of research into such methods is done in part by the same institutions that ratified the ICPD Program of Action: the Human Reproduction Programme of the World Health Organization (WHO/HRP) in Geneva and the Population Council in New York. The other major player in the development of anti-fertility vaccines is the National Institute of Immunology (NII) in New Delhi. Moreover, while contraceptive developers have increasingly become committed to the concerns of women's health organizations and potential users, women's health groups have tried to influence the development of anti-fertility vaccines. Many women's health advocates have raised serious questions about the safety and efficacy of this new method for fertility regulation. They have also been concerned that anti-fertility vaccines could be administered coercively in demographically driven family planning programs (Richter 1993, 1996). A number of groups have begun to campaign to call for a stop to this research and development. Precisely because it has been a disputed area, anti-fertility vaccine development has been at the forefront of experimenting and learning about different ways to accomplish "the integration of users' perspectives" into early stages of technological development.

How can the involvement of users in the early stages of technology development be studied? According to the science and technology scholar Madeleine Akrich (1992, 1995), innovators from the beginning are very interested in their future users and inscribe their hypotheses about users into the technical content of the new object. As a consequence, technologies contain a "script": technologies prescribe a specific usage, invite certain practices and make other practices impossible, and distribute responsibility and power in various kinds of social relations. For example, contraceptive implants that have to be inserted and removed by a trained health worker in aseptic circumstances entail a prescription for use that is incorporated in the technology. Embedded in the script of such a method are a dependency relation between users and health-care personnel and a need for a considerable health-care infrastructure. Clearly, the potential future users of anti-fertility vaccines exist in a wide variety of social, cultural, and personal settings. In what ways are hypotheses about who might use a new technology constructed, and how do these representations come to bear upon technological development? Technology developers and their associates do a lot of work to represent users in such a way that they can properly guide the innovators in the process of developing new technologies. Akrich (1995) has described different techniques by which representations of users are generated, and has distinguished explicit techniques, legitimized by a formal scientific and conceptual basis, from implicit techniques of a more empirical kind, which lack such a basis. For the successful development of a new technology, disparate representations should be combined and superimposed to achieve alignment. One way to reconcile the various facets of users is endowing the artifact with a number of features that would enable it to cope with different situations, expectations, and necessities of users. The reconciliation work is, as it were, delegated to the technology. Another way is to designate the task of bringing together users' requirements and the characteristics of the technology to intermediaries, such as installers, instructors, or counselors. Also, the need for reconciliation between a new technology and images of users can be omitted altogether by creating a new user to fit the artifact.

First, I analyze the way in which attempts to integrate users' perspectives into contraceptive R&D initially evolved. I identify the various actors and ask how women's health advocates became recognized as the political representatives of users. Next, I examine what the integration of users' perspectives meant for the development of anti-fertility vaccines. To what extent could these differing perspectives be integrated into the technological development process? This analysis is based upon an examination of published articles, interviews with a number of the involved scientists and women's health advocates, and reports of meetings between scientists and women's health advocates that have taken place in Geneva at the initiative of the WHO.

Women's Health Advocates as Spokespersons for Users?

Can women's health advocates be considered as spokespersons for the future end users of anti-fertility vaccines, and if so, how? Women's health advocates are a diverse group of individuals, organizations, and informal groups which operate all over the world and share the common goal of empowering women to control their own fertility and sexuality with maximum choice and minimum health problems (WHO/HRP/ITT 1991; Hardon 1992). Reproductive technologies profoundly affect women's lives, and it is therefore not surprising that the women's health movement is concerned about them. The women's health movement developed in response to recurrent problems in the field of family planning. Since the 1970s, women's health groups documented repeated instances of the coercive and not fully informed administration of contraceptive methods, especially in Third World countries.[2] The other major critique made by the women's health movement has been the neglect of women's health, and the biomedical standards for assessing the safety of such methods as the early female contraceptive pill, Depo Provera, and Norplant. The women's health movement has used strategies such as forming watchdog organizations, drawing in the press and lobbying, and setting up alternative services.[3]

Women's health groups, however, are not necessarily well equipped to speak for the practical needs and interests of contraceptive users world-wide. They are organized around political-strategic goals. The issue of representing the enormous diversity of contraceptive users is politically complex. I will not try to address the normative question of who should be allowed to speak on behalf of users. Instead, I will analyze how this question has been handled by the actors involved. In particular, in what way did women's health advocates gain acknowledgment to represent the users' perspectives? Who else could claim to speak on behalf of users?

In 1990, the director of the WHO/HRP, Mahmoud Fathalla, invited one major women's health advocacy group, the International Women's Health Coalition in New York, to assess whether the activities of the WHO/HRP were oriented toward women. The International Women's Health Coalition in turn proposed co-organizing a meeting between scientists working in the WHO/HRP and women's health groups from different parts of the world to discuss specific aspects of the WHO/HRP's work (Cottingham 1995). The resulting meeting, organized in February 1991, was called the Creating Common Ground meeting. The way in which women's health advocates relate to users of fertility-regulating methods was among the central themes debated at this meeting. According to the report:

While there was consensus about the need to bring women's perspectives and experiences to bear on the development, selection, and introduction of fertility regulation technologies, the participants debated the question of who can legitimately and effectively articulate those perspectives. A number of scientists questioned whether, for instance, women's health advocates such as those at the meeting represent the views of poor and rural women (WHO/HRP/ITT 1991: 13).

Specifically in the area of immunological contraceptives, the representativity of concerned women's health advocates and the extent to which they could speak on behalf of users was questioned. Vernon Stevens, the principal investigator of the WHO/HRP Task Force on Immunological Fertility Regulation, commented: "The number of women who have expressed these objections to anti-fertility vaccines is very small and there are no data available to suggest that these views represent those of a significant proportion of women from any country or region in the world." (Stevens 1996: 149)[4]

In fact, women's health advocates had not claimed to speak in the name of users (Stemerding 1998).[5] But they did make it clear that the issue of politically representing a diffuse group such as potential contraceptive users could not easily be resolved. When questioned about their representativity, some women would turn the question around to show the complexity of this issue. For example, at a Conference on Anti-Fertility "Vaccines" convened by women's health advocates in Bielefeld in 1993, the manager of the WHO/HRP Task Force on Immunological Fertility Regulation, David Griffin, posed the question of how representative the opinion of the conference participants was of women in general. Richter (1993: 122) reports that one of the participants "advised [Griffin] to go back to his Geneva office, look into the faces of his colleagues and ask them 'Let's be honest friends, how representative are we to take decisions for the world's female population?'"

But if women's health advocates did not represent the users, on what basis should contraceptive researchers and policy makers take their voices into account? The women's health groups considered that their perspectives were relevant to contraceptive development. The Technical Officer at the Women's Desk of the WHO/HRP, Jane Cottingham, said: "Women who are working in women's health groups and women's health projects have an understanding of women's situation and perhaps an analysis of the situation that has not necessarily been taken account of and that should be represented. And I think that is valid. . . . You cannot have someone from, say Bangladesh, representing 'women in Bangladesh.' We are never going to get to that situation, so let us just forget it." (Cottingham 1995)

But Jane Cottingham represents a particular kind of experience focused on women's health and women's rights, where there has been a lot of reflection and action. That is why her experience and her viewpoint are valuable.

In addition, the contribution of women's health groups was seen as potentially beneficial to the WHO/HRP's work. This in turn contributed to their credits. "This democratization of the research process," said Task Force manager David Griffin (1996: 144), "is not only a welcome addition in its own right, but is likely to lead to greater success of the research effort by ensuring that the methods developed meet the expressed needs and preferences of individuals and couples."

In other words, a mixture of political and instrumental arguments was mobilized to account for the legitimacy of taking into account these different perspectives. Neither the women's health advocates nor the scientists involved assumed, that women's health advocates would be representative spokespersons for users.

Eventually, the varying perspectives of both the scientists and the women's health advocates were specified. In the report of the meeting Creating Common Ground, it was concluded that "both scientists and women's health advocates emphasized that there is neither one monolithic 'scientists' perspective' nor one 'women's perspective,' but rather a broad spectrum of opinion within each community" (WHO/HRP/ITT 1991: 10).

But the consensus reached at this meeting could not prevent questions on the legitimacy of different voices from being raised periodically.

Various Meanings of "Users' Perspectives"

The policy agreements had created room for the integration of users' perspectives, but had not been explicit about exactly what aspects of users would be relevant for integration into contraceptive research and development. Would "integrating users' perspectives" mean learning more about users' contraceptive needs, or would it stand for taking into account women's reproductive health and their rights? This issue needed to be elaborated in practice.

After the first Creating Common Ground meeting, Jane Cottingham was appointed as a special Technical Officer at the WHO/HRP to put into effect the recommendations for action that had been formulated. Her task was verbalized as "to help integrate women's perspectives into the activities of the HRP" (WHO/HRP/ATR 1995). But "immediately when I came here," she later recalled, "I came up against the problem of what is meant by 'women's perspectives.' And I think some women's health advocates tacitly understood that 'women's perspectives' meant 'feminist perspectives.' But it was easier to talk about 'women's perspectives.'" (Cottingham 1995) A feminist women's perspective would mean taking into account women's reproductive health and rights situation, and not to restrict the concern to the practical needs that users express in surveys.

To refer to the envisioned contraceptive needs of users was a widely practiced convention among reproductive researchers and policy makers as they attempted to develop new methods. "Integrating the users' perspective" or "the women's perspectives" thus did not appear to be a drastic new issue for them; they had understood it simply to mean taking into account women's needs and gathering information on what kind of attributes of contraceptive methods were appreciated. Already at its creation in 1972 the WHO/HRP had established special Task Forces on Psychosocial Research in Family Planning as well as on Services for this purpose (Kessler 1992: 48). Social scientists had been appointed by the WHO/HRP to present the needs of users to the reproductive scientists. Since the 1970s, numerous studies on mainly women's/users' needs, preferences, opinions, experiences, and understandings of contraceptives had been carried out by social scientists.[6] However, these representations of users' needs had not played a major role in contraceptive development.[7]

The social scientist Igbal Shah, from what now was called the Social Science Task Force of the WHO/HRP, prepared an extensive overview of the social scientific literature on contraceptive usage (Shah 1995). When he presented this study at the 1995 WHO/HRP Meeting on Women's and Men's Perspectives on Fertility Regulation Methods and Services (WHO/HRP 1996), Shah explicitly positioned the social scientific data that his Task Force had generated as work on users' perspectives: "The understanding of the perspectives of people toward methods of fertility regulation has been an integral part of the activities undertaken by the Programme since its inception in 1972. The bulk of this work has been carried out by the Social Science component of the Special Programme. More recently, the Task Force on Research on the Introduction and Transfer of Technology for Fertility Regulation, Women's Issues Desk and the Unit of Resources for Research have also considered the perspectives of users." (Shah 1995: 2)

By labeling this body of social scientific literature as studies on users' perspectives, the social scientists of the WHO/HRP could claim that their work would meet the recommendations of the Cairo Program of Action that contraceptive development should be guided by women's/users' needs. This would make any further involvement of women's health advocates superfluous. Women's health advocates agreed that social scientific research could indeed be important to gain insight into the situation of users, but they found the representations of users provided by mainstream social scientists too limited. These social scientists had studied and configured users mainly in terms of having unmet needs. The recurrent finding of these studies was that users would prefer a method that was very reliable, safe, and free from side effects. Further, users' needs were found to vary widely, and to change throughout the life cycle (Shah 1995; Cottingham 1997). But what if the contexts in which people plan their families were taken into account? What if such an ideally safe and reliable method was not available, or not accessible? This led the women's health advocates to recommend a different type of social science research that corresponded to what they meant by the integration of users' perspectives. "Well, they may be women," said WHO/HRP Technical Officer Jane Cottingham (1995), "they may be men, they are all in a particular context and they may be aware of their rights or not. And so the problem is that we cannot necessarily get a clear picture of users' perspectives or arrive at a conclusion that, for instance, 10,000 women all say: 'yes we like X.'"

From a feminist perspective on social scientific research, it was important to broaden the scope of these studies to include the users' reproductive health and rights.

Feminist social scientists proposed to place the user in the middle of a contextual and relational analysis (Hardon 1995).[8] In their view, the integration of users' perspectives would also mean a more active role for users as subjects in social scientific research.[9] They asked for more qualitative studies, focus group discussions, in-depth interviews and contraceptive life histories. Following the feminist tradition, they were more explicit than other social scientists about the politics of doing research and they argued that the respondents should benefit from the findings. "Gaining a better understanding of how women make choices and negotiate trade-offs among methods," said the American women's health advocate Lori Heise (1997: 6), "will undoubtedly yield insights that are useful to policy makers and program managers, as well as to women themselves."

In Akrich's (1995) terms, these members of women's health groups proposed to extend the diversity of techniques for constructing representations of users beyond market surveys. But the aim of using a wider variety of techniques was not merely to provide the contraceptive developers with more information or better specifications about the types of methods that users prefer, they wanted to change the focus toward how to improve users' reproductive health, and make them participants in the research process.

Integration of Users' Perspectives into What?

The policy makers and reproductive scientists at the WHO/HRP looked upon women's health advocates in the same way as the social scientists had been regarded, as a source of information on users' needs for contraceptives. They hoped that members of the women's health movement would act as spokespersons and advice them on the needs of users for certain contraceptive products. But women's health advocates did not conform to the role of providing policy makers and reproductive scientists with answers to the question of what kind of contraceptives users want. They proposed a more interactive approach. After the first Creating Common Ground meeting, in February 1991, the Task Force manager asked Faye Schrater to assist the Steering Committee of the Task Force for Immunological Contraceptives. As an immunologist with feminist sensibilities, she was identified as the person who could bridge the gap between scientists and women's health advocates. Schrater (1996) said: "I went to the Steering Committee meeting and we were talking about something, and all of a sudden one of the scientists would turn to me and say: 'Well, Faye, what are the women going to say about this?' Good grief, how many women in the world are there?! I cannot answer. I cannot answer the question in the way they want it answered."

Women's health advocates suggested that the questions on users should be posed differently to include the broader issue of women's reproductive lives. They therefore not only aimed at broadening the scope of the social scientific research, but also sought access to scientific committees and policy-making boards. In their view, the integration of users' perspectives into contraceptive research and development required engaging in dialogue with scientists and policy makers.

There were a number of locations at which members of the women's health movement could engage in this dialogue with the contraceptive developers. For instance, the recommendations of the first Creating Common Ground meeting included a statement about the incorporation of women's health advocates into the policy and research work of the WHO/HRP. This issue was energetically taken up by the Women's Desk. The appointment of Faye Schrater to the Steering Committee of the Task Force of Immunological Methods for Fertility Regulation followed directly from this, which was important, because the Steering Committee was in charge of generating and appraising research proposals, reviewing the scientific work, and deciding on future research needs. In addition, a Gender Advisory Panel (GAP) was formed in 1996 (Benangiano 1995).[10] Largely due to the existence of the Campaign to Call for a Stop to the Research on Anti-Fertility "Vaccines," the work on immunocontraceptives was reviewed at the first meeting of the GAP, in February 1996. In correspondence with its broader mandate, the Panel recommended that representatives of women's health groups should be included in an advisory capacity in the design, monitoring, and evaluation of clinical trials with anti-fertility vaccines (Gender Advisory Panel 1996). This was an important recommendation for the women's health advocates, since it created conditions under which their alternative perspectives on users could actually be taken into account in the research and technology development process.

In sum, the discourse on "integration of users' perspectives" that evolved in the area of contraceptive development did not initially alarm reproductive scientists and policy makers. On the contrary, they hoped that "the integration of users' perspectives" would enable them to get advice on users' preferences and to develop suitable contraceptive methods. But to the women's health advocates, "integration of users' perspectives" was the Trojan horse by which they gained access to reproductive and social scientists at WHO/HRP. Backed by the favorable international climate, women's health advocates introduced a different understanding of integration of users' perspectives. This understanding went beyond the question of what kinds of contraceptive products users want and meant taking into account the users in their contexts, including women's health and their rights, and recognizing the perspectives of women's health groups as valid. It remained to be seen whether and how these perspectives could indeed be taken into account in technological development, and what role women's health advocates would have in this process.

Barriers to Integrating Users' Perspectives into Technological Development

The attempts to integrate users' perspectives into technological development in the way that the women's health advocates had proposed was a whole new experience for all the actors involved. Schrater (1996) commented: "It is complicated; I mean how do we get users' perspectives into this when we do not even have a product yet? We have a potential product. How are we going to get users' perspectives into that? And which users' perspectives are you going to take? . . . I think that that is one of the things that makes the scientists crazy, women's health advocates coming to these meetings saying: 'you don't have users' perspectives, you got to take into account users' perspectives.' Give me a users' perspective."

How did the reproductive scientists involved in the development of anti-fertility vaccines react to the changing meaning of integrating the users' perspectives? As we shall see, they did make various attempts to align the developing product with the user in her context. But at the same time they wanted to maintain their scientific autonomy and to continue their research. Policy makers at the WHO/HRP also faced a complex new situation having to preserve good relationships not only with the women's health movement but also with member states, donors, the pharmaceutical industry, and all kinds of scientists. Below I will analyze which issues the actors involved agreed to discuss, and how certain issues came to fall outside the area in which alternative perspectives could be taken into account.

One strategy that scientists have traditionally used to pursue and maintain their scientific authority is to construct boundaries between science and various forms of non-science (Gieryn 1983; Jasanoff 1987). In the words of Gieryn (1995: 405): "Boundary work occurs as people contend for, legitimate, or challenge the cognitive authority of science—and the credibility, prestige, power, and material resources that attend such a privileged position." In the encounters between the scientists and women's health advocates, such boundary work has played an important role.

Mobilization of the Prototype

Unlike hormonal methods, the immunological methods for fertility regulation are envisioned to cause temporary infertility by stimulating the production of antibodies against substances necessary for human reproduction, such as certain hormones and surface molecules of the sperm and the ovum. These methods were meant to be long acting, low cost, and easy to administer. Worldwide, the developmental stage of anti-fertility vaccines varies, but there was no product ready for introduction into family planning programs at the time this study was carried out. Women's health advocates argued that the safety and efficacy of anti-fertility vaccines were insufficient for use in a less than perfect health-care context. Some of them campaigned to call for a stop to this research and development. In their 1993 petition titled Call for a Stop to Research on Anti-Fertility "Vaccines" (Immunological Contraceptives), they wrote:

Because they use the immune system, they are inherently unreliable. Individuals can react completely differently to the same kind of immunological contraceptive. . . . In addition, stress, malnutrition and disease will cause unpredictable failures of the contraceptive. . . . Immunological contraceptives are unlikely to be ever harmless. . . . Interference with the immune system for contraceptive purposes is indefensible at a time when primary health care systems in many countries are being dismantled, when the incidence of many infectious diseases is increasing, and when we have become acutely aware of the preciousness and complexity of our immune system.

The scientists, on the other hand, considered that the assessment of safety and efficacy belonged firmly within their own domain. According

to them, the safety concerns of the women's health advocates were superfluous. In addition, they stressed that what they had developed was not
the final product but only a prototype. In August 1992, the women's
health advocates met with the scientists at the WHO/HRP in Geneva to
review the current status of the development. According to the report of
this meeting: "It was recognized [also] that the fertility regulating vaccines under development and in clinical trials were still at an early stage
of development and that many of the concerns raised were applicable to
these prototype vaccines which are unlikely to be the ones to proceed to
final product development." (WHO/HRP 1993: 31)

   This labeling of the developing product as a prototype and as relatively
unrelated to the foreseen final product had far-reaching consequences.
The scientists emphasized that the prototype was not meant to be appropriate for the context of use, [11] and therefore the critique of women's
health advocates was not applicable to the prototype. At the same time,
the prototype played a key role in the assessment of the safety and efficacy of the method. One strand of research involves developing a contraceptive vaccine against human chorionic gonadotropin (hCG), a
hormone produced by the fertilized ovum that is necessary for the maintenance of early pregnancy. And, for example, the researchers at the
National Institute of Immunology and those at The Population Council
referred to safety concerns in research on prototype anti-hCG vaccines
in order to lay the basis for further development:

This prototype vaccine was effective in inducing in women the formation of antibodies against hCG. . . . The antibody response was reversible and phase I studies
conducted in six centers located in five countries showed the safety and lack of
side effects of immunization with this vaccine. (Om Singh et al. 1989: 739)

Our study further confirms the safety of this vaccine. . . . The promise of the development of an anti-fertility vaccine, which emerged almost 20 years ago, still holds
true. The process has been slow and we may still be far from the final product. . . .
(Brache et al. 1992: 10–11)

Similarly, the originator of the anti-fertility vaccine developed under the
auspices of the WHO/HRP, Vernon Stevens, pointed to evidence
obtained with a prototype to account for the efficacy of anti-fertility vaccines. Referring to a phase II clinical trial by the Indian team, he wrote:

This vaccine, while not representing a product acceptable for general use, did suf-
fice to demonstrate that immunization against hCG can be effective in preventing pregnancy. This milestone was very important for justifying further research
and development of hCG anti-fertility vaccines. (Stevens 1996: 149)

Thus, the prototype vaccine was not considered susceptible to criticism, but at the same time it was centrally important in safeguarding the continuation of the research. Avril Mitchison (1990: 612), an American researcher involved in the development of the WHO/HRP vaccine, stated: ". . . finding funding is competitive, and the earlier we have something to show for our efforts, the more likely we are to secure further support. In this sense, a prototype vaccine is needed, even though we know that it may not be the optimal choice and may never enter into widespread use."

The researchers assigned safety and efficacy problems of the developing vaccine to its prototype status, while other research findings made with this same unfinished product were characterized as milestone achievements. As a consequence of selectively bringing to the fore the prototype status of the technology, the progress of their research would continue to get the benefit of the doubt.

Compartmentalization of the Research Process

Boundary work was also involved in distinguishing "the vaccine itself" from "the application of the vaccine." The consequence of this distinction was, again, that the issues that women's health advocates raised were framed as irrelevant for the scientific work. A whole range of potential topics to be addressed from a users' perspective was postponed to later stages of development. This can be illustrated by the discussions and events surrounding three elements in the way the vaccine worked. First, the occurrence of a lag period before the antibody response to the vaccine reaches a protective level. Second, the difficulty of predicting the duration of effective immune response. Third, the impossibility of switching off an immune response once it has been set in motion. According to women's health advocates, these problems were intrinsic to the immunological approach, and would lead to all kinds of practical and safety problems for individual women trying to plan their families.

Firstly, primary immunization inevitably entails a period of 3–6 weeks before the requisite immune response is built up. Women's health advocates wrote about the occurrence of a lag period: "Should pregnancy occur during the lag period or occur later due to fluctuations in immune response, the fetus will be exposed to ongoing immune reactions as the contraceptive cannot be switched off. Because of unknown risks for the fetus, this situation is unacceptable for any pharmaceutical product, but in particular for a contraceptive with a lag period inherent

in its design." (Wieringa 1994: 4)[12] WHO/HRP Task Force researcher
Vernon Stevens considered the occurrence of a lag period as a problem
that was relevant only in the application stage and not a problem of the
vaccine itself. He wrote: "Once a safe, effective and acceptable hCG-vac-
cine has been formulated . . . , still other problems must be overcome
before practical application to family planning is feasible. First, administration of the vaccine will not provide "instant" protection against pregnancy." (Stevens 1996: 154)[13] Subsequently, the scientists involved
pointed to the possibility of dual application with a currently available
method to bridge the lag period. The head of the Indian research team
at the NII, G. P. Talwar, wrote: "During the 'lag' period, unless they
either abstain from sex or use an alternate contraceptive with strict discipline, they will be vulnerable to pregnancy. It is, therefore, necessary
to develop 'companion' methods for assuring protection during the initial period." (Talwar et al. 1994: 701)

Secondly, there was the problem of variability in the duration of effective response following injection. Again, women's health advocates considered this phenomenon to be in the nature of immunological
responses, and therefore problematic to the safety and efficacy of the
evolving technology. In Schrater's words (1995: 665): "Some basic biological concerns are related to the nature of the immune response. . . .
Because the degree and duration of immunological responses vary
among individuals, it will be difficult to predict the time span of protective immunity for each person. And because immunity is cryptic, the
body gives no immunological signal that the response has fallen to non-
protective levels."[14]

Women's health advocates assumed that individual users would need
to know the time span of protection. The contraceptive developers
recognized the problem, but rated it as external to the vaccine as well.
For Vernon Stevens (1992: 139), it entailed the need to combine the
vaccine with other means of birth control: "While not technically a part
of new vaccine design, the probable use of anti-fertility vaccines in combination with other means of birth control will surely be a reality and is
worthy of mention in regard to new vaccine development. . . . At the
point in time when immunological birth control methods are ready to
enter family planning programs, this issue will need to be seriously
addressed."

The third issue to illustrate the different categorizations of problems
by the women's health advocates and by the reproductive scientists was
that the effect of the vaccine cannot easily be stopped. Stevens (1996)

considered this an application problem. Again, the women's health advocates viewed the impossibility of switching off an immune response as a safety problem related to the immunological nature of the method. For example, Hardon (1990: 23) wrote: ". . . the method stops working when antibodies to hCG are secreted from the women's body. If side effects occur within that period, the drug cannot be 'switched off.'"[15]

Whether application problems or intrinsic problems of immunological contraceptives, these issues needed to be addressed. The Indian team devoted part of their research to the development of a companion method to overcome the lag period (Upadhyay, Kaushic, and Talwar 1990; Talwar et al. 1993). As Akrich (1995) suggested, delegating the problem to this additional technology can be viewed as an attempt by these researchers to adjust anti-fertility vaccines to the users.

But the women's health advocates did not consider an additional method an appropriate answer. At the August 1992 meeting between women's health advocates and scientists, the need for a companion method to bridge the lag period was discussed: "The women's health advocates were particularly concerned about this aspect of the vaccine, because little is known about the interaction between the vaccine and some of the additional methods that would need to be used during the lag period or about duration and variability of the lag period. Furthermore, the need to use an additional method during this lag period was seen to be a disadvantage." (WHO/HRP 1993: 20)

As a means to address the issue of the unpredictable duration of effective response, researchers under the auspices of the WHO/HRP initiated investigations to develop a test kit to monitor the level of immunity on an individual basis by means of a finger prick blood sample, for home or clinic use (Gupta et al. 1991). Talwar wrote: "[Antibodies] must be present at titres above a threshold if the vaccine is to be efficacious. Titres must, therefore, be monitored on a continual basis each month. Easy to perform 'user friendly' color tests are needed and are currently being developed. The availability of these tests is a prerequisite for the introduction of contraceptive vaccines for family planning." (Talwar et al. 199b: 702)

The test kit was also an attempt by the contraceptive developers to align the emerging technology with the future users in the sense described by Akrich (1995). Again, the problem was relegated to an additional technology. But from their way of viewing the user in her context, a test kit to monitor a woman's antibody level continually appeared problematic to women's health advocates. Schrater (1992: 45) asked:

"Without adequate distribution, rural and poor women may need to return to the clinics for blood tests. If so, how will they get to the clinics? How long must they wait for the results? Who will pay for the tests?"[16] The Forum for Women's Health (1995: 4) asked: "From the point of view of women and demands of women's groups . . . would it not be safer and better to evolve simple user-friendly kits for detection of occurrence of ovulation?"

The third issue was the impossibility of switching off an immune response. The concern of women's health advocates that this might be undesirable to users was taken very seriously. The scientists did not consider reversal on demand technically feasible at short notice (Stevens 1992: 142; Griffin 1994: 93; WHO/HRP 1993: 21). Schrater (1995: 666) put it as follows: "Although scientists say the immune response to beta hCG can be thwarted by injecting large doses of progesterone or the hCG hormone itself, the method would be prohibitively expensive and would probably require hospitalization to monitor for and treat any untoward effects of 'the cure.' . . . The fact that reversal is possible by no means insures that such reversal would be available to all women."

Consequently, a second way to reconcile the anti-fertility vaccines with the users was envisaged. The contraceptive developers proposed that the service providers take charge. Immunological contraceptives were discussed in the newsletter of the WHO/HRP, *Progress in Human Reproduction Research* (1997, p. 6): "One aspect of this method will, however, require special attention on the part of service providers. Since the contraceptive protection offered by the hCG immunocontraceptive will be longer-lasting than the current injectables, users will need counseling to ensure that they understand fully the implications of using a long-acting method that is not reversible before the end of its expected duration of action."[17]

In their encounters with women's health advocates, the reproductive scientists maintained a distinction between their scientific work to develop immunocontraception, and the application of the method. This was an effective boundary work strategy: the perspective of women's health advocates was not to play a part on the technical characteristics of the contraceptive. At the same time, the classification of issues such as the lag period, the unpredictable duration, and the reversibility on demand as external to the vaccine itself had consequences for the kinds of solutions that the contraceptive developers devised to align the contraceptive with users. Problems were delegated either to other, additional technologies or to the service providers.

Forecasting Acceptability

The contraceptive developers repeatedly emphasized that the vaccine would be an attractive method because it would be long acting, easy to administer, and low in cost (Jones 1982: 10, 196; Hjort and Griffin 1985: 272; Thau et al. 1989: 237; Stevens 1990: 344; Griffin and Jones 1991: 190; Griffin 1992: 112; Brache et al. 1992: 1).[18] Significantly, the features that the reproductive scientists mentioned in relation to acceptability were those of the proposed artifact. Immunological contraceptives were not yet long-acting, nor were they easy to administer. They foresaw that the method would be especially appealing in developing countries. For example: "In the developing countries, a family planning preparation that needs to be administered at infrequent intervals and that requires little active participation by the user to remain effective, would have distinct advantages for both the providers and users of family planning services." (Hjort and Griffin 1985: 271) Such announcements aggravated the concerns of women's health advocates that these technical features, together with the impossibility of switching off an immune response once it has been triggered, facilitated possible abuse (Richter 1996).

The women's health advocates successfully put their concerns surrounding the abuse potential of contraceptives on the agenda of the WHO/HRP. The director, Giuseppe Benangiano, and the Technical Officer for Women's Perspectives and Gender Issues, Jane Cottingham, dedicated an article to the subject (Benangiano and Cottingham 1997). The Scientific and Ethical Review Group of the WHO/HRP invited women's health advocates to a special meeting where the issue was discussed. According to the report of the meeting:

There was disagreement as to whether the vaccine has a higher abuse potential than other existing methods. Some people felt that the vaccine is no more open to abuse than currently available methods. . . . On the question of whether research should be stopped because of abuse potential, again sentiments diverged. Stopping immunological research in the field of human reproduction, some felt, would interfere with some of the most exciting leads currently emerging, which hold promise for a whole host of new approaches in the future. (SERG 1994: 5–6)

The assertion that all methods could be abused effectively safeguarded the continuity of the research on anti-fertility vaccines. If all contraceptives were potentially open to abuse, this could not possibly be a reason to stop the research on anti-fertility vaccines. Subsequently, the researchers repeatedly stressed that any contraceptive method could be abused (Griffin, Jones, and Stevens 1994: 113; Stevens 1995; Griffin 1996: 145).

The policy makers at the WHO/HRP also stated that technical objects do not define any specific framework of action. Instead, they viewed contraceptive abuse as a result of political, socio-cultural, and economic situations, and asserted that abuse should be prevented by all means except technological design.[19] "Our position," wrote Benangiano and Cottingham (1997: 43), "is that eliminating research on methods which might be abused will not, in fact, address the problem of abuse. . . . The problem of abuse needs to be tackled where it is happening, by unveiling abusive practices, by informing and educating all levels of the public about ethical requirements, and by extending and strengthening existing safeguards."

Since the contraceptive developers rejected the view that abuse could be forestalled on the basis of the proposed characteristics of the artifact, it became increasingly difficult to maintain the promise of high accept-ability on the basis of these projected features. Therefore, to discuss either acceptability or abuse potential on the basis of the proposed prod-uct was deferred to later stages. In the words of Stevens (1996: 149): "Despite the opposition to further development of hCG vaccines, it is my view that this research should be continued until suitable methods have been obtained before judgment of their acceptability is made."

At the same time, one of the projected technical features of anti-fertility vaccines was actually modified. In May 1995, Judy Norsigian of the Boston Women's Health Book Collective met with Philip Gevas of Aphton Corporation, a small US-based pharmaceutical company that worked with the WHO/HRP on the development of an anti-hCG vaccine. They discussed the critique by women's health advocates of the vaccine design. Subsequently, Gevas (1995) wrote in a letter to Norsigian: "I have already made a, perhaps, profound change regarding the duration of 'protective period,' which I am confident the WHO people will concur with. . . . I believe that the 12 to 18 months originally specified was a sin-cere attempt to determine what might be the best for people with limited access to physicians (e.g. the developing countries) However, the 'abuse potential' . . . considerations convinced me that 6 months, instead, is far better."

Indeed, the WHO/HRP (1993) changed its objectives of developing a vaccine with a duration of effect of 12–24 months to one with a duration of 6–12 months. According to the manager of the Task Force for Immunological Methods for Fertility Regulation of the WHO/HRP,

David Griffin, this change can be attributed to the efforts of women's health advocates (Griffin 1995, 1996). Stevens (1995) also said in an interview that such changes were made because the company wanted to make a product that would be broadly welcomed: "I have been sort of told to make a short acting as well as a long-acting thing. It was a loose kind of talking, not a formal meeting or group. It was centered on women's objections, their fear of losing control over their bodies. Company people considering taking a license said: 'Could you make it shorter?' I said: 'How short?' They said: 'Six months?' I said: 'No problem.' They wanted to make something more acceptable to those women who do not want to lose control over their fertility, their reproductive lives. We will make a longer acting one too, but also short acting so that they will not complain." Similarly, the Aphton Corporation's description of products asserted: "The vaccine is designed to prevent pregnancy for one or two years (being modified to provide six-month protection to be more widely accepted)." (Lyles 1996: 5)

Once the continuity of their research was safeguarded, the contraceptive developers were prepared to admit some of the concerns that women's health advocates had formulated from their alternative perspective on users.

Conclusions

In this chapter I have examined the ways in which the women's health movement has been involved with contraceptive technology, and its relation to users. The women's health advocates did not represent contraceptive users in the sense of articulating their needs, or speaking on behalf of users. Instead, they gave voice to users' perspectives. They mobilized their experiences of working on women's health and rights issues to gain acknowledgment. Ultimately, their differing perspectives on contraceptive users were regarded as a worthwhile addition to the process of technology development.

The strategies that political representatives of end users of a medical technology employ to achieve participation in the making of scientific knowledge have been analyzed by Epstein (1996) in the case of AIDS. AIDS activists were able to influence the course of research by acquiring credibility in the eyes of scientists. They learnt the language and culture of medicine, and by such means they got a foot in the door of the institutions of biomedicine. Moreover, the AIDS activists dispose of a vital source of power: they importantly overlap with the subjects that the scientists wish to recruit for their clinical research: patients. Therefore, the AIDS researchers ultimately depend on cooperation with activists in order to carry on with their research, and AIDS activists could constitute themselves as an obligatory passage point (Epstein 1996). Women's health advocates lack this resource and therefore developed another mix of strategies. Some have indeed acquired cultural competence on immunocontraceptives, but they have also chosen to criticize the language and culture of biomedicine as the only valid perspective to assess contraceptive development.

There were distinctive advantages for women's health advocates to be gained by speaking in the name of users' perspectives rather than engaging in the essentially impossible task of voicing the needs and preferences of the users of the world. Crucially, this strategy enabled them to relate to contraceptive technologies in capacities other than that of potential future users: as researchers or as advocates. Room was created for women's health advocates to introduce different frames of meaning. For example, their critique of a contraceptive that would be long acting and easy to administer was based not on their perception of the needs and preferences of users, but on their concern for the kinds of relations between users and providers that such a technology might constitute. Voicing their concerns in this way was effective: it was the abuse-potential argument that ultimately convinced the contraceptive developers to make a shorter acting vaccine. To speak from users' perspectives also reinforced their position as partners in a dialogue with the contraceptive developers. If women' health advocates had presented themselves as voicing the needs and preferences of users, their role would have been similar to that of the social scientists: advisors on what attributes of contraceptive methods would be attractive to users. To take their advice (or not) would have remained the prerogative of the contraceptive developers. Now, instead, they were invited as dialogue partners the activities of the WHO/HRP.

In their representations of potential users of anti-fertility vaccines, the women's health advocates emphasized the need for a broader analysis of how people plan their families instead of the traditional focus on product features. The attempts of the scientists to align the vaccine with such contextualized images of users put a strain on the users-script of anti-fertility vaccines. The contraceptive developers had envisioned the users of anti-fertility vaccines as women who would want to use a long-acting method, who don't have frequent access to specialized health-care services, who don't want to or can't use hormonal methods, and who may have to hide contraception from other members of the household.[20] These women would fit the ideal characteristics of the imagined artifact. But the method that was actually developing needed several additional technologies to "work" for women wanting to plan their families, and thereby got an ever less coherent users-script. For example, the method used to bridge the lag period could be a hormonal injection, a barrier method (condoms, diaphragms), or "natural" family planning (rhythm method, abstinence, withdrawal). A hormonal bridge method would exclude users who do not want to or cannot use hormonal products. Other possible bridge methods might be unsuitable for users who look for personal confidentiality of use. Similarly, a test kit to monitor the level of antibodies in the blood would make anti-fertility vaccines less convenient for those users who seek to avoid storage and disposal problems, or who don't have regular access to health-care services. These problems that a user of the method might encounter had not played a role in the scientists' work. My analysis suggests that bringing alternative perspectives of users into the technological design and innovation processes might provide one way to anticipate and resolve these difficulties.

8

Inclusion, Diversity, and Biomedical Knowledge Making: The Multiple Politics of Representation
Steven Epstein

Under what circumstances do the "consumers" of biomedicine transform the practices of medical knowledge making? Many abstract discussions of "science and the citizen" conjure up an image of isolated, individual citizens who either do, or do not, feel qualified or empowered to participate in scientific debates. In contrast, a growing body of case studies suggests that meaningful lay participation in science is rarely the business of individuals: more typically and more consequentially, these are projects undertaken by organized social collectivities.[1] In the field of biomedicine, for example, certainly the patient who "does her homework" and confronts the doctor with alternative perspectives about her own condition is making a foray of sorts into the domain of lay participation. But when whole groups of patients suffering from the same disease establish new organizations, elaborate a collective sense of self, and then act in concert to challenge the medical conceptualization of their condition and its treatment, then the intervention is potentially both more radical in character and more transformative in its consequences.

In an essay that tries to give shape to the elusive phrase "democratization of science," Daniel Kleinman (2000a: 140) enumerates some of the prerequisites needed by laypeople before they can participate in scientific debate. These prerequisites include "adequate time and other resources, opportunity to examine deeply held assumptions, and mechanisms that weaken the effects of socially significant forms of inequality." While endorsing Kleinman's analysis, I would supplement his list. Participation becomes more likely when groups build effective organizations, construct new collective identities, and promote groundswells of mobilization and collective action.[2] For example, in the late 1980s people with HIV/AIDS who served as research subjects in clinical trials transcended the status of being merely the passive "stuff" of biomedical research—the raw material out of which medical findings are generated—when they created new organizations that combined militant protest tactics with acquired expertise in biomedical science and research methodology. As AIDS activists amply demonstrated, clinical research can be a domain where experimental subjects who organize collectively may successfully "talk back"—demanding a say in how the experiment itself is conceptualized and conducted (Epstein 1991, 1995, 1996, 1997a,b; Treichler 1991, 1999).

However, the fact that medical knowledge and practice may be transformed by organized political activity on the part of "objects" of research or "users" of medical services ought to inspire closer attention to the sometimes peculiar patterns by which "groupness" is established. What are the politics of representation by which individuals, organizations, or coalitions invoke the interests of social groups and speak in their name when calling for the reform of expert practices? Who, for example, is positioned to speak on behalf of the health needs of "people with AIDS"—or more problematically still, broad social entities such as "women" or "children" or "people of color"? What kinds of "symbolic power," and what practices of social classification, must be exercised by spokespersons who seek plausibly to represent abstract classes of people (Bourdieu 1985; Bourdieu 1991, chapters 8–10; Bourdieu 1998, chapter 3).

Recent controversial changes in biomedical research policies and practices in the United States provide a revealing case study of the dynamics of representational politics. Over the course of the late 1980s and the 1990s, in the face of intense political pressure, agencies of the US Department of Health and Human Services (HHS), particularly the National Institutes of Health (NIH) and the Food and Drug Administration (FDA), have issued policies calling for a more systematic and substantial inclusion of women, racial and ethnic minorities, children, and (to a lesser extent) the elderly as subjects in biomedical research and new drug development. From the standpoint of various proponents of these changes—including scientists, doctors, members of the US Congress, and activists working with health advocacy organizations—no longer would it be acceptable for biomedicine to privilege the health-research needs of white men. Neither could biomedical researchers continue to take white males as the norm, then extrapolate findings from studies done on the bodies of white men to other social groups.

I will argue that these substantial changes in policies governing biomedical knowledge making emerge at the intersection of multiple varieties of representational politics. In order to call for the greater representation of previously underrepresented groups as subjects in biomedical research, various actors had to position themselves successfully as legitimate representatives of social interests and collectivities, invoking the needs, wishes, and interests of groups such as "women" and "African-Americans." At the same time, these representatives speaking for the group had to frame their demands by making claims about the nature of the group—"representation" in the sense of symbolic depiction of fundamental group characteristics. Thus the new research policies emerge out of what might be called "multi-representational politics" that fuses concerns relating simultaneously to numerical inclusion, spokesperson-ship, and symbolic imagery.[3]

The complex configurations of power and knowledge in such disputes become further evident when we explore the lineups of social actors who play roles within them. This is no simple story of a dichotomous and uneven opposition between powerful "producers" of medical knowledge and technologies and disenfranchised "consumers" or "users" of bio-medicine. In considering the coalition of forces that brought the NIH and FDA policy changes into being, it becomes clear that the outcome reflects the efforts of a diverse set of actors whose alliances cut across conventional distinctions between elites and masses. The dynamics of the case therefore reflect the extraordinarily muddy terrain on which struggles over biomedical principles and priorities may often get fought.

A New Policy Regime

In 1993, President Bill Clinton signing the NIH Revitalization Act, the primary purpose of which was to "re-authorize" funding for the National Institutes of Health. However, in addition to mandating an Office of Research on Women's Health and an Office of Research on Minority Health at the NIH, the act contained two important and controversial provisions. First, the NIH, the world's largest funder of medical research, was required to ensure that women and members of racial and ethnic minority groups be "included as subjects" in each clinical study funded by the agency from 1995 onward. Second, the legislation stipulated that every NIH-funded clinical trial be "designed and carried out in a manner sufficient to provide for a valid analysis of whether the variables being studied in the trial affect women or members of minority groups, as the case may be, differently than other subjects in the trial" (US Congress 1993). That is, the prescriptions of this legislation invoked tacit scientific theories about how people differ from one another and when those differences may be medically relevant. And the prescriptions affected not only the composition of the study population, but, conceivably, the framing of the research questions, the logic of data analysis, and the kinds of statistical tests that might have to be employed.

The effect of this legislation was to strengthen policies on inclusion that the NIH had officially (but ineffectually) been promoting since 1986, and to require the agency to enforce what previously had been recommendations to its grant recipients. Congress gave the NIH some leeway in interpreting these unprecedented requirements and in deciding when they might not apply. But, strikingly, the legislation ruled out any consideration of increased medical research costs as grounds for granting exemptions to the policy. The NIH developed guidelines to implement the legislation the following year (interview with Schroeder; US Congress 1993; Nechas and Foley 1994; "NIH guidelines on the inclusion of women and minorities as subjects in clinical research; notice," *Federal Register* 59, 1994, no. 59, March 28: 14508–14513; Rosser 1994; Schroeder and Snowe 1994; Auerbach and Figert 1995; Narrigan et al. 1997; Primmer 1997; Schroeder 1998; Weisman 1998, 2000; Baird 1999). A few years later, in 1998, the agency extended its inclusionary policy making to encompass the variable of age, at least at the lower end of the age spectrum. In this case, the NIH effectively preempted congressional scrutiny of NIH research on children by voluntarily developing and implementing a new policy mandating the inclusion of pediatric populations in many clinical studies funded by the agency (interview with Alexander; "NIH Policy and Guidelines on the Inclusion of Children as Participants in Research Involving Human Subjects," in NIH Guide for Grants and Contracts, March 6, 1998).

The same year in which the NIH Revitalization Act was signed into law, the Food and Drug Administration, the government agency responsible for the licensing of pharmaceutical drugs and medical devices, issued new guidelines governing the participation of women in the clinical trials sponsored by drug companies to test the safety and efficacy of new drugs. Since 1977, women "of childbearing potential" had been routinely excluded from many such trials, whether they were pregnant or not, or had any intention of becoming so, out of concern that an experimental drug might bring harm to a fetus. (In intent, this restriction applied only to early, so-called Phase 1 and Phase 2 trials of new drugs, whose potential for causing birth defects was still unknown; and it was not supposed to apply to trials of drugs for life-threatening conditions. In practice, the broad and automatic exclusion of pre-menopausal women from new drug development had become commonplace.) After members of Congress expressed concern about the exclusion of women from drug testing, and after the HIV Law Project, an activist legal organization in New York, filed a "citizen's petition" against the FDA, the agency removed the 1977 restriction and issued new guidelines. Published in 1993, these guidelines permitted the inclusion of women even in the early phases of drug testing, provided that female subjects used some form of birth control, and also called for drug companies to submit data on the effects of new drugs in both men and women.

In subsequent years—in contrast to the agency's general recent emphasis on expediting the drug-approval process—the FDA increasingly has put the burden on drug manufacturers to present safety and efficacy data by age (including both pediatric and geriatric populations) and by race and ethnicity, in addition to gender.[4] The FDA added teeth to these expectations in 1998, in response to complaints from a national task force on AIDS drug development, warning drug companies that a "clinical hold" could be placed on any licensing application that failed to provide such data (interview with McGovern; interview with Merkatz; interview with Toigo and Klein; "Guidelines for the study and evaluation of gender differences in the clinical evaluation of drugs," *Federal Register* 58, 1993, no. 139, July 22: 39406–39416; "Investigational new drug applications; Proposed amendment to clinical hold regulations for products intended for life-threatening diseases," *Federal Register* 62, 1997, no. 185, September 24: 49946–49954). In passing the FDA Modernization Act of 1997, Congress then took a dramatic step toward promotion of better knowledge about the effects of medications on children: The legislation offered a huge financial incentive in the form of a 6-month patent extension to companies willing to go back and conduct studies of their drugs in pediatric populations (interview with Alexander; US Congress 1997).[5]

How did these various laws and policies promoting representation of diverse subjects in research come about? Although the story is too complex to relate here in detail, an abbreviated version reveals the pathways of "multi-representational politics."

Tracing the History of Multi-Representation at the NIH: Equity, Generalizability, and Identity Politics

In the late 1980s and the early 1990s, the "underrepresentation" of women and racial and ethnic minorities as subjects in clinical research emerged as a recognized public problem (Gusfield 1981) in the United States (Dresser 1992; Mastroianni et al. 1994; Hamilton 1996). In fact there continues to be debate about whether or to what degree women or minorities had been underrepresented previously, and even what "under-representation" means. In the absence of any systematic record-keeping or reporting of the demographics of research subjects by the NIH, the FDA, or medical journals, a range of empirical studies of research demo-graphics, using different samples and methodologies, have arrived at opposing conclusions.[6] Reviewing this issue for the case of women, the report of an Institute of Medicine committee concluded in 1994 that "the available evidence is insufficient to determine whether women have par-ticipated in the whole of clinical studies to the same extent as men. . . . Some studies found that an appropriate number of women were included in specific study populations and that more female-only studies were being conducted than male-only studies. Others found that women were 'overrepresented' or 'underrepresented' in certain types of studies. Others found that women—especially elderly or poor women of diverse racial and ethnic groups—are less likely to be included in studies than men." (Mastroianni et al. 1994: 49)

However, at least in some research domains, such as AIDS research and heart disease, as well as in new drug development, it was not difficult to make a claim that women had been understudied. In addition to the FDA's restrictions, the reliance of many researchers on Veterans Administration hospital patients as research subjects had lessened the likelihood that women would be included in certain studies. Many anec-dotal reports further suggested that biomedical researchers sometimes considered women to be "complicated" research subjects because of monthly fluctuations in hormone levels that could confound the effects of the medical regimes or therapies under investigation. Men's bodies, by this reasoning, were simpler to study: there were fewer "variables" to control for (Mastroianni 1994: 80). Implicitly, men were conceived of as prototypical humans; women were perceived as opposite, deviant, or other (Tavris 1992: 17–20; see also Waldby 1996), and thus as problem-atic objects of biomedical research.

In the case of racial and ethnic minorities, no one had specifically argued that they should be kept out of the subject populations of clinical research. However—despite many findings of racial and ethnic differ-ences in the extent and course of disease, and despite reports in the med-ical literature of "racial differences" or "ethnic differences" in the effects of treatments such as anti-hypertension drugs and antidepressants (Polednak 1989; Cotton 1990a,b; Levy 1993)—few experts had called for ensuring racial and ethnic representation in subject populations across the board. Although some NIH institutes, such as the National Cancer Institute, appeared to have a good record in enrolling racial and ethnic minorities in its treatment trials (Tejeda et al. 1996), other institutes appeared to recruit them in proportions less than their percentage in the population, or less than the percentage they contributed to those suffering the disease in question.

Racial minorities were often considered to be "hard to recruit"—especially African-Americans, who were said to reject the role of medical "guinea pig" out of suspicion of the long history of medical experimentation on black people that dates to slavery and includes the infamous Tuskegee Syphilis Study (Jones [1981] 1993; Thomas and Quinn 1991). No doubt many researchers simply lacked interest in the health needs of minorities, while some researchers may have felt it was best to avoid trying to recruit members of minority groups, lest they be accused of exploiting the socially disadvantaged (interview with Jackson). Furthermore, some experts on clinical trials argued that "homogeneous" subject populations made for "cleaner" trials with less "noise" in the data (although other experts disagreed, arguing that the best research mirrored the complexities of the real world) (Feinstein 1983). Thus many researchers may have felt that "simpler" trials were both more scientifically elegant and more likely to pass muster with the peer reviewers of medical journals. Given the difficulties involved in recruiting minorities as subjects, the predictable consequence of these arguments about homogeneity and simplicity was to solidify the notion that white people were the obvious choice as research subjects.

Beginning in the late 1980s, research policies and practices that had seemed uncontroversial and even ethically advisable suddenly began to appear ludicrous, offensive, and unscientific. In short order, a new "common sense" emerged and replaced a prior, discredited one. Certainly a crucial player in the forging of a new common sense was the women's health movement, which, by the late 1980s, enjoyed support not only at the grassroots but also at elite levels within medicine and government. Women scientists at NIH, such as Florence Haseltine, the director of the Center for Population Research at the National Institute of Child Health and Human Development, worked behind the scenes to call attention to the low profile of women's health issues at the agency (interview with Haseltine). (One of her favorite "sound bites" was that the NIH had 39 full-time veterinarians but only three gynecologists.) At the same time, the broad issue of women's health became a galvanizing one for women in Congress, even as the topic of health reform moved to the top of the policy agenda in Washington (Weisman 1998, 2000). "Every time you picked up the paper, there was something," Pat Schroeder recalled in an interview, thinking of the news reports in the 1980s that trumpeted the findings of medical researchers conducting clinical studies—reports about "men eating fish, men riding bikes, men drinking coffee, men taking aspirin. And we were just wondering whether 'men' was an all-encompassing word, or whether it was truly just men."

Schroeder, a Democrat, along with Olympia Snowe, a Republican, and the other members of the Congressional Caucus on Women's Issues, took their concerns about women's health to Henry Waxman, chair of the Health and Environment Subcommittee in the House of Representatives. Ruth Katz, who served as counsel to the subcommittee, working along with Leslie Primmer, a staffer for the caucus, then devised a "hook" to draw congressional attention to the broader issue of women's health. In an interview, Katz recalled: "I said, 'I wonder if they have any rules about making sure that women are included in clinical trials?'" Upon discovering that, indeed, NIH had already implemented a policy encouraging inclusion of women and minorities in 1986, Katz recalled proposing: "Why don't we get GAO [the General Accounting Office, Congress's own investigative agency] to take a look at the simple question of to what extent NIH is following its own rules?"

Waxman sent the request to the GAO, and the results of its investigation confirmed the suspicions of Katz and the caucus members: The investigators found that the 1986 policy had been poorly communicated even within NIH and had been applied inconsistently (Nadel 1990). "American women have been put at risk," Schroeder declared at the June 1990 House subcommittee meeting at which the GAO report was presented. Schroeder cited the NIH-funded Physician's Health Study, begun in 1981, which had investigated the role of aspirin use in preventing heart attacks. The study had enrolled 22,000 male doctors. "[NIH] officials told us women were not included in the study, because to do so would have increased the cost," commented Mark Nadel (1990: 2), who presented the GAO's findings. "However, we now have the dilemma of not knowing whether this preventive strategy would help women, harm them, or have no effect."[7] Olympia Snowe, whose mother died of breast cancer, described for reporters a federally funded study on the relation between obesity and cancer of the breast and the uterus; the pilot study had used only men. "Somehow I find it hard to imagine," Snowe commented, "that the male-dominated medical community would tolerate a study of prostate cancer that used only women as research subjects."[8]

The release of the GAO report "was carefully orchestrated for maximum public impact" (Weisman 1998: 83), with representatives of the media well in attendance. Still, "never in a million years did I think we'd end up on the front page of the *New York Times*," recalled Katz. What may have caused reporters to sit up and take note was the testimony of the acting director of the NIH, William Raub, who in essence acknowledged that the congressional criticism had merit (interview with Raub). "He flat out admitted it," Katz recalled in his interview. "He did not even try to defend the institution." Members of the caucus began pressing for legislation that would force the NIH to change its ways.[9] That the NIH budget was due for re-authorization provided them with a perfect opportunity to inject their concerns into an existing bill.

Meanwhile, NIH's Florence Haseltine worked with a lobbying agency called Bass and Howes that specialized in women's issues to found a Washington-based advocacy group called the Society for the Advancement of Women's Health Research. The SAWHR explicitly took up the cause of inclusion of women in clinical research as its priority issue and began pressing for passage of the NIH Revitalization Act (interview with Haseltine; interview with Bass). As opposed to the grassroots and national organizations that grew out of the women's health movement of the 1970s, the SAWHR is a more professionally based organization, in which lobbyists and female scientists have played a significant role. The SAWHR now publishes an academic journal, the *Journal of Women's Health*, a senior editor of which was Bernadine Healy, the first female director of the NIH. The SAWHR also receives dues from pharmaceutical companies that belong to its Corporate Advisory Council. By the latter part of the 1990s, the SAWHR had become key proponents of "gender-specific medicine" or "gender-based biology," which (in contrast to the views of other women's health advocacy groups) emphasizes vast, fundamental, biological differences between men's and women's bodies, from the heart, to the brain, to the immune system. Despite the fact that they prefer the term "gender" to "sex," advocates of this movement believe that women (and men) deserve separate medical scrutiny because they are biologically different at the level of the cell, the organ, the system, and the organism (interview with Marts; Haseltine 1997; "10 Differences between Men and Women That Make a Difference in Women's Health," www.womens-health.org).[10]

As legislators and their staffs began work on new language to be added to the NIH Revitalization Act, African-American members of Congress called for a further extension of the legislative mandate. In response, the phrase "and minorities" was added to the wording about inclusion of women in research. This seemed to its sponsors and to others in Congress to be a logical, and politically desirable, extension of the legislative intent: After all, if the NIH was parceling out federal tax dollars, then the research that it sponsored should be of benefit to the whole population. And if whites were already reaping the benefits of better health while men and women of color suffered from higher levels of morbidity and mortality, then it seemed particularly problematic for the government to be investing at higher rates in the health issues affecting white people. Inclusion of minorities also allayed fears expressed publicly about the credibility of medical treatments that had been tested only in white populations. In 1990 Vivian Pinn, the president of the National Medical Association (the African-American physicians' organization) who would go on to become the director of NIH's Office of Research on Women's Health, told a reporter from the *Journal of the American Medical Association*: "Some of our physicians are a little leery [of some drugs because] we can't be certain whether minorities have been participants [in clinical trials]" (Cotton 1990a: 1049).

While congressional and public attention in the late 1980s focused primarily on the goal of gender and racial diversification in biomedical research, attention to the needs of children followed a few years later. Of course, even more starkly than in the case of other social groups, children lack a public voice and do not "speak for themselves" in health policy arenas: their interests are always represented by others. In this case, physicians who were prominent within the American Academy of Pediatrics played a key role as "moral entrepreneurs" who pressed for policy change. Members of the Academy's Council on Pediatric Research, pointing to evidence that showed that the vast majority of medications used by children had never been tested on children, began clamoring for legislation that would extend the policies on inclusion of women and minorities to pediatric populations as well. Pediatricians argued that children are not simply "miniature adults," and that it was crucial to study differences between adult and pediatric populations rather than simply extrapolating from the former to the latter. However, officials at the National Institute of Child Health and Human Development (one of the institutes within the NIH) succeeded in convincing the Academy that legislation was not needed—that the agency was prepared to voluntarily institute guidelines on inclusion of children. These guidelines were published by the NIH in 1998 (interview with Alexander; NIH Policy and Guidelines on the Inclusion of Children as

Participants in Research Involving Human Subjects," in NIH Guide for Grants and Contracts, March 6, 1998).

The new emphasis on inclusion and the distrust of extrapolation across social categories were not without opponents. NIH officials worried about the loss of autonomy and control over the peer review process in their distribution of funds. Conservatives in Congress objected to the "micro-managing" of the NIH. Prominent statisticians and authorities on clinical trial methodology complained that "political correctness" was winning out over common sense and good scientific judgment: In particular, they claimed that the subgroup comparisons called for in the NIH Revitalization Act had the potential to bankrupt research and were not medically necessary, since most of the time men and women, adults and children, and people of different races and ethnicities respond similarly to medical interventions (Piantadosi and Wittes 1993; Wittes and Wittes 1993; Meinert 1995a,b; Piantadosi 1995).

Concerns were also raised about the problematic business of defining medically meaningful racial and ethnic categories. In its implementation of the NIH Revitalization Act's directive concerning "minorities," the NIH following the path of other government agencies by adopting "Statistical Policy Directive No. 15" of the Office of Management and the Budget ("Race and Ethnic Standards for Federal Statistics and Administrative Reporting"). Published in 1977 (and recently revised), Directive 15 specifies the racial and ethnic categories used in the US census. In addition, it suggests that the determination of any individual's racial or ethnic status is best made by the individual: people are what they say they are. NIH's adoption of this directive yielded one "majority" category—"White, not Hispanic"—as well as four "minority" categories: "Asian or Pacific Islander," "American Indian or Alaskan Native," "Hispanic origin," and "Black (not of Hispanic origin)." This way of operationalizing race and ethnicity has provided the basis for the coding scheme used by the NIH in determining compliance with the act. In its own words, "NIH has chosen to use these definitions because they allow comparisons to many national databases, especially national health databases" (Grant Application Instructions, PHS 398, US Public Health Service).[11] It is important to note, however, that census categories are determined in response to a particular set of political needs and pressures, and they have changed with regular frequency since the initiation of the US census in 1790 (Wright 1994; Goldberg 1997, chapter 3). In 1890, the US census included racial categories such as "quadroon" and "octoroon" to designate people who were one-fourth and one-eighth (or less) black (Goldberg 1997: 36–37). More recently, a heated debate about whether the category "multiracial" should be added to the census in the year 2000 ended with the decision to instead allow people to check more than one box in the list of races.

The vexing question of how to define racial and ethnic categories in clinical research was compounded by the concern voiced in some quarters about the reification of race as a form of biological difference. Indeed, some researchers who are heavily invested in promoting the health needs of racial minority groups have nonetheless suggested that the NIH Revitalization Act has pernicious, perhaps even racist, effects. For example, Otis Brawley, an African-American who heads the National Cancer Institute's Office of Special Populations Research, argued in the journal *Controlled Clinical Trials* that "the legislation's emphasis on potential racial differences fosters the racism that its creators want to abrogate by establishing government-sponsored research on the basis of the belief that there are significant biological differences among the races" (Brawley 1995: 293). Brawley, and the cluster of African-American oncologists who share his views, are acutely conscious of the invidious history of racialized thinking in medicine (Brawley 1998; see Proctor 1988; Duster 1990; McBride 1991; Tucker 1994; Wailoo 1997; Tapper 1999). They tend to emphasize the fundamental biological sameness of human beings across racial categories and to regret the lack of "understanding that discoveries about disease in one race are applicable to persons of other races" (Brawley, quoted in Freeman 1998: 220). Furthermore, they tend to attribute differences in health outcomes to social, cultural, and lifestyle factors, including poverty, diet, differences in the consumption of medications, and access to state-of-the-art care (interview with Brawley; interview with Streeter; Roach 1998). In general, however, such opposition to biological reductionism has been a minority position in these debates. Advocates of inclusion have found it more convenient to invoke biological difference as an argument for the diversification of clinical research.

Inclusionary Pressures at the FDA: The Revolt against Paternalism

While arguments about NIH-funded research emphasized the equitable and responsible use of tax dollars and the problem of generalizability, public attention to FDA policies crystallized around issues of autonomy and risk. Here, however, the emphasis on inclusion marks a partial break with a particular way of thinking about medical and research ethics that is itself of fairly recent invention. Although the Nuremberg trials after the Second World War had provided graphic evidence of the horrific uses to which medical experimentation could be put, only in the 1960s, with the publication of reports of widespread abuses of patients in high-profile US medical experiments (Beecher 1966), did many policy makers begin to assert that stricter measures were needed to safeguard human subjects in the United States (Rothman 1991: 70–84). Bolstered by publicity surrounding the Tuskegee study of "untreated syphilis in the Negro male" (Jones [1981] 1993), this wave of concern culminated in the enactment of formal, legal protection of the rights of experimental subjects, along with a new conception of participation in research as a burden which, therefore, must be distributed as equitably as possible in society. Researchers were now obliged to comply with procedures established by the NIH's new Office for Protection from Research Risks; to submit their protocols beforehand to local "institutional review boards" that would ensure that human subjects were not placed at undue risk; and to document the process of obtaining informed consent from their subjects. A distinguishing feature of this regime of regulation was its emphasis on the protection of "vulnerable populations"—children, fetuses, prisoners, the poor, and the mentally infirm—from harm at the hands of the research enterprise (interview with Ellis; National Commission 1979; Edgar and Rothman 1990: 119; Rothman 1991).

As Harold Edgar and David Rothman (1990) have pointed out, one of the most curious aspects of the emphasis on protectionism that arose in the 1970s was that it existed in ironic counterpoint to dominant trends in medical politics of the time: "In a period when autonomy and rights were the highest values in almost every aspect of medical and health care delivery, this was one particular area in which heavy-handed paternalism flourished" (Edgar and Rothman 1990: 121). It is perhaps not surprising, then, that the 1980s saw a "sea change" in attitudes: a shift from viewing participation in research as risky, to viewing it as desirable even if it carried risk (interview with Ellis). Increasingly, patients began to decry governmental paternalism and insist on their right to assume risks—indeed, their right to serve as "guinea pigs" (Edgar and Rothman 1990; Feenberg 1992; Epstein 1996). Some of the same groups that had been singled out for protection in the earlier era, including women and children, were now portrayed as victims of substandard care, stemming from researcher indifference to the particular manifestations of illness in those groups and inadequate access to potentially lifesaving drugs (Corea 1992). Drawing explicit comparisons with recent legal debates about whether women could be excluded from occupations that presented a risk of fetal exposure to environmental hazards, advocates for women emphasized how "protectionism" often served to consign women to second-class citizenship.

AIDS activism proved to be an especially significant source of pressure for change—both away from the "white male model" and away from a protectionist or paternalistic emphasis in the approach toward human subjects and research risks (Epstein 1995, 1996, 1997a). Activists demanded the inclusion of more women and racial minorities in clinical trials of experimental drugs, arguing that clinical trials served as an important means of access to otherwise unobtainable and theoretically helpful new therapies. If, as activists claimed, access to experimental drugs should be considered a social good (rather than simply as a risk from which vulnerable populations should be protected), then it was only right to distribute such access fairly across the population. In practice, AIDS trials were populated primarily by white gay men (Mueller 1998), and as Terry McGovern, the director of the HIV Law Project, discovered, women who sought entry into trials faced extraordinary obstacles. McGovern described the not atypical case of a homeless woman who tried to enroll in a clinical trial in New Jersey in 1991 for an anti-viral drug she saw as her last chance. The woman was told she would be eligible only if she obtained an IUD, but because of her history of AIDS-related gynecological problems, "there was no way that the doctor was going to give her an IUD." Noting that sex was "the last thing [she] was even thinking about" given her state of health, the woman showed up at McGovern's office in a rage (interview with McGovern).

At the same time as activists stressed the ethical principle of equal access to experimental therapies, they also put forward a scientific argument. They noted that, since AIDS had different clinical manifestations in women—something that women with AIDS themselves had observed to be the case—then it made good scientific sense to study the disease separately in women and not to assume that therapies would have the same efficacy or toxicity across groups.[12] Similar arguments were launched about the manifestations of AIDS and the efficacy of AIDS treatments in people of color; and, although the hypothesis was later rejected, well-publicized preliminary findings from one large study of the drug AZT had suggested in 1991 that the drug was less efficacious in "non-whites" than in "whites" (Smith 1991).

These sorts of concerns carried weight in the private sector as well, for while the corporate officers of pharmaceutical companies worried about anything that might add to the often astronomical costs of drug development, they also sensed potential profit to be made from emerging "niche markets" in women's health and minority health (interview with Levy). Many leading pharmaceutical companies have moved to position themselves in the vanguard of women's health research, and many of them have become dues-paying sponsors of the Society for the Advancement of Women's Health Research. In addition, the lucrative incentives offered by the FDA Modernization Act of 1997 have induced pharmaceutical companies to study differences between adults and children in the safety and efficacy of recently developed drugs. Because companies that perform such studies receive, under the legislation, a six-month extension of their patent protection, the advantages to drug companies can be measured in the millions of dollars.

Representation, Classification, and Identity Politics

What are the implications of these complicated histories of reform at the NIH and the FDA?[13] Here, important policy changes that affect how biomedical knowledge is produced were brought about not by research subjects themselves, nor, in any simple sense, by the "downstream" users of medical knowledge and services (that is, patients). Nor, of course, was change brought about by any sort of comprehensive, collective effort on the part of the groups whose interests were constantly invoked, such as women, people of color, and children. Rather, these changes emerged out of the explicit and tacit alliances among an array of diverse actors, including: grassroots activists and health advocacy groups; activist lawyers; sympathetic researchers and doctors working with women, children, and racial and ethnic minorities; the American Academy of Pediatrics; sympathetic "insiders" within HHS; pharmaceutical company scientists and marketers interested in expanding to diverse markets; and members of Congress and their staffers. Thus the coalition that helped bring about lay participation depended upon complicated social relationships that cut across the domains of elites and masses, the powerful and the disenfranchised, and the experts and the laity, even while they served to reconfigure the membership and nature of these categories.

This heterogeneous set of actors both competed and collaborated to speak on behalf of socio-demographic constituencies that do not in fact speak in any single, discernible voice. What are the biomedical interests of "women"? Who is to say? Does Pat Schroeder speak for "women"? Does the Society for the Advancement of Women's Health Research? Do the opponents of biological reductionism most legitimately represent the interests of people of color, or is that position held by those who strategically use conceptions of medical difference by race to press for biomedical reforms?

Even more abstractly, which means of social differentiation are deemed to be medically relevant in the first place? Why gender, race or ethnicity, and age, and why not social class, sexual identity, or religion? Policy reform depended on the capacity of individuals to monopolize the symbolic power needed to classify the group, to bring its concerns into the domains of the political and the scientific, and to define and give voice to its purported collective interests (cf. Bourdieu 1985; 1991, chapters 8–10; 1998, chapter 3). But the ability of individuals to accomplish these tasks was heavily constrained. First, the existing map of identity politics in the United States made it likely that mobilization would occur primarily in relation to the most politically salient markers of difference, such as gender and race. Second, the long history of biological reductionism in the conceptualizing of difference in medicine (Gilman 1985; Fausto-Sterling [1985] 1992; Lacquer 1987; Schiebinger 1987, 1993; Jordanova 1989; Duster 1990; McBride 1991; Oudshoorn 1994; Krieger and Fee 1996a,b; Hanson 1997; Haraway 1997) made it easier to argue about the dangers of extrapolating from, say, whites to people of color, or from men to women, or from adults to children than from, say, rich people to poor people—because race, sex, and age are conventionally understood as forms of difference rooted in the body in a way that social class presently is not. Third, the tremendous power of the state to "produce and impose . . . categories of thought that we spontaneously apply to all things of the social world," and to "[mold] mental structures and [impose] common principles of vision and division" (Bourdieu 1998: 35–63, quotes from pp. 35 and 45) was made manifest in the way that state-sanctioned classification systems, such as "Statistical Policy Directive No. 15," were imposed upon biomedical reformers. These various legacies of past scientific and political practice established the bounds within which legitimated spokespersons could plausibly speak for the group and give voice to its interests.

Acting within these powerful constraints, coalitions brought about changes in biomedical knowledge-making practices by successfully fusing different representational strategies. They simultaneously articulated how groups should be imagined (What were their relevant social, political, biological, and medical characteristics?), how groups should put forward demands (Who speaks for them and articulates their collective interest?), and how groups should be numerically included in studies (What numbers of research subjects were required by the dictates of good scientific practice and equitable science policy?). What I have termed "multi-representational politics" served as the glue that held together a complicated project of transforming biomedical science.

Such events point to difficult questions about the situationally specific uses of categories of identity and difference. What is the relation between "medical" and "political" schemata of social classification? What are the practical consequences when we attempt to make the same set of classifying labels serve "double duty" as bodily descriptors and as names for mobilized collective actors? At the same time, controversies such as these should stimulate us to think carefully about our conceptions of how the "users" or "objects" of scientific knowledge and technology may consciously transform the practices of technoscience. In some cases, it may be meaningful to speak of the processes by which politically disenfranchised political actors, acting in their own name, directly take up the task of transforming science "from below." More frequently, I suspect, it may be more accurate and more fruitful to consider the complex practices of representation by which some individuals or entities, located within heterogeneous coalitions, invoke, and speak on behalf of, broader social collectivities.

Acknowledgments

The work on which this paper was based was supported by an Investigator Award in Health Policy Research from the Robert Wood Johnson Foundation. In addition, this material is based upon work supported by the National Science Foundation under Grant No. SRB-9710423. Any opinions, findings, and conclusions or recommendations expressed in this material are those of the author and do not necessarily reflect the views of the National Science Foundation.

I am grateful to Nelly Oudshoorn for her encouragement; to my research assistants, Nielan Barnes, Paul Chamba, Christine DeMaria, Josh Dunsby, Mark Jones, and Marisa Smith, for their help with the research; and to various audiences who have heard related material, for their comments and suggestions.

I thank the following people for granting me interviews: Duane Alexander, Director of the National Institute of Child Health and Human Development, NIH (interviewed in Bethesda, Maryland, on August 8, 2000); Marie Bass of Bass and Howes, interviewed in

Washington on April 12, 1999; Otis Brawley of the Office of Special Populations Research, National Cancer Institute, NIH (interviewed in Rockville, Maryland, on March 8, 1998); Julie Buring of Brigham and Women's Hospital, Harvard Medical School, and Harvard School of Public Health (interviewed in Boston on April 28, 1999); Gary Ellis of the Office for Protection from Research Risks, NIH (interviewed in Bethesda on March 17, 1998); Florence Haseltine of the Center for Population Research at the National Institute of Child Health and Human Development, NIH (interviewed in Rockville on April 19, 1999); James S. Jackson of the Department of Psychology and Michigan Center for Urban African American Aging Research, University of Michigan (interviewed in Ann Arbor on April 9, 1999); Ruth Katz of the Yale University School of Medicine (formerly counsel to Health and Environment Subcommittee, US House of Representatives) (interviewed in New Haven, Connecticut, on April 21, 1999); Richard A. Levy of the National Pharmaceutical Council (interviewed in Reston, Virginia, on March 29, 1998); Sherry Marts, Scientific Director of the Society for the Advancement of Women's Health Research (interviewed in Washington on April 20, 1999); Terry McGovern of the HIV Law Project (interviewed in New York on May 10, 1998); Ruth Merkatz of Pfizer (formerly with Office of Women's Health, FDA) (interviewed in New York on March 9, 1998); William Raub of the US Department of Health and Human Services (formerly Acting Director of NIH) (interviewed in Washington on April 13, 1999); former US Representative Patricia Schroeder (D-Colorado) (interviewed in Washington on March 18, 1998); Oscar Streeter of the University of Southern California (interviewed in Los Angeles on May 1, 1998); Teresa Toigo and Richard Klein of Special Health Issues, FDA (interviewed in Rockville on April 12, 1999).

III

Multiplicity in Locations: Configuring the User during the Design, the Testing, and the Selling of Technologies

9

Materialized Gender: How Shavers Configure the Users' Femininity and Masculinity
Ellen van Oost

The relation between gender and the material environment, the day-to-day objects that surround us, has recently enjoyed growing interest in the fields of cultural studies, gender studies, and social studies of technology (Horowitz and Mohun 1998; Rothschild 1999; Sparke 1996; Cockburn and Ormrod 1993; Cockburn and Fürst-Dilic 1994; Kirkham 1996; Lie and Sørensen 1996; Oudshoorn 1996; van Oost 1995). These studies contribute—each in its own way—to insight into the complexity and the myriad ways in which gender and material objects are related and mutually constitute each other.

In this field there is a considerable body of work focusing on the processes of how material objects acquire gendered meaning. These studies, especially in the domestic sphere, have analyzed the way men and women accommodate and appropriate new technology in their daily life. A study of gender and the telephone, for instance, found that telephone use became for women a way of expressing their femininity (Rakow 1992). But the appropriation of the telephone by female users not only (re)shaped femininity, but also the telephone itself was being reshaped (Martin 1991). The telephone, originally designed and marketed as a business communication tool, was gradually transformed into to a more general instrument of social communication in the private domain. The main insight to be gained from the type of gender and technology studies that focuses on the use of technology, is that the domestication of new technology is a process of mutual adaptation in which both gender and technology are being (re)shaped.

The gendered meaning given to a technical artifact is often diverse and dependent on the specific use context. For example, different groups have been found to attribute different and often conflicting meanings to computers (Lie 1996; van Oost 2000). Some groups of men constructed the computer as a complex and difficult technology. In this case the skills required to control it became a new way of expressing masculinity. Other groups, however, associated computers with routine office work performed principally by women; the groups did not attribute a high status to computers.

The use context is evidently an important locus in which material objects can function as symbols expressing a gendered meaning. The appropriation of material goods into one's daily life is an important way in which individuals construct their gendered identity. However, the use context is not the only place where objects acquire a gendered connotation. Producers, too, develop and market their products bearing in mind the values and symbols they see as central to the targeted consumer group. A Siemens manager (cited in Verbeek 2000: 12) formulated it this way: "We don't sell appliances, but a lifestyle." Producers know—more than anyone —that it is at the "consumption junction" that success or failure of their product will be manifested (Cowan 1987).

Advertising is an important locus for linking an object to a specific consumer group (Hubak 1996; Kirkham 1996). By creating links between the advertised object and (sub)culturally accepted masculine or feminine symbols, advertisers hope to seduce the targeted group to buy the product. At the same time they actually construct gender and the means to perform gender.

This chapter, however, will focus primarily on the analysis of the objects themselves and the way the gender of the envisioned user influenced the material design of the object. Chabaud-Rychter (1994) was one of the first to study the different ways women users were brought into the design process of consumer appliances. She analyzed the construction of women users as the results of a twofold strategy of designers that aims both at learning about them and at shaping them. I will use the concept of gender script to illuminate how gendered user representations are an inextricably part of designing artifacts. As such artifacts are not neutral objects that only acquire a gendered connotation in advertising or in use; to a certain extent they "guide" the process of giving meaning.

The objects that are central to my study are Philips electric shavers. Shavers are interesting for a gender script analysis because the development of shavers dichotomized into shavers for men and shavers for women. The multinational company Philips is the most important player in the market of electric shavers. Philips was founded in The Netherlands in the late nineteenth century as a manufacturer of incandescent lamps. In the first half of the twentieth century, the company grew into one of the largest multinationals in the area of electrical (later electronic) products. Philips was one of the first producers of electric shavers and has been the market leader in electric shaving appliances for more than 40 years.

Gender Script as Analytic Tool

The inscription of the designers' projected user—or more generally formulated the envisioned use situation—has become an important theme in technology studies. Woolgar (1991) analyzed the design process as a struggle to configure—that is, to define, enable, and constrain—the user. Akrich (1992, 1995) introduced the concept of "script" to make visible how designers' representations of users shape technological development, and how subsequently the artifact shapes the users' environment. Designers construct—explicitly or implicitly—images of users "with specific tastes, competences, motives, aspirations, political prejudices, etc." and inscribe these representations in the technical content of the new artifact (Akrich 1992: 208). As a result, artifacts contain a script and this script prescribes (in a more or less coercive manner) what users have to do (or not do) to produce the envisioned functioning of the technological artifact.

Akrich's script approach has been extended to include gender analysis by introducing the concept of "gender script" (van Oost 1995; Oudshoorn 1996; Rommes et al. 1999; Rommes 2002). "Gender script" refers to the representations an artifact's designers have or construct of gender relations and gender identities—representations that they then inscribe into the materiality of that artifact. Like gender itself, which is defined as a multi-level process, gender scripts function on an individual and a symbolic level, reflecting and constructing gender identities, and on a structural level, reflecting and constructing gender differences in the division of labor. An illustrative example of the latter is given by Hofmann (1996), who found the asymmetrical labor relation between female secretaries and their male bosses reflected in the software structure of the early dedicated word processors.

Gender can be an explicit or an implicit element in the design process. When products are designed for either female or male consumers, gendering is often an explicit process. Existing or even stereotyped images of projected gender identities are transformed into design specifications that are in accordance with cultural symbols of masculinity or femininity. Penny Sparke (1996) provided an extensive elaboration on such cultural symbols in *As Long as It's Pink: The Sexual Politics of Taste*.

Gender scripts can also result from implicit processes. Many objects and artifacts are designed for "everybody," with no specific user group in mind. However, recent studies have shown that in those cases in which designers develop artifacts for "everybody" they often unconsciously base their design choices on a one-sided, male user image (Rommes et al. 1999, 2002). Designers and engineers—mostly men[1]—often use the so-called I-methodology, implying that they see themselves as the potential user, thus creating a gender bias toward male-dominated symbols and competencies. Furthermore, designers often test their products in their own—usually male-dominated—environment. In such cases, the user representation that designers generate is one-sided, emphasizing the characteristics of the designers themselves and neglecting the diversity of the envisioned user group. Configuring the user as "everybody" in practice often leads to a product that is biased toward young, white, well-educated male users, reflecting the composition of the designer's own group (Oudshoorn et al., forthcoming).

Objects, thus, can be perceived as actors that can direct meaning themselves (Akrich 1992; Latour 1997). The use of objects with a gender script often implies a maintaining or reinforcing of prevailing gender definitions. The gender scripts of early dedicated word processors tend to reinforce existing gender inequalities in labor relations. The preferences, competencies and interests of the designers themselves, which served as guidelines in design, may inhibit other social groups (elderly, ethnic, female, poorly educated) from using the artifact.

Clearly the impact of the gender script is neither determined by the artifact nor stable. Gender is an analytical category, the content of which is constantly being negotiated, and objects with inscribed gender relations are actors in these negotiation processes. Obviously, scripts cannot determine the behavior of users, their attribution of meaning or the way they use the object to construct their identity, as this would lead to the pitfall of technological determinism. Users don't have to accept the script, it is possible for them to reject or adapt it. Gender scripts do not force users to construct specific gender identities, but scripts surely act invitingly and/or inhibitingly (Verbeek 2000: 191).

In this chapter, a comparison of the design trajectories of two analogous devices (the Philishave for men and the Ladyshave for women) is used as a method to render visible the gender script of shavers—that is, the inscribed representations of the male and the female consumers. The analysis on the level of artifacts is primarily based on the materials (photographs and product information) collected by Bram Porrey, member of the Dutch Philishave Collector Group.[2] Secondary literature on the historical development of shaving technology (Baudet 1986; Derks et al. 1996; Porrey 1998) is used to contextualize the development of the diverse models.

The Shaping of Electric Shavers

The first electric shavers were developed in the 1920s in the context of the search for safe shaving technology (Baudet 1986). With the traditional straight razor, used by barbers, bloodbaths could only be prevented by experienced hands. A number of individuals were instrumental in the development of the safety razor in the second half of the nineteenth century. But it was the invention of disposable razor blades by King Camp Gillette in 1903 that led to the development of the most successful safety razors in the first half of the twentieth century (Derks et al. 1996). The success of the safety razor stimulated the shift of the shaving location from the barbershop to the home. This shift is characteristic for the much wider development of the consumer society in which the home has become the central unit of consumption (Lubar 1998). Although the safety razor was safe in use, the changing of blades was responsible for a considerable number of injuries. In search of a solution for this problem, Jacob Schick was the first to develop and market an electric dry shaver in 1929.[3] As homes were gradually fitted with electricity, electric shaving became an option. Schick's shaving technology was based on the "clipper system" (Derks et al. 1996). This system consisted of two combs to guide the hairs, with an oscillating, indented knife in between. Electric dry shaving became increasingly popular in the 1930s, and Schick's company acquired several competitors.[4]

At the end of the 1930s Philips entered the market for electric shavers. At that time Philips's main products were light bulbs and radios. During the economic recession of the 1930s the company was in search of new products to keep sales up (Derks et al. 1996). Philips sent an employee to the United States to collect ideas and products. He returned with a suitcase full of electrical devices, among them a number of electric shavers. In the research laboratory of Philips, Alexandre Horowitz, an engineer from the Delft University of Technology initiated the development of a new electric shaver (Baudet 1986). Instead of an oscillating system, he designed a rotating system with three chisels rotating at high speed under a grid. The rotating razor system eventually became the successful trademark of Philips electric shavers, which

Figure 1
The first Philips electric shaver (1939).

were marketed under the names Philishave in Europe and Norelco in the United States.

The first Philips shaver had one shaving head. The bar-shaped body (nicknamed the "cigar") was made of black Bakelite and was delivered with a leather holder. It was presented at Philips's 1939 Spring Exhibition. World War II prevented the "cigar" from becoming a clear success (Derks et al. 1996). In 1946 the company marketed an improved version. Nicknamed the "steel beard," it had the same shape and appearance as the "cigar" but was more robust. It had a larger shaving head and steel blades. Philips gained a place in the market with an intensive marketing campaign and numerous demonstrations (Baudet 1986). The advertising strategy was directed at male users and emphasized the discomfort of wet shaving at a time when the average home in the Netherlands did not have running hot water. With the "steel beard," Philips conquered a segment of the shaving market in the Netherlands.

The next model, introduced in 1948, was even more successful. Nicknamed the "egg," it was ergonomically well designed. The two-head model, introduced in 1951, considerably shortened the time needed for

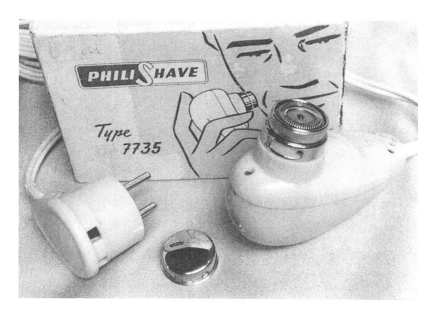

Figure 2
The "egg" model for men (1948).

shaving. The use of new, light-colored synthetic materials was seen as a prerequisite for success, especially in the American market. After World War II, Bakelite was perceived as old-fashioned; ivory-colored plastic was considered modern (Derks et al. 1996).

Philips managed to link this new, modern device to a masculine image. Anthony Quinn used a Philips doubleheader in one of his film roles (Derks et al. 1996). Another marketing technique that proved effective was to provide airline companies with battery-powered shavers imprinted with the airline company's logo. In this way Philips not only linked its shavers with the modern symbol of speed (the airplane) but also saw to it that well-to-do male travelers became acquainted with them. The two-head model became Philips's most successful shaver in the 1950s and the 1960s. The system was improved by the addition of the "fliptop cleaning system." With one press of a little button, the shaving head opened, allowing fast cleaning of the chamber that collected shaven hairs. This innovation was employed in the successor to the doubleheader "egg," launched in 1957. In 1959 this shaver acquired yet another new feature: "floating heads." The two heads were suspended on springs so as to better follow curved facial areas. In 1966 Philips introduced a tripleheader.

Figure 3
The "egg" model for women (1951).

Shavers for Men and Shavers for Women

Although the "cigar" and the "steel beard" were marketed primarily for men, women were also seen as potential users. The market for shaving devices for women had already grown in the first decades of the twentieth century (Derks et al. 1996). Changing fashion, uncovering more parts of the female body, contributed to the growing female practice of removing hair from the armpits, the neck, and the legs. In the 1910s and the 1920s producers of safety razors like Gillette put models specifically meant for women on the market, most of them smaller and rounder to better fit the armpit. In 1939, when Philips introduced its first electric shaver, it already saw women as potential users; indeed, the manual for that shaver included a substantial explanation addressed to women.

With its second generation of shavers, Philips began to differentiate between male and female users. In 1950, Philips's first shaver for women only, the Beautiphil, was introduced. The Beautiphil was a version of the "egg" designed to deal with a different type of hair: there was more space between the slots in the head, and the hair chamber was larger. However, the Beautiphil looked very similar to the men's shaver. Only the storage

Figure 4
The "lipstick" shaver for women (1959).

case was given a feminine touch. The later double- and tripleheaders too had versions for women. The ladies' shavers were nearly the same as the men's; they differed only in being pink and in having slightly different heads.

The main design strategy used in the 1950s and the 1960s to tailor a shaver for female users was to give it recognizable female-coded features, such as a pink housing or a round red storage case. Philips's competitor Braun followed a similar strategy, decorating its women's shavers with little imitation diamonds.

However, in the late 1950s Philips also produced a few women's shavers that cannot be characterized as mere shallow adaptations of men's models. The design was fundamentally distinct from all prevailing models. It can be labeled as a new gender dimension in Philips's practice of designing for women: the masking of technology and shaving. An illuminating example from that period is the "lipstick," whose design emphasized an association with cosmetics instead of with an electrical device and shaving. The masking of the technology was completed by a little pad saturated with perfume to conceal the smell of motor oil. The avoidance of the association with "male" shaving in the case of women using shaving devices was not new.

The design of the "lipstick" fitted into an already existing tradition of avoiding the connotation of male shaving in the case of women's use of shaving devices. Gillette, for instance, had already in the 1920s and the 1930s cautiously avoided the term "shaving" when advertising its products for women. Their first safety razor for women was called Milady Décolletée.[5] The masking of the technical elements for women, however, was a new dimension in Philips shavers design. This dimension would gain importance in designing for male and female users, as we will see.

In the early 1960s, Philips decided to discard the characteristic rotating system for the Ladyshaves (this name had been used since 1956) and replace it with the so-called oscillating system. In this system a clipper head with slots or a foil with small holes catches the hairs that are subsequently cut by oscillating knives. This system—that was standard in all shavers of competitors Remington and Braun—was more effective in meeting specific demands for shaving both legs and armpits. Moreover, the change fitted with the Philips marketing strategy of visually separating the products for female depilation from the male segment. In 1967, Philips also decided to establish a separate production line for Ladyshaves in Austria (Klagenfurst). The production of Philishaves remained in The Netherlands (Drachten). From now on, not only the outside design was different but also the technology inside became specialized. In fact, the whole design, development and production trajectory of Philishaves and Ladyshaves became separated. This segregation on the organizational level was reflected in a widening of the design differences between the two shaving devices. Two different design cultures came into being.

On the one hand, one can see this segregation as a kind of emancipation of the Ladyshave from the Philishave. The Ladyshave was no longer just a derivative of the successful Philishave, like Eve made out of Adam's rib. Now it had its own design and production environment. On the other hand, it also meant that the gender differences could become a more basic and integral part of the design cultures, resulting in a more explicit stereotypical gender script in both Philishave and Ladyshave. The masking of technology and its detachment from male "shaving" gradually would become a core issue in the design culture of Ladyshaves. The premise of the Ladyshave design philosophy became— and still is—that women dislike the association with technology. As a consequence, the Ladyshave was designed and marketed as a cosmetic device, not as an electric appliance. This development is in strong contrast to the design philosophy of Philishaves which can be characterized

Figure 5
The "telephone hook model" for men (1982).

by an emphasis on technology, as we will see by looking at subsequent developments.

In the early 1970s the tripleheader became standard in the Philishaves and improvements were primarily gradual ones (more groves and more/sharper chisels). In 1975 a new so-called TH-design (Telephone Hook) was introduced (a standing model, but with the triple head in a slanting position at the top). This type of model is standard for tripleheaders to this day. In the second half of the 1970s the ivory color and round shapes of the Philishaves disappeared. Instead, black and metallic colors were used and the shapes became bolder. New too, were adjustable shaving heads that allowed the user to adapt the apparatus to his own preference necessitating an extra regulating button.

No basic changes were made to the Ladyshave in the 1970s. The more expensive models were not sold as shavers but as beauty sets and were equipped with several accessories. Within the new design culture of the Ladyshave, the assembly of parts was not done with screws (as was the Philishave) but instead by a so-called clicking system: once clicked together it was impossible—especially for users—to open the device. The clicking system fitted into the design strategy of masking technology:

Figure 6
The Ladyshave beauty set (1979).

visible screws would only enhance the undesired association with the technological content instead of cosmetics.

In the mid 1980s, Philips brought a remarkable innovation on the market: a washable Ladyshave using batteries. Some years later this principle was further developed in the Wet & Dry Ladyshave (1994) which women could use dry as well as while showering. The Wet & Dry Ladyshave was elegantly curved and pastel colored. It had one large button with icons for on/off and for armpit or leg. The advantage of using the shavers during showering was that the softened skin would then be caused less irritation by the shaving. This development again fits seamlessly in the design strategy of enhanced the association of the Ladyshave with cosmetics and body care and not with technology.

The developmental trajectory of the Philishave was quite different. Changes were introduced in three dimensions. The first aimed to improve the quality of the shaving process. With respect to smoothness of the skin, there was still a wide gap between dry and wet shaving. In 1980 Philips introduced a new system called Double Action: the first knife lifts the bristle a little and the second knife cuts the beard hair deeper. To emphasize this innovation the phrase "double action" was imprinted on the outside of the apparatus, a usage that

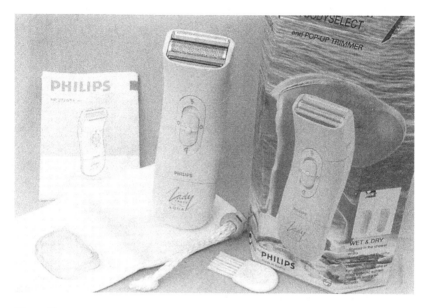

Figure 7
The Ladyshave Wet and Dry (1994).

still exists today. The Double Action system soon became standard in all Philishaves.

The second dimension of change was another weak point of the electric shaver compared to wet shaving: the need for a power point. Cordless shaving became the ideal. Philips had produced some battery-powered shavers, but the battery capacity was too low. In the mid 1980s the technology of rechargeable batteries and the reduction of transformer size was sufficiently developed to produce a usable rechargeable shaver. The rise of micro-electronics in the 1980s was the third dimension to put its stamp on the design of the Philishave. Micro-electronics were used to develop all kinds of monitoring in control features, such as information displays with charge control, number of personal shaving minutes left, etc. The more expensive the model, the more electronics was incorporated in the Philishave. This development continued until the late 1990s.

In January 1998 Philips introduced a new product to tempt the remaining wet shavers to go "electrical": the Cool Skin. This device brings an emulsion onto the skin during shaving that produces an effect resembling the sensation of wet shaving. The Cool Skin proved successful—especially in the United States, which still had 75 percent wet shavers at that time (Wollerich 2000). Also new in the Philishave design

Figure 8
A Philishave with information and control display (1996).

trajectory is that the shaving head of Cool Skin is washable. Both innovations, the use of an emulsion and the washable shaving head, however,
had already been incorporated in the Ladyshaves for some years. Men
using the latest Philishave Quadra can shave under the shower and even
in the bath—possibilities that Ladyshave users had already for at least 5
years.

The Gender Script of Shavers: Constructing Technological Incompetence as Feminine

Masking the technology was a systematic element of the gender script of
the Ladyshave. The methods used included using perfume to mask the
smell of oil, linking the shaver to lipstick, transforming the shaver into a
beauty set, and eliminating visible screws. The script of the Ladyshaves
hides the technology for its users both in a symbolic way (by presenting
itself as a beauty set) and in a physical way (by not having screws that
would allow the device to be opened). The Ladyshave's design trajectory
was based on a representation of female users as technophobic. Whereas
the script of the Ladyshave aimed to conceal the technology inside for

the user, the Philishave design principle was the other way round. The new internal technology of Double Action was made visible on the outside to emphasize the latest technology. In the 1970s metallic and silver-colored materials functioned as tokens representing technological innovation. Slide controls on the outside offered the male user the possibility to relate to the technology within. From the mid 1980s electronic displays fulfilled the same function. The display provided the user with information about and control over what was inside. In the design culture of shavers certain elements were preserved only for men's devices, including black and metallic materials, displays with information and control possibilities, and references on the outside to the technology inside. The new domain of electronics was put fully in the service of developing the gender script of masculine control and technological competence. These types of interfaces and materials were unthinkable in the design culture of the Ladyshave.

The comparison of developmental trajectories of two similar products for women and for men proved to be fruitful for tracing gender scripts in artifacts. The analysis of gender scripts in shavers showed that the bond between men and technological competence has been inscribed firmly in the design of consumer appliances such as shavers. In this respect the shavers reflected the dominant gender symbols and identities, captured in the phrase "To feel technical competent is to feel manly." This quote from the British sociologist Cynthia Cockburn formulates concisely the intimate bond between technological competence and masculine identity that became widespread in the course of the twentieth century western society (Cockburn 1985: 12; Oldenziel 1999; Connell 1987).[6] The Philips shavers not only reflected this gendering of technological competence, they too constructed and strengthened the prevailing gendering of technological competence. The script of the Ladyshave not only "told" women about cultural norms with respect to the armpit and leg hair but also that they "ought to" dislike technology. Clearly, the gender script of the shavers cannot force users to invoke these gendered identities: women can reject the script (e.g. by shaving with a men's shaver or not shaving) or even modify the script (e.g. see it as a technological challenge to open the clicked Ladyshave). But the gender script of the Ladyshave inhibits (symbolic as well as material) the ability of women to see themselves as interested in technology and as technologically competent, whereas the gender script of the Philishaves invites men to see themselves that way. In other words: Philips not only produces shavers but also gender.

Concluding Remarks

This study primarily gave voice to the artifacts—and indirectly to its designers—but not to the female and male users. The question how such gender scripts actually were "read" and what roles the different involved actors had cannot be answered within the limited character of the scope and the empirical materials. A study of the users influence, both in designing and in using shaving devices, certainly will provide an even richer insight in the processes of how gender and technology shape each other mutually.

The content of gender—and thus also of gender scripts—is always situated in time and place. The design strategy of the latest Philishave products seems to converge toward the Ladyshave design. The Cool Skin and other new designs, the Quadra and the smaller two-head Philishave 400 Micro Action, lack an advanced display interface with information and control features. The control features—if present—are unobtrusive. The shape is full of round lines and the material has more (although subdued) colors. The Philishave 400 resembles the curved shape of the Ladyshave Wet & Dry and the Cool Skin allows for cosmetic connoted use under the shower. The Philishave design strategy for men that emphasizes control and technological competence, may well be in retreat at the start of the 21st century.

Acknowledgments

I thank Nancy Brouwer for her valuable master's thesis on the gender script of shavers. I also thank Bram Porrey, who kindly gave permission to use the photographs from his database of Philips shaving devices.

10

Clinical Trials as a Cultural Niche in Which to Configure the Gender Identities of Users: The Case of Male Contraceptive Development

Nelly Oudshoorn

Studies in the sociology of science and technology have emphasized that users play an important role in technological development. Traditionally, users have been considered important actors in the diffusion and acceptance of new technologies. Technologies work only if they become accepted by users and embedded in society (Von Hippel 1976, 1988). More recently, the attention in science and technology studies has shifted from the analysis of users in the sociological sense (i.e., identifiable persons as such involved in the diffusion of technologies) toward users in the semiotic sense. As Madeleine Akrich and Steve Woolgar have suggested, scientists and engineers configure users and contexts of use as integrated parts of the processes of technological development. In the development phase of a new technology, designers anticipate and define the preferences, motives, tastes, and competencies of potential users and inscribe these views into the technical design of the new product (Akrich 1992: 208; Woolgar 1991; Clarke and Montini 1993; Rommes et al. 1999). This semiotic approach challenges the view that users enter the picture only after a new technology has been introduced to the market. Innovators actively draw new configurations of users into the very heart of technological innovation: the development phase.

Configuring the user is thus conceptualized as a major aspect of technological innovation. A closer look at the science and technology literature reveals, however, that these studies address only one aspect of configuring the user. The focus is very much on how innovators anticipate the technical competencies and actions of users. Although Steve Woolgar has described processes of configuring the user as "defining the identity of putative users and setting constraints on their likely future actions," he describes identity only in terms of "who the user might be." His main concern is to reconceptualize human agency in relation to technology. The analysis is restricted to showing how computers are designed in such a way that they define and delimit the user's actions and behavior (Woolgar 1991: 59, 61). Like Woolgar, Akrich analyzes competencies and actions of users rather than user identities. This scholarship thus reflects a rather narrow view of technological development in which user-technology relations are restricted to technical interactions with the artifacts, thus neglecting the broader cultural dimensions of human agency.[1] My conceptualization of the relationships between technologies and users is broader. I suggest that the articulation and performance of gender identities of users is an important aspect of the development of technological artifacts. Technologies will only become successful if technological innovators configure gender identities of users and if future users perform the gender identities articulated by technological innovators.

My point of departure is that technological development requires the mutual adjustment of technologies and gender identities.[2] To study the adjustment of technologies and gender identities, Judith Butler's conceptualization of gender as performance provides a useful approach. Like technology, gender has no intrinsic qualities. As Butler and many other feminist scholars have argued, gender is not something that we are, but something we do.[3] Inspired by Austin's theory of speech acts,[4] Butler has developed a radical critique of the notion of fixed gender identities rooted in nature or bodies. In her poststructuralist theory of gender, gender is considered as the result of discursive practices with the potential to produce what they name.[5] Butler emphasized the role of reiteration in producing and sustaining the norms that constitute gender, which she refers to as a performative process. In this view, gender is not pre-given or fixed but produced as a "ritualized repetition of conventions" (Butler 1995: 31). Each performance of gender may reproduce existing meanings of gender or represent new, subversive readings of gender that produce the possibility of change. In this performative theory of gender, the seemingly universal dichotomy of gender is the result of a constant maintenance of particular conventions of gender, most notably those of compulsory heterosexuality (Butler 1990: 25).

Although Butler's work underscored the constraints on the performativity of gender, she has not reflected on the question of how technologies may contribute to the maintenance or transformation of particular gender performances. Rooted in a semiotic and psychoanalytical tradition, her work primarily addresses the forces of prohibition and taboo in sanctioning and unsanctioning particular sexual practices and gender performances (Butler 1993, 1995). I suggest that it is important to address the role of technologies as non-human actors to understand the processes involved in producing and sustaining particular forms of gender. Technologies may play an important role in stabilizing or destabilizing particular conventions of gender, creating new ones or reinforcing or transforming the existing performances of gender.

Contraceptive technologies serve as a specific case in point to illustrate my argument. Before the introduction of new contraceptives for women in the 1960s, no stabilized conventions existed concerning the relationships between gender identities and contraceptive use. Since only a limited number of contraceptives were available (condoms, diaphragms, natural methods, spermicides, sponges, sterilization), neither men nor women had many options for contraception.[6] This situation changed drastically when new contraceptives for women became available: high-tech methods that intervene in the physiological processes that regulate ovulation and conception in female bodies. The introduction of a much wider variety of modern contraceptives for women has disciplined women and men to consider the use of contraceptives a woman's responsibility. Because of the innovation in female contraceptive methods— including the hormonal contraceptive pill, the intra-uterine device (IUD), and hormonal methods such as Norplant—women's methods have come to predominate practices of family planning. Only about 17 percent of contraceptive users rely on so-called male methods, i.e., condoms and male sterilization (United Nations 1994: 4). Female sterilization, oral contraceptives, and IUDs are the methods used most frequently (Robey et al. 1992: 11). Contraceptives thus function as important tools in delegating and distributing responsibility and control over procreation. In Foucauldian terms, contraceptives are "disciplinary technologies": "They are part of the 'socialization of reproductive behavior,' that can discipline such behavior in multiple ways" (Clarke 1998: 165). The predominance of modern contraceptive drugs for women has disciplined men and women to delegate responsibilities for contraception largely to women. Contraceptive technologies thus constituted strong alignments between femininity and taking responsibility for reproduction.

Another illustration of the performative and integrative capacity of technologies to create and sustain gender identities is the emergence of the women's health reproductive movement in the late 1960s and the early 1970s. One major reason for the establishment of this social movement was concern about the health risks of the first generation of contraceptive pills and IUDs (Oudshoorn 1994). Since then women's health groups have been important actors in lobbying against the introduction of contraceptives considered as unsafe or having the potential for abuse, and simultaneously advocating the development of better contraceptives for women (van Kammen 2000b). In contrast, no men's reproductive health movement exists. The difference in emergence of social movements concerning the reproductive health of women and men can be understood in terms of a "technosociality": people construct collective identities based on a shared experience with specific technologies, in this case contraceptive technologies.[7] In the second half of the twentieth century, the idea of woman as the sex responsible for contraception thus came to be the dominant cultural narrative materialized in contraceptive technologies, social movements, and the gender identities of women and men.

Consequently, contraceptive use became excluded from hegemonic masculinity.[8] Inspired by the Gramscian notion of hegemony, Robert Connell has introduced this concept to refer to the cultural dominance of particular forms of masculinity. Like Butler, Connell conceptualized gender as a cultural construct, emphasizing the diversity in masculinities and femininities. He explicitly included power as an important aspect of the relationships between genders and within genders.[9] Anticipating, in a way, Butler's performative theory of gender, Connell described gender as "something that does not precede but is constituted in human actions" (Demetriou 2001: 340), emphasizing that we should use "gender" as a verb (Connell 1987: 140). According to Connell, hegemonic masculinity implies the subordination of women and subordinated masculinities (ibid.: 185, 186). Hegemony emerges from "preventing alternatives from gaining cultural definition and recognition as alternatives, confining them to ghettos, to privacy, and to unconsciousness" (ibid.: 186). Connell identified heterosexuality as the most important aspect of contemporary hegemonic masculinity.

Connell's theory of gender is important because it enables me to differentiate between different performances of gender, including hegemonic masculinity and non-hegemonic masculinities.[10] However, as Butler, Connell does not theorize the role of technologies in creating and sustaining particular forms of masculinities. Although Connell occasionally refers to the role of technologies in constituting masculinity,[11] he does not classify technologies as a "gender regime," a concept he confines to the labor market, the state, the family, and, more recently, the "structure of symbolism" (Connell 1995: 357).[12] My point of departure is that in contemporary societies the production and the use of contraceptive technologies are crucial to an understanding of how particular forms of gender gain cultural dominance whereas others remain marginalized. In the last two decades, feminist studies of technology have suggested that the development and use of technologies are very significant sites to understand the social and cultural aligning of technologies and masculinities. Feminist historians and sociologists of technology have shown the strong alignments of technology and hegemonic masculinity in technological practices, particularly in the field of engineering, and the enduring and pervasive cultural equations drawn between hegemonic masculinity and technology. As Cynthia Cockburn (1992) and others have argued, we can never fully understand technology without masculinity and vice versa. In this view, gender and technology are seen as mutually constitutive or co-produced (Berg 1996; Cockburn 1983; Faulkner 2000; Lie 1995; van Oost 2000; Oldenziel 1999; Oudshoorn et al. 2002; Wajcman 1991). These studies, however, do not focus on technologies that have weak social and cultural alignments with masculinity, such as contraceptive technologies and reproductive technologies in general.[13]

As illustrated in my reflection on the history of contraceptives above, the predominance of contraceptives for women has contributed to a stabilization of performances of gender which constituted a strong alignment between femininity and contraceptive use. Masculinities that ask men to take responsibility for their reproductive bodies became excluded from hegemonic masculinity and were constituted as a subordinate form of masculinity. Equally important, physiological means of contraception that separate sexual from reproductive functions challenge hegemonic views of masculinity that emphasize the intertwinement of the male sexual and reproductive body. The development of new contraceptives that enable men to perform sexually without being fertile thus conflicts with two aspects of hegemonic masculinity: delegating the responsibility for contraception to women and safeguarding the unity of male sexual and reproductive bodies (Scale 2002: 1). The "feminization" of contraceptive technologies created a strong cultural and social alignment of contraceptive technologies with women and femininity and not with men and masculinity, which brings the development of new contraceptives for men into conflict with hegemonic masculinity. The development of new contraceptives for men thus requires the destabilization of these conventionalized performances of masculinity.

From this perspective, the developmental phase of a technology becomes an intriguing location for understanding the co-construction of users and technologies. I view the testing phase of a technology as a cultural niche in which experts, potential users, and other people participating in the testing of the technology articulate and perform nonhegemonic identities of users to create and produce the cultural

feasibility of a technology.[14] Based on an analysis of two large-scale clinical trials of hormonal contraceptives for men organized by the World
Health Organization, one in the late 1980s and one in the early 1990s,[15]
I will show how the development of new contraceptives for men required
specific procedures to discipline men as reliable test users, including a
renegotiation of male identities. Adopting the concepts of projected and
subjective identities,[16] I will describe the configuration of the identities of
test users as involving both the projection of male identities of the trial
participants as articulated by reproductive scientists and other actors,
and the articulation and performance of identities as created and experienced by men participating in the clinical trials. We will see how male
contraceptive researchers have configured trial participants by articulating specific representations of masculine identities. Men participating in
contraceptive trials articulated and performed male identities that
largely matched the researchers' projected identities. The chapter first
describes the specific procedures introduced to discipline men as reliable test users of the contraceptive on trial and continues by analyzing
the renegotiation of male identities as a crucial part of the clinical testing
of hormonal contraceptives for men.

Disciplining Men as Reliable Test Users

Clinical trials are a peculiar type of testing. Not only do they require
material resources, such as the availability of drugs, instruments to measure the effects of the drugs, and forms and statistical procedures to register and produce data; they also depend on the collaboration of human
beings, known more formally as trial participants. Clinical trials thus represent a very specific practice of configuring the user. Whereas most
other configuring processes take place in the absence of users, clinical
trials, like other user tests, require the presence and the cooperation of
potential users.[17] As Stephen Epstein (1997a: 691) has suggested, "clinical
trials are a form of experimentation that requires the consistent and per-
sistent cooperation of tens, hundreds, or thousands of human beings—
'subjects' in both senses of the word—who must ingest substances on
schedule, present their bodies on a regular base for invasive laboratory
procedures, and otherwise play by the rules." Thus, one of the major
aspects of clinical testing is ensuring the cooperation of trial participants.
For researchers, this is a complicated endeavor because test subjects talk
back, may decide to discontinue their participation, and may fail to comply with the procedures of the trial (ibid.: 693).[18]

In the trials under discussion here, ensuring the cooperation of subjects proved even more complicated than usual. The researchers and clinicians involved in male contraceptive trials faced a new situation. For decades, contraceptive testing had been focused on women. In the twentieth century most attention in reproductive medicine had been focused on women. Since the late nineteenth century the female reproductive body had become firmly entrenched in the infrastructures of the medical world and beyond. Knowledge, diagnostics, and therapies concerning the female reproductive body had been made robust by alignments across laboratories, gynecological clinics, pharmaceutical companies, family planning policies, family planning clinics, and social movements, particularly the women's health movement. As I noted above, since the introduction of the female contraceptive pill in the early 1960s, collective actors had focused almost exclusively on women, neglecting men as potential subjects of research, users and clients. Contraceptive researchers, predominantly men, who became involved in male contraceptive research were therefore not used to experimenting on men, as is illustrated by the following quotation from one of my interviews:

None of us have ever seen a male method introduced. There is no track record for male methods since the condom, going from basic studies to market place. It's not like an antibiotic where there are fifty antibiotics introduced and the next one is small change really. So this is a truly revolutionary event, introducing a new male method. I mean that is just unheard of really in anybody's experience. . . . (interview with William Bremner, 1994).

Moreover, men were not used to being subjected to contraceptive trials or any other form of medical experimentation or examination relating to their reproductive organs. Whereas women had been and still are subjected to widespread experimentation and testing practices, such as screening programs for breast and cervical cancer, procedures for assisted fertility, physical examinations related to pregnancy and the use of contraceptives, and clinical testing of new contraceptives, such routine practice had been virtually absent for men. Consequently, noncompliance was a serious problem in the male contraceptive clinical trials that were initiated in the early 1970s.[19] In that period, several academic clinical centers in the United States and Europe initiated the testing of predominantly, but not exclusively, hormonal compounds as contraceptives among small groups of men, ranging from four to thirty volunteers.[20] The first large-scale clinical trials involving several hundred men took place in the late 1980s and the early 1990s when the WHO initiated two so-called multicenter clinical trials in which several clinical centers in Asia, Australia, Europe, and the United States participated (WHO 1996: 125). Through the years, many researchers experienced problems in ensuring compliance, reporting high dropout rates, sometimes of half of the test population (Foegh et al. 1980: 631; Foegh 1983: 25; Lobel et al. 1989: 123). In a trial published in *Contraception* in 1979, researchers described how few men followed the instructions recommending the use of alternative contraceptive methods to prevent their female partners from becoming pregnant if the method on trial happened to fail (Barfield et al. 1979: 123). In the late 1970s, a French research group decided to leave the field of male contraceptive research because trial participants failed to comply with treatment (Lobel et al. 1989: 123).

In view of the newness of the situation, the organization of clinical trials for the testing of hormonal contraceptives for men required a lot of extra work compared to the testing of other drugs. Researchers had to create specific tools to ensure the cooperation of the test subjects. To configure men as reliable test subjects, clinicians selected men in stable relationships. This selection criteria reflected one of the major worries of the male contraceptive community, that is that trial participants, and future users, would be unreliable in using contraceptive methods. In "Birth Control after 1984," published in *Science* in 1970, Carl Djerassi, one of the developers of the pill for women, was among the first to articulate the problem of the unreliability of men in matters of contraception. Djerassi (1970: 948) cited "the male's generally lesser interest in, and greater reservations about, procedures that are aimed at decreasing his fertility." He continued: "If the agent were to be administered orally, men would probably be even less reliable about taking a tablet regularly than women have proven to be, and efficacy could probably be determined on a large scale only though long-term studies of married couples."

In the last three decades, the notion that men cannot be trusted in matters of contraception has become familiar in family planning, feminist, and journalistic discourse (Oudshoorn 1999, 2000). To solve anticipated problems with noncompliance, researchers have put their faith in the female partners of the trial participants. Procedures for clinical testing, both in the United States and the United Kingdom, illustrate the crucial role of women in transforming their husbands and partners into reliable test subjects. First, female partners functioned as key actors in motivating men to participate in the trial. Men who participated in the two WHO multicenter clinical trials reported that one major reason for participation was that they were encouraged by their female partners

(Ringheim 1995: 76). Researchers actively exploited the positive attitude of women toward male contraceptive trials. They usually asked women to come along for the first visit to the clinic and to participate in informed consent procedures. Second, women are enrolled as agents to monitor changes in their partners. In the trials in the United Kingdom, female partners were asked to keep records of any changes in the sexual activity or behavior of the trial participants (interview with Fred Wu, 1994).

Practices during the trial show how procedures to discipline men into reliable test subjects are not restricted to the selection phase of clinical testing. To ensure compliance during the entire period of the trial, researchers offer trial participants special services and treatments. Sessions to take blood and semen samples, and other laboratory tests, are usually organized in the evening to accommodate the men's work schedules (interview with Alvin Matsumoto, 1994). Moreover, researchers spend quite some time in sharing the results of the medical examinations and laboratory tests with trial participants and offer medical care. The practices of clinical trials show how these tests have a dual function. For researchers, clinical tests function as tools to investigate contraceptive efficacy and the side effects of contraceptive compounds. For the participants, clinical trials function as a health check that keeps them motivated to visit the clinic on a regular basis.[21] Medical examinations during the trials provide men with attention and health care they do not request or receive in other places, which reflects the growing awareness among men of the importance of health issues which emerged in the 1980s (Bernardes and Cameron 1998; Nahon and Lander 1993). Researchers in Seattle emphasized the importance of free medical care in ensuring compliance:

We are very successful with keeping volunteers going. That has a lot to do with recruitment but it also has to do with the fact that we have a monthly clinic and we set it up so that it is an evening clinic, after hours and so they come, get their exam, their blood drawn, drop off their sperm counts and get to know the investigators pretty well. They get examined very carefully, so they get medical care, if they have a cold, we take care of that, if they have a little acne, we. . . . Yes, that might motivate people to stay and there is some camaraderie in a sense that is built up over the years. They get some feedback about what is happening to the sperm counts, they get laboratory evaluations, so they see what the cholesterol is doing, what the blood count is doing. A lot of the volunteers like to see that feedback. (interview with Alvin Matsumoto, 1994)

Medical care thus played an important role in disciplining men as reliable test subjects, at least according to the researchers.

Practices of clinical testing of contraceptives for men thus show how researchers had to put great effort into disciplining men to be reliable test subjects. As in the testing of other drugs, the reliability of trial participants is a crucial requirement in establishing stable relationships between researchers and trial participants. For male contraceptive trials, however, reliability is a concern that goes beyond the relationship between researchers and test subjects within the secluded and relatively malleable domain of the laboratory. In contrast to the testing of female contraceptives, where any pregnancy can be directly identified as a failure of the contraceptive method (or in the case of contraceptive pills, as indicating noncompliance of the trial participant), contraceptive failures in male contraceptive trials can never be excluded because the untreated partner can become pregnant by having sex with someone else (WHO 1996: 958). The assessment of contraceptive efficacy of hormonal compounds thus required specific procedures to ensure the compliance of the female partners of the trial participants, a highly peculiar practice compared to other drug testing. In this context, the selection of "couples" or "men in stable relationships" as described above can be considered as an adequate, although not 100 percent effective tool to avoid skewing the data due to the extramarital sex of the female partner of the trial participants.

In this process, researchers not only configured the trial participants, but they also configured the future users of male hormonal contraceptives, as is noted by British researchers involved in clinical trials in the early and mid 1990s in Manchester and Edinburgh. Shankland (1993) quoted Fred Wu, an investigator in Manchester, as having said "The men we think this will appeal to are those in stable relationships and those prepared to share the burden and benefits of partnership." Marlin (1998) quoted David Kinniburgh of the Medical Research Council's Reproductive Unit in Edinburgh as having said "The pill will appeal to couples where trust has built up and men want to take responsibility for what happens between the sheets."

The selection of couples in stable relationships was not just introduced as a tool to make men into reliable test subjects. Equally important, it was constructed to create a distinctive niche in the contraceptive market for male hormonal contraceptives. In reaction to skepticism about the acceptability of the new contraceptive voiced by groups of feminists, health-care providers involved in AIDS prevention, and journalists, suggesting that men are unreliable in matters of contraception or that hormonal contraceptives (as non-barrier methods) do not help to prevent

AIDS, researchers configured the potential user of reversible non-barrier contraceptive methods as men in monogamous, stable relationships where partners trust each other.[22] By configuring the user as couples in stable relationships, researchers simultaneously constructed the non-user: men with casual relationships, as was noted, again, by researchers in Edinburgh and Manchester:

It won't work for the 17-year-old at the nightclub looking for a contraceptive but will for men in relationships. (David Kinniburgh, as quoted in Smith 1997)

A woman would be mad to believe a chap she met in a nightclub who said "You're all right, love, I'm on the pill." (A. Bellis of Manchester, as quoted in Sweetenham 1994)

Disciplining men as reliable test subjects and future users thus entailed the construction of users as men with stable, monogamous relationships and the construction of promiscuous men (and women!) as non-users.

The procedures introduced to discipline men as reliable test subjects have been quite successful. As against the complaints about non-compliance articulated in reports of clinical trials in the 1970s and mid 1980s, reports in the 1990s were much more optimistic about compliance. Drop-out rates after the initial screening procedure has been estimated at approximately 10 percent, which is much lower than the drop-out rate in female contraceptive trials which can be as high as 30 percent (interview with Alvin Paulsen, 1994; interview with Fred Wu 1994; WHO 1996: 958).

The Responsible, Caring Man

Disciplining men as reliable test subjects not only required specific selection and test procedures. Intriguingly, it also involved a renegotiation of male identities. As described in the introduction, this renegotiation can be understood as a dual process involving both the projection of male identities of the trial participants as articulated by reproductive scientists and other actors, and the articulation and performance of identities as created and experienced by men participating in the clinical trials. Documents used to communicate with the media and trial participants, such as posters and press bulletins to recruit trial participants and leaflets to inform trial participants about the procedures of the trial, are important sources through which to study the first part of this process. The rhetoric of these texts show how male contraceptive researchers and public relations officials constructed a specific image of the potential trial participant. In Seattle, men who applied to be trial participants received a leaflet titled "Questions and Answers," first introduced in 1994, which opens with the section "Why is a male contraceptive needed?" (Paulsen et al. 1994: 6). After a short description of the contribution male contraceptives can make to reducing the "exponential population growth," the document continues to highlight the importance of male contraceptives for enhancing "equality between men and women":

> . . . the primary value of a male contraceptive may be that it will allow couples to share not only the benefits but the responsibilities and risks of contraception. While the development of contraceptive agents has allowed women to control their fertility and thus has been an important factor in freeing them from most traditional roles, the responsibility for contraception has remained almost exclusively a female role. (Paulsen et al. 1994: 6)

Documents used to recruit men in the United Kingdom contain a similar emphasis on sharing responsibility between the sexes as a major reason why it is worthwhile for men to participate in the trials. A press bulletin launched by the University of Manchester's Communications Office on July 9, 1993, to recruit male trial participants for the second large-scale WHO clinical trial, articulated the need for new male contraceptives:

> The move toward providing more options for male contraception is really reflecting social trends that equality between the sexes should extend to Family Planning. Of course, it also has important implications for the Third World, where the population explosion is uncontrolled. (WHO 1996: 2)

In both the documents just quoted, the potential trial participant is configured as a man who wants to contribute to helping his partner as well as to reducing the population growth in Third World countries. The poster used in Edinburgh to recruit men for clinical trials in the mid 1990s exemplifies this altruistic image, although it also adds a third interesting motive. The poster begins with three questions:

> Interested in helping develop a new contraceptive pill for men?
>
> Fed up by the lack of choice for men?
>
> Want to help your partner get off the female pill?

In contrast to most of the documents used to recruit and inform male volunteers, this poster explicitly addresses men in terms of their individualistic interests. Taking part in clinical trials is portrayed as relevant for men because it may increase their choice of contraceptives. Most researchers configure male contraceptive trial participants, however, as men who are willing to share the responsibilities and risks of contraceptives with their partners thus constructing the image of men as responsible, caring partners.

The ways in which men articulated their motives for participating in contraceptive trials shows how the image of the responsible, caring men has become part of the identity of these men.[23] Many men participating in the two large-scale WHO clinical trials portrayed themselves as willing to take responsibility for contraception (Ringheim 1996a: 6). For example:

It's about time fellas start taking responsibility for this kind of thing. I hadn't been wandering around with the burning desire to take part in male contraceptive trials. (ibid.: 7)

I think men have been allowed to be lazy about this. I don't know who decided it, but it always seemed to be pushed on the woman to be responsible. (ibid.)

A man should have 50 percent of the responsibility. This attitude is becoming more common. Women are not objects. They're the same as us. We're equals. To some older guys, women are second-class citizens. In [the United Kingdom], they go to the pubs and leave the women at home. I think it will probably take 20 years before this dies away, but a male contraceptive would appeal to my circle of friends. They are like me and think men should be responsible. (Ringheim 1996b: 87)

Demonstrating prior awareness of the potential for problems, the majority of men who participated in the WHO trials (61 percent) articulated their motivation in terms of helping their partners who experienced problems with the female pill. (Ringheim 1996a: 6)

It's got to do with the fact that my wife gets depressed when she takes the pill, and I saw this on the telly and I just rang up. That's the main reason I came on the trial. (ibid.)

If she goes on the pill again there is always the risk, isn't it? And my way of thinking is, once she's taken the risk for a few years, I'll take the risk. Then you halve it. (ibid.: 8)

My wife taking estrogens was like the shrew that couldn't be tamed. She would wake up depressed . . . and after a period of time I said "Honey, it's the pill, stop taking it, I don't care, I'll use condoms, or other forms of birth control, I'll go on the program that my friend is on, but you stop taking the pill right now." (Ringheim 1995: 76)

Participants in the WHO trial in Thailand also explained their motivation by referring to problems with the female pill, although they articulated concerns about their partners forgetting to take the pill (Ringheim 1995: 77). Incentives to participate in the trials were not expressed only in terms of problems with the female pill, but also with dissatisfaction with the use of condoms or vasectomy as means of contraception in stable relationships (ibid.; Ringheim 1996a: 81). The motives to participate in the trials thus also contained non-altruistic components: the trials could help men to avoid the use of condoms or vasectomy. Another motive which shows the self-interest of men participating in the trials is the argument that the trial enabled them to be in control of their own fertility (Ringheim 1996b: 86; Ringheim 1995: 77). The dominant image articulated by male trial participants, however, was their interest in sharing responsibility for contraception with their partners.

The language used by these trial participants reflects how they considered taking responsibility for contraception as a largely unfamiliar and exceptional activity for men in long-lasting relationships or marriage. By taking part in contraceptive trials, men thus actively performed non-hegemonic male identities, which unmistakably reflected the researchers' projected identities of responsible, caring men. Participants of the trials in Sydney, Australia, constructed a self-image portraying themselves as different from other men:

We all know that at this stage of time, it's not socially acceptable for men to use male contraception. We are doing this because we are different. (Ringheim 1993: 22)

I figure that the people who are doing this program are a different kind of guy anyway, we're not SNAGS (sensitive New Age guys). I hate SNAGS. . . . I don't think we are typical of white Australian middle class society. (ibid.: 24)

Some of the Australian trial participants also explicitly articulated their new role in terms of masculine identities:

I think that men have always had soft sides, gentle sides, nurturing sides, but for a long time they have been repressed. To a certain extent all these norms, morals, and values are raised into prominence because we are precisely in that period of change so people are forced to think about "do men have to do things a certain way," and "what's a typical male?" (Ringheim 1993: 11)

In assuming non-hegemonic identities, male trial participants did not receive much support. Most male colleagues and friends considered their decisions to participate in a contraceptive trial as rather peculiar, as shown by the experiences of trial participants in Sydney:

You still get people who would say "What are you doing that for, can't your wife take the pill or something?" It seems like the abnormal rather than the normal, the idea that the bloke, apart from condoms, would actually take any part of sexual responsibility for contraception, particularly not one which involved needles. (ibid.: 23)

I told a lot of males about it because . . . I felt quite proud about the fact that I was on it. I thought it was a great thing to do. Probably out of the maybe 50 guys I told,

X [another men participating in the trial] was the only one who considered it. . . .
I thought a lot more people would have said—that sounds great. (ibid.)

[They] weren't particularly interested in the contraceptive side effects, they were
more interested in the anabolic effects. (ibid.)

They worry for us most of the time. My boss does. (ibid.)

Trial participants thus had to defend and negotiate their new identities.
Interestingly, these men received much more encouraging reactions
from women, particularly their female partners (ibid.: 25). As I have
described above, women played a crucial role in encouraging their partner to participate in the trials (Ringheim 1995: 76; Ringheim 1993: 13).
A significant share of the participants in the WHO clinical trials (23 percent) mentioned the encouraging role of their partner as main reason
for participation (Ringheim 1996b: 76). One of the men expressed this
as follows:

Quite honestly, I never would have volunteered if my wife hadn't complained. My
motto is "if it isn't broken, don't fix it." I think most men are only too happy to
have women use contraception. We know they have problems sometimes. Why
would we want to share that? But when the wife says "I've had it. Use a condom
or get the snip [vasectomy], then we begin to look around and realize, there isn't
much else for men, is there? (Ringheim 1996b: 86)

The reasons why women adopt this role is quite obvious: the participation
of their male partners in the trial frees them, although only temporarily,
from the use of contraceptives, at least if they are monogamous. In many
studies investigating the experiences of women with the pill, a substantial
number of women have expressed their dissatisfaction with oral hormonal
contraceptives or other current methods, as is reflected in the previously
quoted remarks by men participating in the male contraceptive trials.[25] To
quote two female partners of the WHO trial participants:

I thought it was absolutely brilliant. I loved it. The break from the pill really gave
me a chance to get my head straight. I've always suffered from depression. I didn't
always know it was the pill until I went off of it. (Ringheim 1996b: 84)

The trial was an interesting experience for him. We'd do it again. I found it great.
I didn't have to do anything. Nice not to have to think about it. I wasn't worried
about pregnancy. I was relaxed. We definitely had more sex, but I was also more
receptive. I felt happy that he was taking responsibility. (ibid.)

Women thus used the clinical trials as a location to renegotiate responsibility for contraception with their male partners. By doing this, they
actively engaged in the construction of non-hegemonic male identities:
caring, responsible masculinities of various types.

"Astronauts in the Sperm World"

To articulate this male identity, male contraceptive researchers and trial participants relied on hegemonic representations of masculinity. The illustration on the poster used in Edinburgh as described above exemplifies this imagery in a nutshell. The upper half of the poster shows a picture of an astronaut, standing on the moon with a flag in his hand, with the word "Exclusive" in a balloon near his head. The left side of the picture says "First Man on the Pill." In a funny and clever way, the poster suggests that men who decide to become volunteers are performing a heroic act like the man who first set foot on the moon. Participation in a male contraceptive trial is thus portrayed as an exciting new endeavor. Potential trial participants are addressed as adventurous men who want to explore a territory where no one has gone before.

Space metaphors were also adopted by trial participants. One participant in the second large-scale WHO clinical trial in Sydney described himself and his colleagues as "astronauts of the sperm world" (Ringheim 1993: 10). Other male volunteers constructed images with similar connotations. They identified themselves as pioneers in the development of a new male contraceptive method for men which they felt was important to them (ibid.: 7). Others described "the excitement of trying something new and possibly risky" as the most important feeling of being a trial participant. Researchers and trial participants thus transformed participation in a clinical trial into a brave, pioneering act.

The way in which male contraceptive researchers and the female partners of trial participants described men participating in contraceptive trials also adds to this image of the brave man. In reports of the trials, male volunteers were praised for their commitment to the trial and their perseverance in enduring the demands of testing. In the report of a French clinical trial published in 1983, trial participants were given credit for their compliance: "The authors wish to thank the 6 men for their strict adherence to the protocol's requirement in spite of the constraint of their professional lives." (Glander 1987: 631)

Including trial participants in the acknowledgments of a clinical trial report is rather exceptional: usually only funding agencies, pharmaceutical firms providing drugs, technical assistants and secretaries, or laboratories that have performed specific tests are acknowledged. Other reports of male contraceptive trials included credits for the trial participants in their preface. For example: "The volunteers took a keen interest in the research and felt very responsible for fulfilling their part of the studies, although they were not paid." (Foegh 1983: 7)

In the report of the second large-scale WHO trial and the press bulletin reporting the results of the trial released by the WHO in April 1996, the trial participants were portrayed similarly. In the press bulletin, Dr. Benagiano, the director of the WHO's Human Reproduction Program, praised all the men who ever volunteered in a WHO trial:

> The willingness of men to volunteer for the recently completed study, and other similar WHO-supported studies in the past, as well as their motivation and commitment to continue with the protocol of weekly injections, demonstrates the interest in—and demand for—a reversible male contraceptive of this type. (WHO 1996)

Benagiano not only praised the volunteers for their commitment, he also used them as examples to articulate the need for the new method. Male volunteers thus have a dual role in these reports: they figure as trial participants and as "prototypes" of future users. The rhetoric of publications in scientific journals exemplifies this transformation of trial participants into future users. In abstracts and method sections, these male trial participants are portrayed as active agents rather than passive test subjects. Instead of the usual phrases such as "the subjects were given an intramuscular injection" (Schurmeyer et al. 1984: 417), or "experiments performed on 10 normal volunteers" (Skoglund and Paulsen 1973: 358), or "a male contraceptive trial was undertaken in 23 men" (Bain et al. 1980: 365), trial participants are described as "men requesting contraceptives" (Guerin 1988: 187; Foegh 1983: 7; WHO 1996: 821; Soufir et al. 1983: 625; Lobel et al. 1989: 118). This subtle shift in discourse in which agency is attributed to the trial participants suggests that they have taken the initiative or asked for the trial, transforming trial participants into initiators of the new technology.

Finally, the female partners and friends of men participating in the contraceptive trials also contributed to highlighting the special role these men have played:

It's absolutely noble. The man's so brave. (Ringheim 1996b: 82)

I thought it was very noble of him to have injections. I go hysterical with needles. I wouldn't have been able to do that. (ibid.)

Researchers, trial participants and their female partners and friends thus actively constructed the image of the brave, pioneering man.

Conclusions

We can conclude that the organization of clinical trials for the testing of hormonal contraceptives for men required a lot of extra work compared to the testing of other drugs. Researchers had to create specific tools to ensure the cooperation of the test subjects. Since men were not accustomed to being subjected to contraceptive testing, and researchers were not used to having men as test subjects in contraceptive trials, the testing required specific procedures. To configure men as reliable test subjects, clinicians selected men in stable relationships. Test subjects and future users of the new technology became represented as monogamous men. Women also played a crucial role in configuring men as reliable test subjects. Female partners were important actors in encouraging men to take part in the trials and to ensure their cooperation during the clinical testing. Last, but not least, these women were creative agents in articulating male identities. Women used the clinical trials as a location to renegotiate responsibility for contraception with their male partners. By doing this, they actively engaged in articulating the image of the caring, responsible man. We thus can conclude that clinical trials functioned as an important location to configure the identities of the test users. This configuration work was not restricted to technologists, as has been suggested by Akrich and Woolgar, but included the work of family and friends of the test users.

Most important, men taking part in the clinical trials of male hormonal contraceptives have in turn performed this projected identity. By participating in the clinical tests, men consciously or unconsciously performed an aspect of male identities that conflicted with hegemonic representations of masculinity, that men are not inclined to take responsibility for contraception. As we have seen, a majority of the men participating in contraceptive clinical trials portrayed themselves as altruistic men who wanted to help their female partners who had experienced problems with the female pill. The dominance of altruistic images can be understood in the context of the contested nature of male contraceptives. An articulation of the users of the new technology in terms of incentives of self-interest would run the risk of providing critics and opponents of new male contraceptives with arguments to reject the new technology. An image in which male contraceptives are portrayed as drugs that serve the interests of men, particularly if it emphasizes men's control over contraception, conflicts with feminists' advocacy of women's autonomy in reproductive matters.[26] Being in control of reproduction is thus not included in configuring the identities of users of male contraceptives. This is in sharp contrast to other technologies recently introduced for male reproductive bodies, most notably Viagra. The discourses of Viagra are dominated by modernist rhetoric that portrays the capacity to be in control of one's body as "the proper and appropriate order of masculine things" (Mamo and Fishman 1999: 16). The debates on Viagra and the Male Pill thus show a reification of hegemonic masculinity which emphasizes men's mastery and control of sexuality rather than reproduction as essential aspects of masculinity (Mamo and Fisherman 1999: 17; Connell 1995).

In sum, we can conclude that the clinical trials have functioned as a cultural niche for the co-construction of a new technology and a new male identity: the caring, responsible man. This image has dominated male contraceptive discourse in the scientific community and in policy circles since the late 1960s and was also adopted by participants in clinical trials. This does not imply that hegemonic masculinities were completely absent from these narratives. To negotiate this new male identity, clinicians and trial participants relied on dominant cultural representations of masculinity which represent men as brave and pioneering subjects.[27] Although the long and winding road of the development of hormonal contraceptives for men has not yet come to an end, the quest for new male contraceptives has had a definite impact. Activities in laboratories and clinics and the ongoing debates in the news media have transformed male reproductive bodies from invisible bodies into public bodies, thus breaking with the practices and traditions that have long dominated medical and bodily discourses. Most importantly, technological innovation in male contraceptives technologies has brought gendered routines and conventions concerning contraception into the headlines. Technologies have thus the capacity to make visible and to destabilize dominant cultural narratives on gender.

Acknowledgments

I thank William Bremner, Alvin Matsumoto, Alvin Paulsen, and Fred Wu for granting me interviews, and Adele Clarke for her useful comments to improve an earlier version of this chapter. A more extended version will be published in my book *The Male Pill: A Biography of a Technology in the Making* (Duke University Press, 2003). The second section of this chapter has been accepted for publication in *Men and Masculinities*, a special issue of *Technology and Masculinity* to be published in spring 2004.

11

The Mediated Design of Products, Consumption, and Consumers in the Twentieth Century

Johan Schot and Adri Albert de la Bruheze

The historian of technology Thomas Hughes called the twentieth century a century of invention and technological enthusiasm. Technology became subject to conscious organization (for instance, in large technical system building), policy, and reflection (Hughes 1989). In the twentieth century, people for the first time referred to the concept of technology in the singular. Technological development became an independent and abstract phenomenon that was far beyond all the specialties of the many fields of application.[1] Technology in an abstract sense became the symbol of modern society.

In social economic history and in the history of technology, engineers, planners, producers, and managers working in trade, industry, and government often are portrayed as the masterminds, initiators, designers, and makers of modern industrial society. It is as if the twentieth-century world has been made in the small world (the network) of large companies, bureaucracies, and professional organizations. Other social groups, including consumers, laborers, women, and children, are considered passive bystanders in such a history; they adapt, although sometimes against their will and after some resistance. The consumers' role is reduced to that of purchasers of new products. Moreover, consumers do not have a face in many studies—in contrast to the "masterminds," the planners, and the "makers," they are often anonymous.

This image of a modern society created by technology push, (large) technical system building and its momentum, and production has been nuanced, corrected, and complemented by various historians of technology, in particularly by Ruth Schwartz Cowan and David Nye. Cowan has argued for focusing on the actual or potential consumer and for viewing the networks surrounding artifacts and systems from the consumer's point of view. She introduced the notion of the consumption junction, "the place and time at which the consumer makes choices between competing technologies," and she tried "to ascertain how the network may have looked when viewed from the inside out, which elements stood out as being more important, more determinative of choices" (Cowan 1987: 263).[2] Nye's work on electricity, as Nye explains, examines "the process of electrifying America from the general public's point of view. It shifts attention away from inventors and captains of industry to ordinary people: consumers, workers, reformers, housewives and farmers." (Nye 1990: xi)[3] In another book, Nye contends that "it would be illogical to consider consumers as passive recipients of corporate messages. . . . They have long been actively involved in a process of defining themselves through the acquisition and display of goods." For him, choices made by Americans have decided the course of American history (Nye 1998: 10).[4] By emphasizing developments in daily life and in the use of new products, these historians (and others) have made users visible as co-designers and co-makers of modern technological society.[5] Pushing users to the fore not only nuances traditional histories of designers and makers but also makes visible the often-hidden role of gender at an early stage.[6]

The complementation of production-oriented studies by consumer-oriented studies was a necessary and productive development, and a lot of work remains to be done in this regard.[7] Consumption-oriented studies, for instance, take the existence of consumers for granted, and focus on their choices. As a result, they tend to de-emphasize production and system building. In doing so they largely obscure the ways in which consumers and consumer images constructed in laboratories, factories, and marketing departments influence actual consumption. To remedy these flaws, we would like to contribute to the development of a new perspective which aims at reconnecting production and consumption. Central to our perspective is a focus on the mediation process between production (supply) and consumption (demand).[8] We will characterize this mediation process as a process of mutual articulation and alignment of product characteristics and user requirements. In the process of mutual articulation and alignment (or mediation), product characteristics, the use, the user, and the user's demands become defined, constructed, and linked. Mediation as a process of mutual articulation and alignment is influenced not only by the work of producers and users but also by the work of mediators and by the existence of institutional loci and arenas for mediation work. We call such a locus a mediation junction.

In modern societies the mediation process became structured by a tendency to concentrate design activities in laboratories and in design agencies and firms. As a consequence, the identification of consumers and their needs became increasingly difficult—especially in the twentieth century, when the rise of mass production and the upscaling of production in most industries widened the gap between production and consumption. For producers the consumer became a white spot, hard to locate and hard to get a grip on when products are under development. It is revealing, perhaps, that in management studies an accepted lesson is that most product innovations fail because of a lack understanding of users' needs. At the same time, it is also clear that such an understanding is hard to get; a merely increasing emphasis on market research itself does not lead to a better understanding and a higher probability of product success. Designers do not seem to seek relevant market information; even when they do, they often neglect relevant market information. What seems crucial is a process of mutual articulation and alignment of product characteristics and user requirements. Yet it is precisely because of a felt uneasiness about where the market is that marketing, advertising, and branding thrived in the twentieth century: they held the promise of conquering the unknown land of the consumer. These activities alone could not, however, sell the overwhelming array of new products flooding the market in the twentieth century. In this chapter, we will focus on the mediation process between production and consumption—a process that, in addition to being crucial for overcoming the distance between production and consumption, resulted in the construction of products, production, consumers, and consumption in one movement. As a result—as we will argue in our concluding section—this mediation process has been constitutive for the building of the consumer society defined as a "system in which consumption is dominated by consumption of commodities, and in which cultural reproduction is largely understood to be carried out through exercise of free personal choice in the private sphere of everyday life" (Slater 1997: 8).[9]

The case material we use is drawn from Dutch case studies. Although some would argue that mediation is a typically Dutch style of producing and consuming, we hold that our perspective is valid for research in other Western countries. Comparative research yet to be undertaken will have to determine to what extent the Dutch style is a special one.

Mediation Junction and the Quality of Mediation

Innovation studies based on the work of Schumpeter often distinguish three phases of technical change: invention, innovation, and diffusion.

In Schumpeter's analysis, invention is exogenous to the economic process; its character is mainly technical, engineers focusing on design. Technology is made endogenous in the innovation phase by entrepreneurs who find and develop markets. Finally diffusion is a process of imitation, of which the causes of failure to adopt an innovation are sought in the characteristics of the users and the nature of the communication process, for the existence of a profitable technology is presumed. Note that in this model the continued existence of an old technology cannot be explained, other than by conservatism and defective means of communication. Nathan Rosenberg was one of the first economists to criticize this highly influential model. Rosenberg believes that innovation and diffusion are also processes in which technical change reigns. New technologies are not automatically superior to old technologies, they have to be made better in a learning process. In this context, Rosenberg (1976, 1982) introduced the notions of "learning by using" and "learning by doing."[10] Dorothy Leonard-Barton and other authors building on his work have argued that it is not only a matter of adopting technology to its environment, but also of adapting the environment to the technology: "implementation of technical innovations is best viewed as a process of mutual adaptation, i.e. the re-invention of technology and simultaneous adaptation of the organization" (Leonard-Barton 1988: 253). Leonard-Barton proposes to enrich the invention-innovation-and diffusion model by introducing large and small cycles of redefinition of technologies and their contexts. Figure 1 illustrates these dynamics, in which continuing feedback guides the redesign of product characteristics and user requirements and their (ultimate) alignment.

While Leonard-Barton discusses implementation of new technologies within an organization and as a "normal" development, Lundvall among others has argued that similar processes of mutual shaping characterize innovation and diffusion processes in situations in which the producer and the user of a technology are separated by the market. He introduced the notion of "learning by interacting," suggesting that market transactions do not result in "satisfactory innovations." Not enough feedback is possible on the user requirements. Consequently all kind of mismatches arise. Lundvall argues for the importance of networks and inter-organizational relationships to carry the learning processes between producers and users.[11] This argument can also be found in various other diffusion and implementation studies (Lundvall 1988; Von Hippel 1988; Habermeyer 1990; Slaughter 1993; Fleck 1994). In these studies, users are not seen as passive recipients; instead they are often the source of

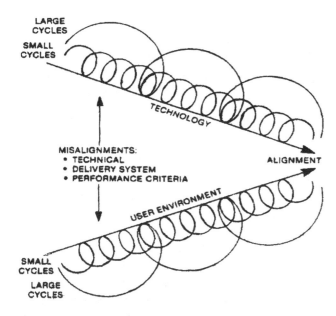

Figure 1
The process of mutual articulation of technology and user environments
(source: Leonard-Barton 1988).

new innovations (Leonard 1998). Rosenberg had already argued that
actual use is necessary to generate the knowledge required to improve a
product. Such knowledge cannot be generated in any other way, he
argued, because interactions between products and their use environments are too complex to be predicted. This means that designers can
invest in fault-anticipation strategies (such as simulation, using analytical
procedures, and incorporation of already successful field-tested subsys-
tems in the design) depending on the costs and benefits that they expect,
but learning by doing is the default strategy (Von Hippel and Tyre 1995).
This line of argument implies the impossibility of designing a perfect
technology in house, an observation that will be dealt with in our second
case study. According to Lundvall, new technologies need to be tested
and tried in practice, which is the only way to specify and articulate precise requirements (both technical and contextual) for their implementation. In the same vein, Von Hippel's work on innovation stressed the
importance of lead users. According to Von Hippel, lead users have three
main characteristics. First, they are competent users: they can define the
problems with which they are confronted in the implementation of the
new technology, distinguish trivial problems from more fundamental

development problems, actively assist the developers with technical knowledge, and formulate their experience so that it can be incorporated into the ongoing development work. Second, they are resourceful users: they have good access to economic, personnel, and know-how resources that can be of particular benefit in development and implementation work. The third characteristic of lead users is their incentive for innovation.

Hoogma and Schot (2001) have argued that user (and producer) innovativeness does not only follow from actual use or from a set of specific user characteristics, as Rosenberg stressed, but also from the nature (and quality) of the interaction process between producers and users. One of the important points to make is that the consumer does not yet have "precise demand requirements" and a clear view of relevant product attributes. Producers and consumers have to develop the product— and, thus, so-called user needs—at the same time the product is introduced and diffused. User needs and possible alignments with design options cannot be discovered ex ante, as Rosenberg stressed, but have to be constructed and negotiated in a process of mutual articulation and alignment of demand and supply. Koch and Stemerding (1994: 1212) describe this process as follows: "This attunement involves on the one hand processes of technical specification, in which the content and meaning of a particular technical option (like a genetic test) becomes further specified in practice, and on the other hand processes of articulation of demand, in which producers as well as consumers further specify needs and demands in relation to available technical options."[12]

The process of mutual articulation of demand and supply depends crucially on the work of mediators. Mediators are actors who focus on articulating and aligning demand (consumption and user requirements) and supply (production and product characteristics). They include home economists, fashion intermediaries, consumer organizations, auto clubs, marketing and testing agencies, retailers, and supermarkets.[13] For mediation purposes, often a specific institutional locus—an agency or a platform for example, will be created. We call such a locus a mediation junction (or perhaps a technology junction). The mediation junction is, thus, the place at which consumers, mediators, and producers meet to negotiate, articulate, and align specific technical choices and user needs. It is an arena where agenda building and technology development become connected.[14] Our assumption is that the nature of a mediation junction will influence the mediation process. This assumption will be explored in two brief case studies.

Since the 1990s, in many fields, such as history of technology, science and technology studies, gender studies, and media studies, the active consumer and (to a much lesser extent) the importance of the mediation process have been identified and explored.[15] The main thrust of this work is to show the co-construction work involved in use of new (and old) technologies. For our purposes, it is important to refer to a central concept introduced and discussed in this literature: the concept of script, first explored by Madeleine Akrich. In Akrich's view, designers "define actors with specific tastes, competences, motives, aspirations, political prejudices, and the rest, and they assume that morality, technology, science and economy will evolve in particular ways." A large part of the work of innovators is "that of '*inscribing*' this vision of (or prediction about) the world in the technical content of the new object. I will call the end product of this work a 'script' or a 'scenario.'" (Akrich 1992: 208)[16]

The concept of script is useful because it makes visible a new kind of user: a projected user. The mediation process can now also be perceived as a process of articulation and negotiation of projected users, as we emphasized at the beginning of this chapter.[17] In this process, we would like to argue, a third kind of user participates: the represented user, brought into the process by mediators who often claim to represent specific users. In the mediation process a number of users proliferate: projected users, real users, and represented users. Implicated in the process of mediation is the contestation of these users and their demands, the quality of the projections (by designers), the legitimacy of the representations and the representers, and the acceptance of real users

We now turn to two brief case studies to explore the nature of the mediation process. Both cases primarily deal with product innovations that implicate a range of changes on the consumer side. The case studies have been selected because they illustrate important features of two modes of mediation. The first case is the introduction of the disposable milk carton in the Netherlands. In this case the mediation junction was partly located outside the firms supplying milk and packaging, allowing also consumer organizations and other parties to influence the mediation process. In the second case, the introduction of snacks by Unilever, the mediation junction was incorporated inside the firm, and the mediators called upon were market research and consumer research departments. In these cases we will explore our assumption that the nature of a mediation junction is of crucial importance in the mediated design of products, consumers, and consumption.

The Disposable Milk Carton[18]

Until the 1960s, milk distribution in the Netherlands was dominated by the system of door-to-door delivery by licensed milk and dairy retailers. This system became problematic with the increasing building of blocks of flats after the Second World War. Because of this, daily delivery of bottled milk (and the return of empties) became a virtually impossible task for the milkmen. Therefore, the dairy industry advocated the installation of elevators, door phones, shopping safes, and cold storage in apartment buildings. Because milk stayed fresh longer in cold storage, it would no longer have to be delivered every day. As a result of protests and delivery strikes by milkmen, emotions flared in discussions of milk distribution. In 1958 in Schiedam there were even "skirmishes" between milkmen and apartment dwellers. The apartment dwellers demanded daily door-to-door delivery of dairy products and threatened to establish user cooperatives if this demand was ignored.

The conflict over milk delivery resulted in a growing and "broad" interest in new types of milk packaging and milk distribution. Questions on the subject were asked in the Dutch Parliament and attracted the attention of Minister of Agriculture Mansholt and of Minister of Housing and Building Witte. Supermarkets seemed a promising and "modern" distribution alternative to the delivery monopoly of licensed milk and dairy retailers. As far as packaging was concerned, the lighter milk carton seemed to be the best substitute for the heavy glass milk bottle.[19] Moreover, the two alternatives were linked together in the postwar policy directed at an increase of productivity and efficiency. In 1957 Minister Mansholt argued for instance that "by way of paper milk packaging cheaper production will be possible, provided that the entire distribution system is adapted to this type of packaging."[20]

Food packaging in the Netherlands started receiving serious attention after the Second World War within the context of efforts to increase efficiency in production and business. In 1946 the Institute for Packaging of the Dutch Organization for Technological and Scientific Research (TNO) was established as the research institute of the packaging industry. With the establishment of the Dutch Packaging Center (Nederlands Verpakkingscentrum) in 1953, the Dutch packaging industry got its first umbrella organization and spokesperson. The TNO Institute of packaging and the Dutch Packaging Center closely cooperated in the development of new packaging materials, packaging methods and packaging machines. In the 1950s and the 1960s, when industrial attention to consumption was increasing,[21] food packaging became an important subject.

In the Dutch door-to-door distribution of milk, unpacked milk remained important till the 1960s mainly because this milk was cheaper than glass-bottled milk. Many Dutch housewives therefore kept preferring unpacked milk even if this (pasteurized) milk had to be cooked and stored in pots and pans. Owing mainly to urban regulations requiring glass bottles and to "modern" views about food hygiene and health, the glass milk bottle began its advance in the Netherlands in the 1950s.[22] Despite this advance, particularly the dairy industry did not find the glass milk bottle satisfactory. The milk bottle was laborious. Delivery, return, storage, cleaning, filling, capping, pasteurization, and cooling required a lot of space, labor, and expense. Already in the 1950s the dairy industry looked for different and particularly cheaper alternatives for milk packaging. Attention was focused on America and Scandinavia, where the milk carton had a considerable market share. This was especially due to the method of distribution. Unlike the Netherlands, those countries allowed milk to be sold in grocery stores and supermarkets. In the Netherlands only licensed milk and dairy retailers were allowed to sell milk. Each day, dairy products were delivered to the door by milkmen. In 1955 the cooperative dairy industry began to explore the market viability of the milk carton.

The Koninklijke Nederlandse Zuivelbond (Royal Dutch Dairy Union) carried out an extensive investigation into the marketability of the milk carton from 1955 to 1957. That body concluded that the paraffin-wax layer on the inside of paper packaging might crumble, that the taste of milk packed in paper deteriorated rapidly at higher temperatures, that milk cartons could start to leak, that the strange form of the milk cartons caused aversion, that the milk was not visible, and that opening a milk carton was not really a simple task. Moreover, it concluded that the Dutch housewife did not benefit from the lower weight and the easy transport of paper-packaged milk, because her milk was after all delivered at home, and the empty bottles were collected. The milk carton would have better chances in countries where milk was sold in grocery stores, according to the committee (FNZ 1957). A "field test" was done, and it appeared that the housewives declined to buy milk in cartons because it was more expensive (by 2–5 cents) than glass-bottled milk and because there were numerous problems of habituation and usage with milk cartons (opening, pouring, and conserving). In 1959 the Nederlandse Verpakkingscentrum (Dutch Packaging Center), the umbrella organization of the Dutch packaging industry, concluded: "All the same, [the dairy industry] is convinced that there is a future for paper and that milk will be packaged in paper in due course, but for the time being the Dutch consumers are not ready for it. In 5 to 10 years they possibly are." (NVC 1959: 23)

In this period, and within the context of debates about milk distribution in apartment buildings and possible changes in the door-to-door delivery system, organizations of housewives became involved in the matter of the disposable milk carton. Because food and nutrition were generally considered women's responsibilities, milk packaging was discussed within associations of housewives and information services, and it received increasing attention in the press (particularly in women's columns). Consumer organizations like the Dutch Household Council (Huishoudraad), established in 1950 as an umbrella organization of women's organizations and (household) advisory organizations, wanted to get involved in the development of the disposable milk package. Representatives and spokesmen of producers and consumers got together on this. In 1958, the Verpakkingscentrum and the Nederlandse Huishoudraad took the initiative for the foundation of a joint working committee, "Housewife and Packaging," which included—besides the Verpakkingscentrum and the Huishoudraad—representatives of the paper and packaging industry, of TNO (the Study Group for Layered and Lined Packaging Materials), of the Centraal Instituut voor Voedselonderzoek, of the Instituut voor Bewaring en Verwerking van Tuinbouwprodukten (Institute for Conservation and Processing of Horticultural Products), of Dutch housewives, of the department store De Bijenkorf, and of the broadcasting companies AVRO and NCRV. The study committee on Housewife and Packaging and its subcommittee on Milk Packaging may be seen as a mediation junction where producers and designers, mediators, and consumers could negotiate about the design of a new product, its consumption, and its consumer. Each actor had his or her own vision of use and user.

According to the cooperative dairy industry—from the perspective of not-yet-recovered investments in the milk bottle system—the milk carton was too expensive and awkward to use. The paper industry, on the other hand, emphasized that housewives would benefit from the lower weight and the ease of use of milk cartons. According to nutritionists, the milk carton might encourage women to buy more milk and have their families drink more milk: "It is after all sufficiently pointed out to her that a certain quantity of milk per day for each member of the family should be incorporated in her menu in some way or other." (NVC 1959: 18) According to the Nederlandse Huishoudraad, it was in the housewife's main interest to have adequate milk delivery and storage facilities in the house. The Huishoudraad concluded that the Dutch housewives had no reason to switch from bottled milk to paper-packaged milk within the existing distribution system of door-to-door delivery. The council's own research among housewives showed that many housewives thought the milk from cartons tasted of paper after a few days. Moreover, the paper was often thought too soft, and many housewives found it inconvenient that they could not see how much milk was left in the carton.[23] This picture was confirmed in a poll by the glass manufacturers: "the" Dutch housewife had considerable problems in using the milk carton. The glass manufacturers did not expect the replacement of the glass milk bottle by the milk carton to be very successful, because in the existing distribution system of door-to-door delivery a new way of milk packaging was of no advantage to the housewives (Verenigde Glasfabrieken 1965: 2). This process of negotiation and articulation was important for developing and constructing product requirements and user demands. It led to the broadly shared conviction that the purchase of paper-packaged milk by the Dutch housewife would not be a success. Practical use seemed to prove this conviction. The diffusion of the milk carton in Dutch cities was slow.[24] As a result of the negotiation and articulation process it had become clear that developments in the existing method of milk distribution would be of crucial importance for diffusion to take off. This broadly shared conviction was being prepared in the work of the milk packaging junction.

Within the governmental policy directed at industrial rationalization and societal modernization, the retail trade was liberalized in the early 1960s. As a result, groceries and supermarkets from then on were allowed to sell packaged milk and milk products. This, in combination with "dairy price stunts" by the supermarkets, the direct delivery of dairy products from the diary industry to supermarkets, and the upscaling in the retail trade as a result of emerging supermarkets, made door-to-door delivery less profitable. By the end of the 1960s it had begun to decline. Nevertheless, door-to-door delivery of milk remained the most important form of distribution until the early 1970s. Not until around 1972 did food companies and supermarkets become competitors of any significance (Nagelsmit 1975; ten Bruggencate 1969). As the Dutch housewife increasingly bought milk in the supermarkets, and also because small retailers were disappearing quickly, the glass milk bottle became a "burden" for both the housewife and the retailer. The disposable milk carton was the solution. It was lighter than the milk bottle, and it could be stacked in a refrigerator. (At the time, most refrigerators were small.) In the gradual replacement of the milk bottle by the milk carton after 1965, milk consumers, consumption practices and consumer representatives played important roles. According to the Centrale Verpakkingscommissie (Central Packaging Committee) of the cooperative diary industry, the advantages of milk cartons and the purchase of paper-packaged milk in the supermarkets became clearer with the increase in recreation and the growing number of women working outside the house. These represented consumer images nicely fitted onto consumption representations described by the Central Packaging Committee as "a modern life style and dietary habits."[25] This might be true, but it conceals the crucial process of mutual articulation of demand and supply that prepared the way (and the actors) for large-scale introduction of the disposable milk carton. Its history can best be understood as a story of negotiated diffusion (de Wit, Albert de la Bruheze, and Berendsen 2001). Producers, mediators, and consumers articulated their views and ideas about the milk carton. In tests, projected users, represented users, and real users were aligned in order to articulate the product and the interests of consumers.

Unilever and Indoor Snacks in the Netherlands[26]

Snacks and snacking have a long tradition in the Netherlands. Herring, sausages, peanuts, candy, and cake were sold and consumed on the streets. A national snack culture developed with the increasing production and consumption of French fries ("patates frites") and hot (uniquely Dutch) meat snacks like *kroketten* and *frikadellen* that were (and still are being) sold by cafeterias and *automatieks*, especially after the Second World War (Shuldener 1998). These snacks contributed greatly to the habit of outdoor eating. In 1958, potato chips were introduced by a British concern called Smiths. Potato chips immediately became a big success, and the Dutch quickly became the largest consumers of potato chips in Western Europe (Hesen 1971).

In the 1960s, the multinational food concern that is now Unilever decided to enter the booming Dutch indoor snack market. It saw as the most promising market option the large-scale industrial production of complex or "composed" snacks and snack food made up of a great number of ingredients that could be combined and varied indefinitely. Nevertheless, Unilever decided to adopt a prudent strategy to get a foothold on the Dutch snack market. With this strategy the production and marketing of simple traditional potato chips was the first step. The second step implied the development of a branded composed snack assortment once a market position had been established. In order to position the corporation on the snack market, the project team and management responsible for snack development addressed the questions what snacks were, who the snackers were, and when snacks were consumed. It claimed to find that snackers consisted mainly of housewives and children, and that snacking was being done during dinner, between meals, and at social events where drinks were served. The corporation decided to concentrate on the development of "between meals" snacks and "social events" snacks. Because in the Netherlands these snacks would be consumed in between the traditional three main meals and after dinner, the characteristics of these snacks were defined as "tasty light and crispy but not filling."[27] As the most promising potato snack meeting these criteria, the corporation considered its reformed potato stick, a more complex and extruded potato snack that contained less potato and thus less fat, and that—according to the corporation—looked and tasted better than the traditional potato chips that dominated the Dutch snack market.

Consumer research was considered necessary to find out whether these consumer and consumption representations fitted with social reality. From a 1965 "caravan test" in which 183 Dutch housewives participated, it appeared that consumers clearly preferred the new reformed (and extruded) potato stick over the traditional potato chip. The housewives particularly liked the pronounced taste and crispiness of the reformed sticks. This test—as did others—also revealed that consumers increasingly minded the consumption of fat and that they judged the reformed sticks as too fatty. Unilever translated this finding into the consumer opinion that "fat makes people sick and thick" and into the consumer requirement that potato snacks not be too fatty. Unilever thereupon decided to reduce the fat content of its reformed sticks and to test the adjusted snack on the Dutch snack market.

In 1967 Unilever optimistically introduced the reformed sticks by means of a test introduction in Rotterdam. Supported by commercials in shops, papers, and magazines, the initial sales looked promising. In the first week, 7,000 units were sold. Within a few weeks, however, sales dropped to fewer than 2,500 units a week. In order to understand what had gone wrong, Unilever marketing employees interviewed 600 housewives who had bought the sticks. The results of the interviews were quite shocking for Unilever. It appeared that the Rotterdam housewives did not consider the sticks to be "snacks" at all but rather "savories." Neither did they considered the sticks a completely new snack. Even worse, many said that the sticks did not look good, were sticky and greasy, and did not taste good. Unilever had great difficulties in accepting the outcomes of the interviews with real consumers because it considered real users and their preferences as difficult to grasp. Unilever therefore preferred to deal with—and to rely on—aggregated consumer and consumption projections mapped by its marketing departments in order to deal with market trends adequately. "It is always tricky to understand from housewives' statements what they really mean," said one Unilever official, reacting to the results of the market test interviews.[28] Notwithstanding these reservations—and after an internal investigation—Unilever officials had to admit (internally) that the fat content and the taste of the reformed sticks had not been improved in the time interval between the first and the second market test, although this had been ordered. Additional investigation revealed that a correction of these failures would require a substantial improvement of the production process, and—even worse—that the reformed sticks turned out to keep far less well than was hoped and claimed. To correct this would require among others a far-reaching and costly improvement of the packaging material. In order to be better prepared the next time, Unilever officials decided to incorporate the articulated user requirement as "user demands" into snack R&D.

The Dutch snack consumer thus did not behave in the way Unilever— based upon consumer and consumption research—expected them to do. Within the company this led to reflection with respect to the question whether Unilever should continue its activities in the snack sector or not. Because the snack market was conceived as a "growth market," it was decided to go on, and to concentrate first on national snack R&D and then on the development of "second-generation snacks"—improved potato chips, nut snacks, etc. Unilever subsidiary Calve–De Betuwe became the R&D center for snacks in the Netherlands. In 1971, after successful market tests, Calve introduced the borrelnoot, a new "social event" snack consisting of a peanut coated with a fried crispy and spicy layer. The borrelnoot became very successful and provided Unilever with a substantial market share. Unilever marketing attributed the success of the borrelnoot to the perfect matching of product innovation and perfectly mapped Dutch consumer preferences: "The basic concept of the fried, coated peanut comes from Indonesia, and many Dutch families would probably have been aware of the concept."[29]

Both the market failure of the reformed potato sticks and the market success of the borrelnoot convinced Unilever that the market—i.e., consumers and consumption—should determine product development. Because the success of the borrelnoot had shown the fruitfulness of working with projected consumer and consumption, Unilever decided to continue working with in-house marketing and user representations. Unilever therefore undertook great efforts to map and represent consumer preferences as accurately as was possible. Represented consumer demand and consumer requirements articulated during in-house product tests and consumer panels became increasingly important for the articulation of technical requirements during snack development. Based upon the results of in-house product tests and consumer panels, professional consumer researchers and Unilever project teams constructed the image in the 1970s that consumers increasingly favored healthy, fresh, and natural products. These projected consumer and consumption images were "built into" several new snack concepts based upon fruit with "crispy and/or multi-textured properties" that should provide an inevitable association with "health" and "nature." In the same vein, in-house Unilever consumer research discovered in the 1970s that the number of daily main meals steadily decreased while the number of "eating occasions" increased. Unilever thereupon anticipated market opportunities for snack food with a higher nutritional value than the existing snacks like potato chips and the borrelnoot. Projected consumer requirements of healthy and natural foods were connected with market expectations concerning new kinds of snacks into the concept of the "wholesome snack."[30]

In its continuing efforts to penetrate the snack market and to master consumers and consumption, Unilever developed all kinds of new snack products through an in-house process of working with corporate mediators (marketing and consumer research departments) and consumers in all kinds of in-house product tests and consumer panels. It also bought snack companies that had developed successful relations with consumers. Nevertheless, the risk remained that the confrontations between projected consumers, represented consumers and real users would not lead to market success. During the introduction of new products consumers often acted differently than was projected and expected.

Disposable Milk Cartons, Snacks, and Beyond: Some Exploratory Conclusions

The examples of the milk carton and snacks show that consumers were co-designers and co-makers of new products and technologies, whether they were represented, projected, or real-life consumers. During product tests, user panels, market experiments, consumer research, conferences and workshops, representatives of producers, and consumers negotiated and articulated the design and the use of snacks and milk in disposable cartons. The cases also show that during the market introduction of new products not only the products are tested, but also demand and supply requirements are articulated and attuned to each other. This process is shot through with difficulties, and actors struggle to identify a productive sociotechnical fix, a product that fulfills constructed user needs and design specifications. Mediators are very important in this process, as both cases testify. Products are designed not only on the design table and in factories but also during a process of market introduction. This market hardly resembles the ideal typical neo-classical market of free supply and demand, but rather consists of a series of forums and arenas where mediators, consumers, and producers meet and negotiate—a mediation junction.

The case materials presented here show interesting patterns in the mediation of consumption. They show the workings of two different kinds of mediation junctions. In the case of the disposable milk carton, several junctions operated next to each other—in particular, the joint working committee "Housewife and Packaging" and the Committee for Milk in Paper Packaging. The latter committee was controlled by the supply side, but the former committee was not. Meetings between mediators and producers in which projected and represented users were discussed led to a process of mutual articulation of demand and supply. A diffuse learning process developed, including field tests with real users in which all actors developed a deeper and shared understanding of productive alignments between user requirements and product characteristics.

In the snack case, the mediation process was fully controlled by Unilever's corporate mediation junction. This resulted in a more haphazard process. The mediation process for snacks was basically a process of market testing and testing of projected users against real users. This can lead to market failure as well as to market success, as the snack case exemplifies. Which factors caused success and which ones caused failures was rather unclear to Unilever, and in the "reformed sticks" example failure was explained by referring to the unpredictable behavior of women and housewives in particular. In an attempt to remain in control of consumers and consumption, Unilever preferred to deal with projected consumers, thereby relying on its corporate mediation junction of marketing and consumer research departments.

Although the two mediation patterns distinguished need additional and comparative research to warrant corroboration, our hypothesis is that the nature of the mediation junction—whether it is fully controlled by the producer or not—influences the mediation process between production and consumption. The case materials suggests at least that out-house junctions create more favorable conditions for the matching of projected, represented, and real users than in-house junctions. Out-house mediation junctions seem to provide better opportunities for more symmetry, and in that sense they have more potential for clarifying confrontations, assumptions, expectations, and scripts of the actors involved. This becomes clear in the milk packaging case, but whether this fits to every product and firm—as well as in what way and to what extent different kinds of mediators and mediation junctions influence the outcomes of mediation processes—remains to be answered.

Our second hypothesis is that the emergence and proliferation of mediators and processes of mediation junction building in the twentieth century were of decisive importance for the shaping of the twentieth-century Dutch consumer society, because they accommodated the selling and buying of a large variety of products. The fast expansion of mass consumption in the twentieth century and especially in the 1950s and the 1960s (the "economic miracle") cannot be ascribed entirely to so-called supply-side factors, as if mass consumption followed the rise of mass production. In this perspective, at some point mass production needs rising standards of living to secure markets and to absorb products, and to secure industrial and political peace. Contrary to this position, in our view, mass production and mass consumption are constructed simultaneously in a mediation process. To study the emergence and development of the consumer society fruitfully, we consider it necessary to pay serious attention to this mediation process, and to the history of this mediation process and its institutions, which has yet to be written.

Acknowledgments

This chapter is one of the results of a collaborative research project on the rise of Dutch consumer society, New Products and New Consumers. A History of a Reciprocity, 1890–1970. funded by the Netherlands Organization for Scientific Research—Humanities (NWO/GW), File 250-29-056. This project is part of the national research program Technology in the Netherlands in the Twentieth Century (TIN-20) coordinated by the Netherlands Foundation for the History of Technology (SHT). We thank our research collaborators Marja Berendsen, Liesbeth Bervoets, Gijs Mom, Ruth Oldenziel, Anneke van Otterloo, Peter Staal, Anton Schuurman, and Onno de Wit for stimulating discussions and a free flow of ideas. We are indebted to Frank Geels and Onno de Wit for detailed and broad-ranging comments on an earlier draft.

12

Giving Birth to New Users: How the Minimoog Was Sold to Rock and Roll

Trevor Pinch

The role of users in technology studies has gained increased attention since Ruth Schwartz Cowan (1987) introduced the notion of the "consumption junction." Rather than conceive of users as passive consumers of technology, scholars have focused on how users interact with artifacts to become agents of technological change (Douglas 1987; Nye 1990; Martin 1991; Woolgar 1991; Fischer 1992; Mackay and Gillespie 1992; Akrich 1992; Cockburn and Ormrod 1993; Kline and Pinch 1996; Kline 2000; Bardini 2000; Mackay et al. 2000). Woolgar (1991) has shown how designers actively "configure" users, and Akrich (1992) has argued that the appropriate user interaction with a technology is "scripted" into its design. Users, in turn, can challenge such scripts and reconfigure technologies (Latour 1987; Akrich 1992). A good example of such a reconfiguration is the use of record turntables for "scratching"—a use inconceivable to the engineers who first developed turntables. Historians have studied how users come up with such completely new uses for technologies. Douglas (1987) drew attention to the role of amateur operators in the history of radio, Fischer (1992) and Martin (1991) have shown how rural women first used the telephone for extended conversations rather than simply short business calls, and Kline and Pinch (1996) documented the use of the automobile as a stationary source of power rural in the rural United States.

Most of the aforementioned studies deal with how technologies are adapted to users or how users adapt technologies. In this chapter I want to consider a more fundamental question: How do users come into being in the first place? Often the answer is that manufacturers "invent" the user to go along with their technology. Bardini (2000), in his study of how Douglas Engelbart developed the first computer mouse, considered the part played by an invented "virtual user." Jenkins (1975) showed how George Eastman's need to create a new user for his roll-film Kodak camera changed photography from a specialized professional realm to a popular hobby. But sometimes the manufacturers have no idea where the users will come from. The case I focus on in this chapter—the development of a new form of electronic music synthesizer, the Minimoog—is such an example. The predominant synthesizer manufacturers were largely unaware of potential new users of their technology. In this case, someone unconnected with the manufacture of synthesizers noticed the importance of the new users, recruited them to the new technology, and in the process developed a completely new way of selling synthesizers.

Technology studies have not paid sufficient attention to mediators such as marketers and salespeople in the development of technology.[1] This reflects a general lacuna in the social sciences whereby selling and salespeople have long been neglected.[2] It is argued here that field sellers—sellers who go on the road selling a product and meeting with customers—because they have access to both manufacturers and users, moving backwards and forwards between the two groups, are a particularly important type of mediator to study.

Field sellers are highly mobile and build up elaborate networks with their customers (whether retail stores or end users).[3] Because of their interaction with users, field sellers often are the first to notice the emergence of a new use for a technology. They also are often the first to hear about deficiencies in current use, how a technology can be improved, and what works and what doesn't. Such information is often passed back to designers and manufacturers. Field sellers are the active agents of how a technology is domesticated.[4] Aside from occupying a strategic position, salespeople provide a methodological resource in the study of users. It is well known that advertisements, sales brochures, and the like provide important clues to how a technology is to be used.[5] Most selling involves persuasion. In their sales activity salespeople often make explicit the implicit uses of a technology. Their sales strategies, including sales patter, rhetorical displays, and demonstrations, may reveal how they and the manufacturers think technologies are to be used. The practices of such salespeople are, I would claim, a key place in which the study of technology and users can be pursued.

In this chapter we will see how salespeople had to devise de novo the social and technical practices to sell a technology and at the same time recruit new users. This case is interesting because the main salesperson who is the topic of this chapter was first and foremost a user of the technology. By drawing on his own experiences as a user, he was able to understand the wants, needs, and problems of other users and to devise effective means to recruit them to the new technology. By recruiting, training, and sustaining new users, he was able to build the new retail market for synthesizers. By following this one salesman we are able to see that what is normally taken to be an economic concept—a market—is built from a series of social and technical practices.[6]

The Synthesizer

Today most portable electronic keyboard synthesizers are made by Japanese multinational companies (notably Korg, Yamaha, Roland, Casio) and are sold mainly through retail music stores via dealer distribution networks. Synthesizers are sold in the same way as other keyboard instruments, such as electric pianos and organs. The major synthesizer manufacturers display their range of instruments at trade shows such as that of the National Association of Music Merchants[7] (NAMM) or its European equivalent, the Frankfurt Music Messe.

But synthesizers have not always been sold in this way. The synthesizers invented by the pioneering American engineers Robert Moog and Don Buchla in the early 1960s were very expensive ($3,000–$10,000) and were almost exclusively sold to independently wealthy musicians and academic studios, either by direct order from Moog or Buchla or (in the case of Moog) through two sales representatives located in New York and in Los Angeles. The early synthesizers—large modular units that looked like telephone exchanges—were extremely difficult to use for live performance. Producing any particular sound was a complex process that involved adjusting a number of knobs and wires. These synthesizers were mainly used in studios to produce works of electronic music with the aid of dubbing, tape splicing, and multi-track recording.

It was not until 1970 that synthesizers were first sold in retail music stores. The first instrument to be sold in significant numbers in this way was a cheaper portable keyboard synthesizer, the Minimoog. This small synthesizer had a built-in keyboard and used switches rather than patch wires to set up and change the different sounds during live performance. The Minimoog was significantly cheaper than the earlier modular synthesizers, retailing for about $1,500. Paul Theberge (1997: 53) notes that "the Minimoog was the first synthesizer designed for the popular music market and the demands of real-time performance" and that "for many it defined the word 'synthesizer' throughout the decade of the 1970s." In 2001 Bob Moog was awarded the Polar Prize (sometimes called "the Nobel Prize of music") by the king of Sweden for his invention of the Minimoog.

Figure 1
The modular Moog synthesizer.

Figure 2
The Minimoog.

The Minimoog went from first prototype to final production model in less than 6 months and in dramatic circumstances (Pinch and Trocco 2002). At the time (late 1969 and early 1970), Moog's company was nearly bankrupt. Always undercapitalized, it had gone through a rapid expansion in 1968 and 1969, when the modular synthesizer became popular, first in rock music (the Byrds, the Doors, and the Beatles used it[8]) and then in classical music (with the surprise crossover hit *Switched-On Bach*). Many more "switched-on" records followed, in many genres. But the gimmick did not last long, and by late 1969 orders for modular synthesizers had slumped.

The first prototype Minimoog (Model A) was made by an engineer named Bill Hemsath, on his own initiative, out of junk that he had found in the attic of Moog's factory (Pinch and Trocco 2002). Bob Moog was convinced of the project's worth only when the prototype was refined into a more elegant design. The final decision to go ahead with a production model of the Minimoog (Model D)—a decision taken while Bob Moog was away on a speaking tour—involved a small insurrection on the part of his engineering team. Here is how Jim Scott, one of the engineers, described the events:

As soon as Moog left we all looked at each other and said, "You know, if we don't get this thing engineered to the point where it can be produced, we're all going to be out of a job anyway. And it's going to take us forever to hand-wire these things. . . ." And so [Moog's business manager] said, "Okay, just go ahead and lay out the [circuit] boards."[9]

When Moog returned, a couple of weeks later, the Minimoog was in production. Jim Scott takes up the story:

[Moog] came back and found ten D Models nearing completion, and he was not pleased. And he called us all into his office, and he let us know in basic Anglo-Saxon exactly what he thought of all of us. He was absolutely livid. And we all just kind of sat there and looked at our fingernails and nobody said anything.[10]

Bob Moog himself had no idea that the instrument would be a success:

. . . I remember thinking, and saying to a lot of people, we're going to make a hundred of these and then we'll stop and see where we are. You know, the funny thing is, we never did stop![11]

What use could there be for a synthesizer that was less versatile than the modular units in the sorts of sound it could make? Able to produce only one note at a time, the instrument was musically far less useful than an electronic organ. With 44 knobs and switches, it was also far more complex to operate. In short, it seemed too limited for serious electronic composers, and both too limited and too complex to appeal to commercial musicians. When it was first developed there was no clear conception within the Moog company who would buy or use this new instrument. Bob Moog believed that session musicians needed something more portable to take from studio to studio. But no one had tried to sell such an instrument, certainly not in retail music stores. Moog: "I don't think I, or anybody else in the company, went into a music store before . . . March of 1971."[12]

Some musicians were interested in the prototype Minimoogs. The very first prototype (Model A) was premiered in a concert at Cornell University on Easter Sunday 1970 by the first-ever live synthesizer ensemble, Mother Mallard's Portable Masterpiece Company, led by David Borden. Borden was pioneering a new form of electronic minimalism. His ensemble mainly used modular synthesizer equipment but found the portable synthesizer a useful addition. The space-fusion jazz player Sun Ra bought another early prototype (Model B) for use in his Solar Arkestra. The production model (Model D) was eventually premiered by the keyboardist Dick Hyman at New York's Museum of Modern Art in the summer of 1970. A *New York Times* reporter described the latter event rather unenthusiastically: "The results were gentle, melodic and rhythmic and relatively free of the blips and bleeps that often turn up in electronic music."[13] In the latter part of the concert, Hyman apparently tried to stage a "free-form happening," which didn't happen: "The happening . . . lasted for 55 minutes, which was more than a large part of the audience did."[14]

The main use of the Minimoog, however, was not to be found in any of these genres of music, but in rock. With its monophonic keyboard, its pitchwheel (which allowed the operator to "bend" notes, like a violinist or a guitarist), and its distinctive sound, it allowed a band's keyboardist to challenge the guitarist for the limelight. Part of the story of how this came about involves the development of "progressive rock," with its new keyboard heroes, the British performers Keith Emerson and Rick Wakeman.

The story of "progressive rock" is well known,[15] but how rock musicians got introduced to the Minimoog is a story that has never been told. The story involves the efforts of many salesmen who worked for synthesizer companies. In this chapter I focus on one such pioneering synthesizer salesman, David Van Koevering. In 2 years, Van Koevering went from a demonstrator of novelty musical instruments to vice-president of the Moog synthesizer company. It was Van Koevering who first saw the new instrument's potential in rock music and who took it on the road in a Cadillac to recruit a network of dealers, first in the United States and later throughout Europe and around the world.

Little David

David Van Koevering was the son of a Florida evangelist whose show featured novelty musical instruments, such as Swiss cowbells, tuned frying pans, a Theremin (an early electronic instrument), and musical stones played with hammers. David was a musical prodigy who was taught by his father to play many musical instruments—for instance, at age 6 he won prizes for playing the ukulele behind his neck. Young David and his brother, sister, and mother took part in his father's show, which played in churches from Michigan to Florida and which even ran on television for a while. (One musician told me that the young Van Koevering was known as "Little David.") Eventually, David took over his father's show and brought it to schools all over the United States.

Van Koevering always kept an eye open for new instruments to add to his show. When he first heard the Moog synthesizer, he was amazed. Frank Trocco and I interviewed Van Koevering on January 30, 1999. When he recalled the moment 30 years earlier when he first recognized the potential of the new instrument, his voice broke with emotion:

I saw something. I visualized performers using the Moog, the power of the sound, the sonic energy, and I believed that it could become common. I imagined it as powerful as electric guitar to the first guys that ever played this thing. . . . I saw something . . . and something happened throughout the cosmos that changed me and music. I knew, I could see it. [very emotional] So I argued with Bob [Moog], this is a performance instrument, and he'd gently give me all the reasons why it was difficult to perform with it.

In early 1970, Van Koevering visited the Moog works in upstate New York with the intention of acquiring a modular synthesizer for use in his show. Moog's engineers warned him of the extreme difficulties of using the Moog for live performance. Not only did the patch wires take an inordinate amount of time to change, but the oscillators were notorious for drifting out of tune. "The Moog," according to Van Koevering, "didn't stay in tune, it was a tough thing to set a pre-set up on, it sounded like no acoustical instrument ever, and it was hard to replicate an acoustical instrument, and everything you did with every one of the knobs gave you a brand-new sonic experience, some of which you never could repeat again. So this was hardly traditional. . . ."[16]

Van Koevering (along with Moog's own sales reps) constantly pushed Moog to produce a performance instrument. Other musicians desiring a performance instrument have told us about Moog's reluctance to consider this market. For Moog the synthesizer was a means to make music in a studio. Keith Emerson, the first major rock musician to take a modular Moog synthesizer on tour (with his band Emerson, Lake and Palmer), recalled Moog's doubts about the instrument's suitability for live performance:

> He said that he'd invented this instrument really for use in the studio, he wasn't too sure how it would stand up in live performance. He didn't really recommend traveling or touring with this instrument. And it soon became apparent to me that the tuning was really problematic in live performance. . . . The first concert we did, it was a bit of a nightmare. You'd get it sort of like tuned up, and then when the audience came in and the temperature of the building went up [laughter indicating that it went out of tune].[17]

But Van Koevering, like Emerson, was determined to use the modular Moog live, and once he did he found that his audience loved its sound. He started to think about new venues where he could perform live with his Moog.

One of Van Koevering's Florida business acquaintances was Glen Bell, the founder of Taco Bell. In the early 1970s Taco Bell was expanding into the southeastern US. Tacos were somewhat of a novelty there, and Bell was on the lookout for new ways to sell them. His meeting with Van Koevering was productive for both men. They struck up a partnership whereby Van Koevering agreed to hand out vouchers during his school shows for kids to obtain free meals at their local Taco Bell. Once there, the children and parents were entertained by Van Koevering playing pop tunes on his Moog synthesizer. The audiences loved the shows, and Van Koevering was paid handsomely. For a couple of years he regularly performed in Taco Bell restaurants throughout the Southeast.

The Island of Electronicus

Van Koevering was well aware of the effect of the Moog on his audience; he was on a personal mission to try and realize the potential of the new electronic sounds. He developed plans for a much more ambitious project than making music to eat tacos by and set out to create a special venue where electronic music could be listened to and experienced. It was on an island off the coast of Florida that Van Koevering's dream of an "Island of Electronicus" came to him:

. . . something happened to me. I saw a room. I saw this huge ceiling, I saw this glass dome at the top, and I realized that the room existed, that I'd seen it in a newspaper story about an island off from St. Petersburg, Florida, called Tierra Verde. It had been built by the Deltona Corporation and it had originally been planned to be a gambling casino . . . this room with these huge, big wooden arches. This thing was about 90 feet across, or 100 feet across, and it had a bowl-shaped floor with ridges in that floor, and it was all glass, on a man-made island. . . . I went to the man that leased that. . . . Glen Bell, from the Taco Bell restaurants. And I told Glen, I want that room, I'm going to do an "Island of Electronicus" at that room.

Van Koevering leased the island, equipped a "Moog Rotunda" with a big sound system and a light show, and had the words "Love and Peace" emblazoned over a "Happening Stage." The arena was furbished with pillows for the audience to lie on. Van Koevering advertised the venue on local TV and radio by showing Atlantis sinking into the ocean and out of it rising the Island of Electronicus. He was an able publicist. It became clear to Van Koevering that much of his audience were young people, and he soon directed his commercials advertising the island specifically to teenagers:

. . . the place filled up. Three dollars a head per show . . . three shows per night . . . and they're lined up, the parking lot's full, and they're lined up all the way around the building.

TP: What sort of people were coming in?

DV: Teenagers, kids, young kids, senior adults. Lenny Denny [a famous American Hammond organ player] showed up and said "My God, are you going to ever serve liquor in this place?" Because he had the Hammond organ show going on over on Treasure Island. I said "No, I'm not going to go for liquor, I can't get teenagers in here, kids in here. . . ." And he said "Well, God, my future's secure, I guess I'll stay in business." He said, "If you ever serve liquor or do a dinner in here, I'm out of business!"

It was just as the Island of Electronicus was up and running that the very first batch of Minimoogs came off the Trumansburg production line. Van Koevering bought six of the first ones ever produced to use on the island. Besides playing pop tunes, he also made soundscapes:

We do the first show—I had two or three musicians working with me—we took these Minimoogs and we layered all this Moog synchronized background stuff from all these patches that I'd learned from all these early synthesists, and we had a killer show, lights and automata. And we'd start a motorcycle up—you'd hear a Minimoog sound look a motorcycle, you'd hear 'em kick it over [imitates motorcycle sound], and then we'd take noise . . . and you'd hear 'em choke it—you could cause that sucker to make it shift gears, and you could hear that filter screech and like a wheel would chirp. And you'd get his motorcycle going around the room [imitates sound], with the Doppler effect . . . now, we did this with two Minimoogs—a four-cylinder sports car would start its engine. And you heard four cylinders running, you had four oscillators going . . . that engine of a sports car, would go the other way around the room. And you'd hear the motorcycle going one way and you'd hear the sports car go the other way, and a horrendous crash would happen . . . over the stage and parts were rolling all over the room. And the audience would go nuts. They'd stand and they'd cheer and they'd clap, and it was an awesome, awesome sonic sound effect event. A sonic picture, we'd painted a picture with synthesizers. . . .

Van Koevering did not abandon his evangelistic leanings. On Sundays he ran a special show for the local churches. "It was a hit," he recalled. "We did three shows a night, six nights a week, Sunday night off. Sunday night we'd rent the whole place to churches to come in and we did a similar kind of a show for the churches in the area."

During the day, Van Koevering kept three local musicians busy making jingles on the synthesizers. He sold these jingles to local TV and radio stations.

Van Koevering's experience with the Island of Electronicus was invaluable. It enabled him to experiment with the new instrument, the Minimoog, in live performance. It also showed him that there was a significant audience for the new sounds among young people. Furthermore, by allowing members of the audience to jam along with the show, it further showed the potential of the Minimoog as a performance instrument with which young musicians could express themselves:

Rock groups would come and sit in the pit. . . . The stage was tiered and down in the pit there was a place you could sit that had a Minimoog in the pit, and you could get into that seat with the Minimoog and you could put headsets on and you could play your Minimoog along with the show, just ad lib along with the show, jam with us. And I had an earplug up on the stage, and I could stick it in my ear, and with a switch I could listen to this Minimoog and *this* Minimoog, and these kids were good, some of these kids were great, and I could patch them to the sound system, and when I patched them to the sound system a spotlight would come on over the kid, over that pit, and you could hear the kid play along with us. . . . They had to stand in line to get to do that.

With Minimoogs now coming off the production line, Van Koevering came up with an even more ambitious plan. He would go on the road selling Minimoogs:

Bob Moog called me and said, I now have—and I forget the number—eight more synthesizers for you, Minimoogs, and he shipped them to me. And they showed up, and the day they showed up I stood up on the stage at the end of the third show . . . and I made the announcement, "You have just experienced the last Island of Electronicus show." And I pulled my bow tie off, because I did all this in a tux. . . . And somebody from the audience says, "Well, what are you going to do?" And I said, "I'm taking the Minimoogs on the road and I'm going to establish a dealer network." And my wife thought I was absolutely insane.

On the Road

Once Van Koevering had chosen to sell Minimoogs, the problem he faced was not the problem that faced the average salesman who needed to sell a product. His problem was much, much bigger: There was no known market for this instrument, and synthesizers had never before been sold to music dealers. Van Koevering boasted of having been thrown out of many music stores: "I've been thrown out of more music stores than any man alive, because before you would condition the market there was no market, and I had to invent the market."

In order to invent the market, Van Koevering first needed a base. With a partner (Les Trubey, the owner of Central Music, whose main music store was in St. Petersburg, Florida), he formed VAKO (after VAn KOevering) Synthesizers. The plan was that Van Koevering, while out on the road, would find music stores to stock Minimoogs, Trubey would then ship them the synthesizers by United Parcel Service, having earlier bought them directly from the Moog factory.

As Van Koevering set out on the road, he was confident: "I owned a brand-new Cadillac, told my wife, I'll be back when the Minimoogs are gone, and I piled all the Minimoogs in the trunk in the back seat of this Cadillac and I'd go to the first city." That first stop was Gainesville, Florida, which had a strong music store (Lipham Music) and a college. Van Koevering recalled:

So I go to Gainesville. . . . I go to Buster Lipham, tell him the story of what Les Trubey's doing, and he laughs at me. He says, "You want me to sell that thing? Show me how to do a violin, show me how to do a flute." Well, the flute was easy, that was a sine wave . . . and there had never been a store in the world that had sold a synthesizer. And he literally laughed at me. And he said, "If you can prove to me that musicians will do this, you come back and do a clinic [a demonstration], and I'll sell them." And I thought, "Man, that's a lot of work. I didn't know what hardly a clinic was." . . . I could do anything with the synthesizer, I could do any sound . . . because I performed with the thing . . . but no one else could, the musician couldn't, the store guy couldn't.

Buster Lipham, on being confronted with what looked like a very complicated instrument, needed to be convinced that keyboardists would be able to play it. Then Van Koevering had a spot of luck. One of the people in the store as he demonstrated the Minimoog was a rock musician, Bob Turner:

Bob saw me demo that thing. And he said, "I'd love to play that in my rock band." And I shoved it across the table at him, I said, "Well, then, do it." And he said, "I can't afford to buy it." I said, "Who said buy? I said do! You put this in your rock band and you start playing this thing in Gainesville. . . . Tell people you got it from Van Koevering, and I'm going to bring that guy, whoever that guy is that wants it, back to Buster Lipham and we're going to get Buster Lipham to see there's a potential for this." So I got musicians psyched up about the instrument.

Bob Turner was soon recruited by Van Koevering to be his first sales rep. Van Koevering realized that music stores were unlikely to take this new instrument, with its 44 dials and knobs—it was just too unfamiliar. The strategy he hit upon was to go directly to the musicians, persuade them to buy the Minimoog, and then take them back to the store. Then, having seen that customers existed, the store might be persuaded to stock the instrument. He concentrated almost exclusively on rock musicians to begin with; later he would widen his net to include all sorts of gigging musicians.

Using the Minimoog

Van Koevering quickly realized that he would have to teach the musicians he met how to use the new instrument. "So," he recalls, "I came up with the five things that you had to know about the Minimoog. You had to learn about the oscillators, number one, and the mixing, number two, and the filters and the timing of the filter and the timing of the amplifier—those were the five basic things they had to learn."

The Minimoog, although much simpler than the modular system and with a smaller range of parameters to adjust, is still a complex instrument. Just changing the setting of one potentiometer by a tiny amount can produce a totally different sonic effect. By working out how to teach musicians to play the instrument, Van Koevering was in effect writing the instruction manual for the Moog company. Van Koevering also came up with an ingenious idea to enable musicians to get their favorite sounds:

. . . I carried rolls of tape in my pockets, colored tape, mystic tape, with a scissors, and I would create a sound that he liked, and we'd put red slivers [shaped in a V] on all the 44 knobs and switch positions that meant all to red is sound red, and all the yellows are sound yellow, and all the blues are sound blue, and all the whites and all the green and—and those were the pre-sets. "On the way from one pre-set to the other, anything you find that's musical, play it, experiment, create a song for the sound, create a mood with the Hammond or the Fender Rhodes to accompany this melody line." And I was teaching them synthesis, and they'd get it.

The technique Van Koevering employed here to teach the musicians was the same one his father had used to teach him how to play novelty instruments—often the instruments were color coded with tape and all young David had to do was pick out the chords with the correct color.

The difficulty of repeating sounds with the Minimoog was legendary. According to Keith Emerson, Rick Wakeman (keyboardist of the British progressive rock group Yes) used to solve the problem by using thirteen different Minimoogs, each with its knobs super-glued into position to make a different sound! Again Van Koevering's efforts were to have a lasting impact on the industry. Eventually the Moog company and other manufacturers of portable keyboards, such as ARP, developed cardboard sound charts that could be placed over the controls, enabling all 44 controls to be set correctly to make the one sound corresponding to the cardboard chart.

The Hard Sell

Most of the rock musicians Van Koevering met did not have the money for one Minimoog, never mind thirteen. Although much cheaper than a modular synthesizer, the Minimoog still cost as much as a group's van. How could these young amateur and semi-professional rock musicians afford such an expensive instrument? Van Koevering found a way of solving this problem too, and in effect he created a financial technology to help make the instrument available:

The financial thing, I worked out a formula that's really funny. I'd go in when they had three sets, and I'd only work with a rock band that had a contract with a club. If they didn't have a contract, they had a problem, I couldn't waste my time with them. . . . They could borrow money through the signature of their girlfriend's mother, or they could get a loan on their girlfriend's car—they never had a car [themselves], they worked in a rock bus, or they worked in a rock van, they didn't own a car that was financeable. . . . And you had to figure out the guy that had the potential of borrowing the money.

Having carefully selected his prospect, Van Koevering would spend the afternoon with the musician helping him (all the musicians he dealt with were men) prepare for his set that night by teaching him new sounds, marking the new sounds on the instrument with colored tape, and so on.

Part of Van Koevering's pitch was to appeal to the impact the musician would make as a rock soloist:

. . . the rock guitarist was the superstar, and the Minimoog could make, because of its sonic energy, it could make the keyboard guy a superstar—he now had something powerful. His Hammond was powerful, and his Fender Rhodes was powerful, but not like the guitar. But a monophonic, piercing electronic sound coming out of four or five . . . amplified speaker stacks, could give him some energy and he could compete with the guitar, and he wanted to do that.

After teaching the musician a few basic sounds, Van Koevering would lend him the synthesizer to use in the first set. He prepared the instrument by displaying the word "MOOG" in large mailbox sticky letters on the back of the walnut cabinet, facing the audience. The only condition he placed on the musician was that he had to ask the audience three times during that first set whether they liked the sound of the Moog. Van Koevering takes up the story:

. . . I'd stand next to the manager, and I'd say to the manager every time the kid would say "How do you like the sound of the Moog?" and the crowd would go nuts, "Yeah, whooee!" . . . I'd say, "This kid needs this thing," and I'd reinforce that the manager has got to know that the kid needs this thing.

At this moment, Van Koevering employed a clever strategy: he left the club:

First set's over, I pull the slow-blow fuse and I split. They couldn't start that thing up if they wanted to. . . . I've even had kids try to short the suckers out while I was gone. Because while he's gone, he's in the second set, now, the audience has been stimulated, they've been told this is the Moog and they, "Do you like it?" and they've said "yes" and a request would come up from the floor, "Play that thing." Now, the manager didn't know that I'd split, and the manager would say to the kid, "Play the Moog." He'd say, "I can't play the Moog." What I'm doing is I'm conditioning the manager to know that this is meaningful for his attendees, this is for his participants, for his customers, they like this—and without exception, they would.

Many sales strategies work in this way, by creating demand.[18] Having created a demand for the instrument, Van Koevering would return just after the second set:

. . . after the . . . set I'd come back, and the kid's mad, he's been bawled out by his boss, the boss is upset with me, "Why the hell did you leave, you're not supposed to leave, you get this kid all going and. . . ." I had another show I had to go do at another club. I'd go and often sit in the car and read a book. And then it would go on that the next day, or that night, I would say to the kid, "You ought to have this." "Oh, I know, I've got to have this." "Well, I can't do it again, this is it. You get your loan together, and here's how you could do this. You go to your girlfriend, you go to her mother, and you get a loan from a signature at a loan company, you call me, I'm staying at this hotel, you do that today—it's 1 in the morning, 2 in the morning—you stay up and you do this before you go to bed, you go down and you talk to this woman, this girlfriend, you find your father, you find this guy who can loan you the money, you talk to your manager, you get an advance against your gig, and you go to a loan company." And I would have been to the loan company before and I would have had the loan papers and I got them all for the person to sign and they'd go in and I'd get three or four of these kids doing this in a city. There's hundreds of cities that I've done this in. I did this until we had a Moog network selling Moog synthesizers coast to coast. I did this all over the country.

Such high-pressure sales tactics entailed risk, and once or twice Van Koevering ran into trouble:

. . . there was resistance to that, and there were times managers got upset with me. There's horror stories that could be told where a manager would get very mad in a club in Underground Atlanta and was willing to beat me up for taking advantage of, of tricking this kid into doing the first set and not doing the second so I could sell an instrument. And those kind of things were very painful.

Anyone who has met David Van Koevering will quickly realize that he is the sort of guy who can sell anything. He has, without doubt, the gift of gab and can gift wrap any object in such a compelling verbal spiel that it starts to look special.[19] On one famous occasion, later in his life, he took a whole bunch of Memorymoog synthesizers that the Moog company could not sell, repackaged them as "Sanctuary Synthesizers," and sold them all to American churches. Salesman or visionary, David Van Koevering was extraordinarily effective at what he did. The introduction of consumer appliances into America was often accompanied by "hard sell" routines.[20] Such methods were typically used for a new product being sold to an entirely new group of users. Van Koevering was simply following in the tradition of such itinerant salesmen. From his perspective, he was not only selling the young musicians instruments but also giving them the means to release their musical energy: "Bob Moog has respectfully called me a great salesman, and I suppose that that's a correct term, but there's a passion that I carry for creativity and it's my job to unlock that in that kid. . . . I take it very personally. . . . And if I can do that I can change that kid's life."

ARP Sales

David Van Koevering was relentless as he traveled from city to city. Like all great salesmen, he refused to take "no" for an answer, and would not leave a city until he had sold at least one Minimoog. Not only was he persuading musicians to buy the instrument; he was also persuading dealers to stock the instruments and recruiting sales representatives. He was, in short, inventing a market. And he soon realized he was not alone. Other companies were also starting to sell portable synthesizers. In particular, the ARP Company of Lexington, Massachusetts entered the portable synthesizer market at just about this time with its ARP Odyssey and ARP 2600 synthesizers. In fact, Van Koevering claims the ARP salesmen were following him around. He often found himself giving clinics for these other instruments as well. Van Koevering's commitment to the nascent industry was such that he saw it as his duty to help dealers in any way that he could:

... I remember hundreds of times, literally, that I would have to do a clinic on synthesis and include in the clinic how to sell the ARP, and demonstrate it, and the Korg, and demonstrate it, because I knew that if that dealer had synthesizers that he couldn't sell he'd throw the Moog out. And I knew that the success of the Moog was dependent upon the industry forming. . . . We weren't competitors, we were building an industry. But it wasn't the size of my ship that I was concerned about, it was the depth of the harbor.

There was a camaraderie among these early pioneering salesmen. ARP too soon realized the importance of appealing to the new market of rock and pop musicians. ARP's founder, Alan Pearlman, was an engineer from a previous generation, but ARP had someone who could better connect with young people. David Friend, a Princeton graduate student, joined the company at age 21. It was Friend, along with a musician named Roger Powell, who first went on the road for the ARP company to establish their dealer network. Powell, a studio engineer and a talented synthesist, had arrived at ARP one day from Atlanta. Powell later recalled: "David and I traveled all over the place in a red Chevy van. We tried to sell the 2600 [one of ARP's first portable synthesizers] in hi-fi outlets as well as music stores. We got thrown out of most of them. . . . The turning point was when [the music retail store] Sam Ash decided to take the 2600, late in 1971. They were really forward looking. They gave us credibility among retailers, and exposed our instruments to all the musicians in New York."[21]

Once the ARP 2600 was picked up by rock musicians at New York's most famous music store, the world of rock and the world of synthesizer sales began to reinforce one another. Pete Townshend of the Who purchased an ARP and used it on the album *Who's Next*. The same synergy between rock stars and sales was to help propel the Minimoog into the stores.

Van Koevering realized the importance of letting store managers and musicians know that they were not alone, that the synthesizer as an instrument was being widely taken up to make music. He would ply managers and musicians with records on which the synthesizer was featured: ". . . I collected all the albums, which I already had anyhow, for my own experience, they were mine, and I compiled an unauthorized, illegal set of copies and had it copied, mass produced. And I could leave these with dealers, let them know that this is legitimate stuff, this is an industry that's being born."

Lucky Man

The uptake of the Moog by high-profile rock keyboardists such as Keith Emerson and Rick Wakeman further confirmed the importance of the instrument for rock soloing. In the early 1970s, Emerson, Lake and Palmer were one of London's leading progressive rock groups. Keith Emerson incorporated a Moog synthesizer in his flamboyant stage act and had a huge hit with his first Moog record, "Lucky Man." It was the distinctive Moog solo at the end of "Lucky Man" that young rock groups wanted to recreate on their Minimoogs. Emerson had seen the potential of the Moog earlier when he had borrowed one to use in a concert with the London Philharmonic Orchestra. It was clear to him that the Moog had enormous potential as a live instrument: "We had a sold-out concert, the orchestra was sitting there. . . . Everybody in the audience was absolutely staggered, they didn't really realize what this was all about. It was like making extraordinary sounds. Just from that one concert I realized the potential that this instrument had."

While Emerson, Lake and Palmer and Yes were touring the United States, van Koevering charted the orders for Minimoogs that came in after their concert appearances:

. . . "Lucky Man" shows up, and then there's a Keith Emerson tour, and then Rick Wakeman, he's got two of them laying on Melotrons. And I mean, we knew where these artists were by the cities that were calling . . . because when the guy did a show they'd go to the music store to find one, and the music stores had to get a Moog—so we knew where Keith was by the phone calls, because we knew he was in Boston because you got 30 phone calls from Boston. We knew he's over in Wilmington, or he's in New York City or he's in Chicago, or whatever. . . .

Images of Keith Emerson were used extensively to advertise the Minimoog, and Emerson soon struck up a close relationship with Bob Moog (Pinch and Trocco 2002). The Minimoog soundcharts included instructions as to how to obtain specific Keith Emerson and Rick Wakeman sounds. By this point the Moog Company had realized the importance of rock musicians as a new sort of user.

Figure 3
A Minimoog advertisement featuring Keith Emerson.

As the success of the Minimoog spread, Van Koevering started to broaden the sorts of musicians he approached. There were plenty of working musicians who could use the instrument. Furthermore, these musicians often had a regular income and could afford the instrument more easily:

The simple thing that we found, there were musicians of every type, not just rock, that bought Minimoogs. We had country musicians that used them very efficiently, very effectively. The big bulk of them was rock and roll, the garage bands, the local bands, the wannabes, the regional bands that were working regularly and making a living at their work. Only the regionals made livings. Everybody else carried an extra job and went in debt. Rock and roll didn't pay the bills. I think that was a part of it.

Van Koevering's success selling Minimoogs did not go unnoticed by the Moog company. Jim Scott recalled: "All of a sudden we'd get an order from Van Koevering for a hundred or so of these things, whoa!"[22] But Bob Moog was facing mounting financial problems. In 1971, rather than go bankrupt, he sold the company to what we would today call a venture capitalist.[23] Bill Waytena paid nothing for the business but subsumed Moog's $250,000 of debts.[24]

It was Waytena who insisted that Moog start attending NAMM shows. Moog: "As crass and as unmusical as he was, he probably, because he was so far out of the mainstream of all this, he was probably able to look at it and see where it was going in two or three years. . . . And what he saw was that in our company in 1971 it was at the cusp, going from one thing into another, and that in a couple of years we were going to be part of the musical instrument business, but we weren't yet."[25] Before Waytena's involvement, Moog had gravitated toward his fellow audio engineers at the biannual Audio Engineering Society meetings, where there was always an exhibition hall where he could demonstrate his new products. But if the synthesizer was to become a musical instrument rather than a specialized piece of audio hardware, this meant reaching music stores and dealers. And the most effective way to reach such people was through NAMM. Today synthesizer companies, including Roland, Yamaha, and Korg, have some of the biggest displays at NAMM shows. But back in 1971 synthesizer manufacturers had never attended such shows. The Minimoog was first demonstrated at NAMM in Chicago in June 1971, but it did not make much of an impact. Van Koevering recalled: ". . .we had a terrible time at the first few NAMMs, because dealers didn't get it. . . . They didn't know how to demonstrate it they couldn't sell it. And nobody was coming in in those early years asking for the Moog until the rock stars started performing. When that happened then it took off. But it all happened simultaneously."

Part of the difficulty the Moog company faced was that musical instrument trade shows were still dominated by organ companies, which displayed their latest products with slick demonstrations. It was ARP that first developed the musical demonstration that was necessary in order for synthesizers to have a similar impact. While Bob Moog at his first NAMM show in 1971 felt like "a fish out of water,"[26] ARP was comfortably swimming around in the new forum. Bob Moog was impressed by ARP's slick demonstration: "ARP was down the aisle from us and they were really hip because they had actually worked out a demonstration. You know, they *played music* on that, cornball, everyday, pop music."[27]

Before Moog was taken over, VAKO Synthesizers had always operated as an independent entity. Waytena's plan from the outset, however, was to bring Van Koevering into the fold. Van Koevering, as always, was out on the road when the transfer of ownership occurred:

He finds me on the road in a music store, and he says, "You're going to meet me in Atlanta tomorrow." And I was someplace a long ways from Atlanta, within driving distance, and I said, "I have no plans to be in Atlanta tomorrow." He said, "If VAKO Synthesizer ever gets another synthesizer you're going to meet me at the airport in Atlanta tomorrow—I just bought the Moog Company. . . ." So I went to Atlanta, and he told me, "You're going to become vice president of Moog Music." I said, "I don't think so." He said, "If you ever sell another synthesizer you will." And I knew that I couldn't argue with him, he had deep financial resources.

Van Koevering had never needed a formal contract with Bob Moog. They had become friends, and anytime Van Koevering needed synthesizers Moog was eager to supply them. He now realized that Waytena held the upper hand. "So I said to Bill Waytena, 'Okay, what do you want me to do?' He said, 'I want you to come to New York and you're going to give up your network and you're going to sell synthesizers to the world.' And he appealed to that part of me that had this mission. I wanted to go to the world."

Thus it was that David Van Koevering, former novelty instrument showman, became a vice-president of the best-known name in the synthesizer business. It truly was meteoric rise in status for Van Koevering, and it was too good an opportunity to miss. As head of Sales and Marketing, he was now in a position to recruit a sales force and repeat on a larger canvas what he had already carried out. He was going to sell synthesizers to the world.

Going to the World

As vice-president of Moog, Van Koevering was in a position to reorganize their sales force and to start the campaign to sell the synthesizer internationally. He immediately put into effect what he had learned on the road, always emphasizing the importance of having dealers and distributors who knew the potential of the instrument and who could show new musicians how to play it. A manual and sound charts soon followed, now made officially by Moog. Van Koevering visited the Frankfurt Music Messe, where ARP also was trying to establish its own dealer networks in Europe:

The ARP company . . . they were at Frankfurt. And of course they were watching what we were doing. . . . And the first night . . . David Friend and Alan Robert Pearlman came over and said, "How many did you sell today?" And I said, "I haven't sold anything." I was interviewing, I'm talking to dealers and I'm talking to distributors and I'm figuring out who can do the job because I truly knew what the job was, and I'm asking for the dealers that had the sensitivities that could train the people to go do what I did in the clubs, and do they have the heart for that—there was a whole series of questions, I'm taking notes. They said, "Man, we sold four dealers today, and we've got one distributor who's going to sell in Spain." The second day they came by and said, "What did you do today?" I said, "I didn't sell anything." They said, "We sold two today." Third day, same answer. Fourth day. Fifth day.[28]

While ARP was setting up dealers, Van Koevering was biding his time. He did not want just any dealers; he wanted dealers who were fully committed to the products and were prepared to sell them in the way that he knew worked:

They had to come over here. They had to take a tour of the factory, they had to see how a Minimoog was made, they had to learn how to adjust the Minimoog on the inside for calibration reasons, how to put circuit boards in them, how to fix them in the field, and go out on the road with me. Now we're building a network over here, now we've got road men over here traveling—the same things that I did in Florida.

Van Koevering demanded a lot of his dealers, but by the last day he had lines of them waiting outside his cubicle:

There must have been 45 or 50 distributors standing in line to find out, after they had given me their information, if they were going to be granted the right to sell the Moog synthesizer. And I sat there, and we sold several hundred synthesizers, and we locked in the distributors for England and we locked in the right distributors for Germany, and we did Italy, and we did Switzerland and we did Greece and we did all of the European nations, the Scandinavian countries. . . . We poured that information into 28 distributors—and Moog took off in Europe.

The list of distributors for Moog performance synthesizers which Van Koevering put together included distributors in Canada, Denmark, Norway, England, Finland, France, Germany, Greece, Hong Kong, Italy, the Netherlands, Portugal, Spain, Sweden, and Japan. In all, 685 different dealers made up this international network. Van Koevering truly had found a way of taking his vision to the world. The stepchild of the modular Moog synthesizer that had first emerged from the Trumansburg attic only 3 years earlier was now on the shelves in the best music stores everywhere.

At this point we leave the story of the Minimoog, the first portable keyboard synthesizer to be sold in music stores and the first synthesizer to be sold in large numbers to rock musicians. The Minimoog paved the way for the massive success of the synthesizer industry in the 1980s when portable synthesizers like the Yamaha DX7 became huge sellers and when every pop band had to have a synthesizer. The rapid growth of the synthesizer industry is evident from the differing sales numbers. Whereas 12,000 Minimoogs were sold in the lifetime of the instrument (12 years), 200,000 DX7s were sold in 3 years and Casio sold 15 million instruments in total between 1980 and 1990.[29]

The Vision and the Mission

David Van Koevering was indeed a man with a vision and a mission. For Van Koevering himself the two things were linked. It is easy to dismiss his "visions" as a salesman's bluster or as hocus-pocus. But this is to miss the point as to how such charismatic salesmen operate. All of Van Koevering's many projects were accompanied by this sort of visionary rhetoric. Whether selling God, selling synthesizers, or selling synthesizers for God, he still had to persuade a group of people to commit to something. Building commitment to a product is something that all effective sales-people do whatever their product, and having a vision that can be shared is a compelling way of building such commitment.[30] To go out and create an entirely new market requires not only the right strategy but also extraordinary energy and dedication. And Van Koevering had the latter in abundance. His vision for the synthesizer as a performance instrument drove him and obsessed him and made all his driving seem worthwhile. The vision, in effect, also served as a scenario, a way of planning the road ahead, a way to assess where he was going and what he had yet to achieve.[31]

His mission developed as his circumstances changed.[32] I doubt that he planned it all out in advance; he responded to the situation he faced and was prepared to make changes, uprooting his family if necessary. His introduction of the Moog into Taco Bell, which sounds now like a dead end, was actually a good way to find out how ordinary people reacted to the Moog, and the Hammond organ was a good model here—it had been for years used for such light entertainment. The Island of Electronicus project was a remarkable way to bring to fruition his vision of the sonic power of the Moog. That project was a bit like running a test laboratory for the new instrument. Big companies that make consumer appliances have in-house test laboratories.[33] Van Koevering built his laboratory out there in the world. It gave him a chance to interact with the new users, especially rock musicians, and to learn from them what their requirements were. It gave him a chance to see the sales potential of the Minimoog among this new group of users.

Van Koevering's last change of direction—from the lab back out into the world—was the most extraordinary of all. He took the lessons from his laboratory and applied them in the real world with great success.[34] In the process, he managed to persuade numerous musicians, mostly in the field of rock and mostly amateur or semi-professional, to buy Minimoogs. To do this he had to devise de novo the social and technical practices to enable this instrument to be sold. The gift of gab was not enough. As we have seen, his sales practices evolved into a combination of material practices (labeling the instruments) and interactional techniques (which he used to close the deal). He also had to have in place the financial apparatus to enable these young musicians to purchase their instruments (loan agreements). He also didn't want just individual sales, he wanted to establish a market. And to do this he had to establish a dealer network, find a way to instill product loyalty, and figure out how to sell the synthesizer so that others could repeat his success. He had, in effect, to identify a new group of users and recruit them to take up the instrument. In short, his own claims of inventing the synthesizer market are not that farfetched. And, crucially, Van Koevering was helped in all these activities by the fact that he was himself a user.

Sellers Are the Missing Masses

Perhaps the key idea of this chapter is that selling as an activity provides us with a new research site to study users. Whether through their interactions with users or by moving from use to sales (and here it would be interesting to study the recruitment of sales forces for selling new technologies—how many, like Van Koevering are actually themselves users?), it is sellers who tie the world of use to the world of design and manufacture. Sellers are "boundary shifters" (Pinch and Trocco 2002). They are the true "missing masses" of technology studies.[35]

Synthesizers would have remained instruments for elite rock musicians and composers if not for the efforts of those who developed the skills, practices, and expertise to sell them. That Moog is the name we remember is due in part to the fact that Van Koevering made it the name to remember by putting mailbox letters on the backs of the Minimoogs that were played in clubs.

Finally, it is important to realize that not only a new industry and a new sort of user but also a new part of our sound culture was being established. Emily Thompson (2002) has recently shown how new soundscapes were built in the earlier part of the twentieth century from new acoustical practices, the new science of acoustics, new architectural practices, and new technologies. In this chapter I have documented some of the material and social practices that brought into being not only a new set of users but also a new soundscape of electronic sounds—sounds which had not been heard before and which became increasingly recognizable and identifiable as the synthesizer revolution unfolded. I leave the last word to David Van Koevering:

The sound was in the culture. They heard it on radio, and they heard it on television, not just as the musical sound, but the sound of the synthesizer.

Acknowledgments

I am grateful to Knut Sørenson and Ronald Kline for helpful comments on an earlier draft. I thank Robert Moog and David Van Koevering for permission to reprint images. I acknowledge the Harvard University Press for permission to reprint material contained in Trevor Pinch and Frank Trocco, *Analog Days: The Invention and Impact of the Moog Synthesizer* (Harvard University Press, 2002).

Notes

Introduction

1. To be sure, the literature on the relationships between users and technology is much more extensive than we discuss in this introduction. We don't intend to present a complete overview of research in this area. We restrict our discussion to the approaches that are central for the authors of these chapters. Research traditions that are not covered here include innovation studies, the design literature, and psychological research on the role of users in scientific research. (Examples of innovation studies: Von Hippel 1976, 1988; Coombs et al. 2001.)

2. The relationship between technologies and male gender identities has been, and still is, a major topic among feminist historians and sociologists of technology who have shown the strong alignments of technology and masculinity, particularly in the field of engineering (Berg 1996; Cockburn 1985; Faulkner 2000; Lie 1995; van Oost 2000; Oldenziel 1999; Wajcman 1991).

3. Feminists have not only sensitized historians to include users in the history of technology; they have also made the case for including different technologies in historical studies, such as household technologies and health technologies (Wajcman 1991). For a more detailed analysis of the ways in which feminists have shaped research on users, see Oldenziel 2001.

4. See e.g. Clarke and Olesen 1998; Franklin and Ragone 1998; Ginsburg and Rapp 1995; Hartouni 1991; Kirejczyk 1993; Lock and Kaufert 1998; Martin 1989 and 1994; Morgan and Michaels 1999; Petchesky 1987; Petchesky and Judd 1998; Saetnan et al. 2000.

5. See e.g. Jacobus et al. 1990; Oudshoorn 1994; van Kammen 1999.

6. Exemplary studies: Turkle 1984; Wright et al. 1987; Rasmussen and Hapnes 1991.

7. See e.g. Bereano et al. 1985; Cowan 1976, 1983.

8. Exemplary studies: Cockburn 1985; Faulkner and Arnold 1985; Kramarae 1988.

9. Friedman (1989: 184, 185) has introduced a typology of users of computer systems that includes six different types: patrons (the initiators of the technology),

clients (for whom the system is intended and designed), design inter-actors (who are involved in the design process), end users (who operate the system), maintenance or enhancement inter-actors (those involved in the further evolution of the technology), and secondary users (individuals who are displaced, de-skilled or otherwise affected). For a discussion of taxonomies of users introduced by other scholars, see Mackay et al. 2000.

10. The concept of the implicated actor thus resembles Friedman's category of secondary users.

11. See e.g. Lohan 2000. Similar criticism has been articulated in STS studies (Mackay and Gillespie 1992).

12. Feminists not only pushed others to include users in the analysis, they also inspired STS to include the work of all traditionally invisible actors in network analyses. Susan Leigh Star (1991: 29) criticized actor-network approaches for erasing the work of actors such as secretaries, laboratory technicians, and wives of scientists and engineers. Star argued that a focus on this invisible work would reveal completely different networks involved in technological development.

13. Although the notion of implicated actor was initially restricted to humans, Adele Clarke (forthcoming) has recently extended the concept to include non-human actors.

14. For detailed discussions of the politics of Haraway's cyborg figure, see Prins 1995 and Moser 2000.

15. See e.g. Casper 1994; Cussins 1998; Downey and Dumit 1997; Gray 1995; Lock 2000.

16. Latour and Akrich (1992) went on to develop an extensive terminology to elaborate their "semiotics of machines."

17. Given the heterogeneity of users, designers will, consciously or unconsciously, privilege certain representations of users and use over others. Studies of the development of information technologies indicate that design practices are dominated by the so-called I-methodology, in which innovators consider their own preferences and skills as representative of the future user (Akrich 1995; Oudshoorn et al., forthcoming; Rommes et al. 1999). Because of the underrepresentation of women in this domain, technological objects become attuned to the interests and skills of young middle-class men, rather than women or other groups that are underrepresented in the world of technology.

18. For an analysis of subject networks that focuses on other domains (an analysis of the attachment of music amateurs and drug users), see Gomart and Hennion 1999.

19. These studies build on feminist-inspired research on how subjectivities and bodies are constructed and performed in the context of biomedical technologies (Cussins 1998; Dugdale 1998; Singleton 1996).

20. Whereas technology studies include a wide variety of technologies, cultural and media studies are restricted to consumer technologies, including food, clothes, music, and the old and new media. According to Morley (1995: 296), "the very development of the field of media studies has been premised on an understanding of the centrality of the process of media consumption in contemporary social and cultural developments."

21. Bourdieu argued that the structure of consumption resulted in a reproduction of class relations. His argument also included a methodological claim: a focus on consumption provided a new way to study social relations in some objectified form—in his study, as a pattern of taste (Miller 1995: 267).

22. The specific relation of women to consumption does not imply, however, that they have been the principal agents in its development. For an overview and discussion of the literature on the involvement of women in consumer culture, see chapter 5 of Lury 1996.

23. For an overview of this literature, see Lerman et al. 1997. Inspired by feminist scholars, historians have extensively studied the history and culture of what is familiarly called consumer society, a concept introduced to identify the emergence of a specific type of market society: the Western capitalist system of exchange. Dutch historians of technology have written detailed accounts of the active role of intermediary organizations such as consumer groups in the emergence of the consumer society in the twentieth century—a role they describe in terms of a co-evolution of new products and new users (Schot and Albert de la Bruheze, this volume). For a discussion and an overview of the historical accounts of the birth and the development of a consumer society, see chapter 1 of Storey 1999.

24. On the work of the Frankfurt School, see Storey 1999: 18–23 and du Gay et al. 1997: 87.

25. Exemplary studies: Ewen 1976; Strasser 1990; Walker 1998.

26. On the different views of the relationship between consumption and identity, see chapter 8 of Lury 1996.

27. For an exemplary study of the symbolic work involved in appropriating consumer technologies, see McCracken 1988.

28. Silverstone (2000: 14) describes the restriction of the domestication approach to studies of the home as an arbitrary limitation.

29. For a similar criticism, see the chapter by Oudshoorn in this volume.

30. For a detailed criticism of the dominance of the linear model of innovation in STS, particularly the literature on the public understanding of science, see Sørenson et al. 2000. Similar criticism of the adoption of a linear model of technological innovation has been addressed to domestication approaches (Lie and Sørensen 1996: 11, 12). As design processes, domestication processes don't

reflect any linear progress, but include conflicts and contests in which practices and routines in the use of technology may be broken again which may result in a drastic re-domestication (Lie and Sørensen 1996: 11).

31. For such criticism, see Sørensen et al. 2000 and Mackay et al. 2000.

32. Marx (1980: 24), who described the overlap between production and consumption in the course of an analysis of mid-nineteenth-century capitalist production, suggested that "production is . . . at the same time consumption, and consumption is at the same time production." British cultural studies scholars, most notably Stuart Hall, have elaborated this view. For further discussions of the artificiality of the distinction between production and consumption, see Silverstone and Haddon 1996: 44 and Lie and Sørensen 1996: 9, 10. As Lie and Sørensen have suggested, this blurring of boundaries between design and use does not imply that production and consumption should be considered as identical processes, or that designers and users have the same roles in technological innovation. They argue that "we have to examine consumption and production to identify their characteristics," and they suggest that "this examination is hampered by an a priori dichotomization of the two processes" (ibid.: 10).

Chapter 1

1. A home television set could be used in place of the visual display unit. The programs and the data were both stored on audio cassette tapes.

2. The movement of the home computer from the garage to the basement, implied by its suggested use in food preparation, is similar to the initial advertisement for the Apple II, introduced at the same time as the TRS-80. The text of the Apple advertisement reads: "Clear the kitchen table. Bring in the color television. Plug in your new Apple II and connect to any standard cassette recorder/player. . . . Only Apple II makes it this easy." (*Byte*, August 1977) However, the Apple II was much more expensive than the TRS-80, and was claimed to be the first "all in one" computer that did not require any special skills to put together.

3. All this with 4 kilobytes of storage, Basic as the programming language, and hardly any available computer programs.

4. For a good introduction to studies on the masculine culture of technology, see Wajcman 1991.

Chapter 2

1. This article is based on material from Kline 2000. I use the terms "rural" and "urban" in accordance with the demographic meanings given to them by the US Census Bureau in this period. The Census Bureau called a community having fewer than 2,500 people "rural," one having more than this population "urban." For

comments on varying uses of the terms "urbanization" and "modernization," see Kline 2000: 6–8.

2. That was one of the main points of Kline and Pinch 1996.

3. My analysis of reciprocal social relations, mediators, and users is also indebted to Berger and Luckmann (1966), Cowan (1987), and Fischer (1992). Bijker (1995) has developed a different reciprocal methodology, the convergence model, in which technological frames link artifacts and social groups.

4. Akrich and Latour (1992: 261) make a similar point about antiprograms.

5. For complaints about playing music on the lines, see e.g. *Telephony*, September 1904: 258; November 1904: 460; October 1907: 260.

6. *Telephony*, December 1907: 385.

7. I use the analytical framework of gender structure, identity, and symbolism involved in social relations. See Scott 1986; Harding 1986, chapter 2; Lerman, Mohun, and Oldenziel 1997.

8. *Rural New Yorker*, December 20, 1902: 841; *Literary Digest*, October 17, 1914: 733; *Telephony*, July 22, 1911: 118.

9. *Telephony*, July 1905: 52; March 4, 1911: 294.

10. See e.g. *Telephony*, August 1907: 103; Deer Creek Co-operative Telephone Company, Constitution and Bylaws, adopted December 31, 1909, Claude Wickard Papers, Franklin Delano Roosevelt Library, Hyde Park, New York, Box 19; Fischer (1987a).

11. *Telephony*, January 1906: 67; April 10, 1909: 446; June 19, 1909: 707; September 11, 1909: 254; July 22, 1911: 118; September 16, 1911: 354; November 11, 1911: 604; December 2, 1911: 683; December 23, 1911: 791; November 16, 1912: 764–765.

12. *Wallaces Farmer*, January 21, 1910: 96.

13. *Outlook*, 86 (1907): 767–768; *Telephony*, April 10, 1909: 433; October 21, 1911: 500; *Literary Digest*, October 17, 1914: 733.

14. Interviews by Suzanne Moon with Eva Watson, February 21, 1995; George Woods, March 18, 1995; Lina Rossbach and Sonia De Frances, January 21, 1995, in the possession of the author.

15. *Telephony*, May 1903: 296–298; March 5, 1910: 286–287, on p. 286.

16. *Telephony*, April 1906: 280.

17. Angus Hibbard to John Sabin, April 25, 1903, AT&T Archives, Warren, New Jersey (hereinafter referred to as ATTA), Box 1342; Miller (1905: 433-434); *Telephony*, March 1905: 277; April 10, 1909: 433; March 25, 1911: 382; *Literary Digest*, October 17, 1914: 733 (reporting on a story in *Telephony*, October 3, 1914).

18. P. L. Spalding to Joseph Davis, November 5, 1903; Davis to Spalding, November 9, 1903; J. Fay, "Induction Coil for Farmers Lines," memo, November 9, 1903, ATTA, location 21-06-02-08.

19. C. E. Paxson to Hammond Hayes, July 3, 1905; Hayes to Paxson, July 6, 1905, ATTA, location 21-05-01-05.

20. John Carmody to Benton Rural Electric Association, August 12, 1938, John Carmody Papers, Franklin Delano Roosevelt Library, Hyde Park, New York (hereinafter referred to as JMC), Box 88.

21. *Rural Electrification News*, October 1935: 25.

22. Margaret Anderson, field report, January 28, 1941, records of the Rural Electrification Administration, National Archives, Washington (hereinafter referred to as REAA), Entry 86, Box 8.

23. William Nivison to John Carmody, November 5, 1937, JMC, Box 84.

24. *Rural Electrification News*, April 1939, special issue on safety.

25. William Nivison to John Carmody, November 5, 1937, JMC, Box 84.

26. See e.g. Mary Willis to John Carmody, November 4, 1937, JMC, Box 84; Robert Tisinger to Loren Jenks, October 5, 1938, REAA, Entry 86, Box 2; Warren Hamilton to J. W. Pyles, April 7, 1938, REAA, Entry 86, Box 11.

27. Elva Bohannan, field report, January 25–28, 1941, REAA, Entry 89, Box 2.

28. For a study that reaches similar conclusions for a region, see Adams 1993.

29. William Nivison to John Carmody, November 5, 1937, JMC, Box 84.

30. See e.g. US Department of Agriculture, Bureau of Agricultural Economics, "Farm Homes Use Wide Variety of Electrical Appliances," attached to E. C. Weitzell to Mr. Haggard, November 29, 1948, REAA, Entry 27, Box 3; US Department of Commerce, Bureau of the Census (1952): 213.

31. Interviews by Suzanne Moon with George Woods (March 18, 1995), Gerald Cornell (May 24, 1995), and Thena Whitehead (February 11, 1995), in possession of the author.

32. See e.g. E. M. Faught to George Munger, July 16, 1940, REAA, Entry 86, Box 21.

Chapter 3

1. I do not want to give the impression of being a neo-Heideggerian. I own several high-technology consumer goods and travel frequently by plane.

2. The Reclaim the Streets! movement is closely associated with the so-called anti-globalization protests which have taken place at various meetings to discuss free trade, such as at the meeting of the World Trade Organization in Seattle,

November-December 1999. One of the ironies, from the perspective of this chapter, is that these groups use the Internet extensively to organize actions and the use of the Internet by bodies such as the World Trade Organization itself becomes the target of such actions.

3. For a more extensive discussion of the use of autobiography in technology studies, see Henwood et al. 2001.

4. For a fuller discussion of the relationship of inequality to the use and production of the Internet, see Thomas and Wyatt 2000.

5. For a critique that highlights the absence of users in SCOT (social construction of technology) approaches, see Mackay and Gillespie 1992. Two good collections that address the position of users are Silverstone and Hirsch 1992 and Lie and Sørenson 1996.

6. Identifying non-users raises again the questions surrounding the dictum "Follow the actors"—questions raised by Martin and Scott (1992), Russell (1986), Bijsterveld (1991), and Star (1991). In particular how can invisible actors be identified?

7. The Internet Society maintains comprehensive links to sites with data on Internet use. See Internet Society 2001.

8. Source: Nua 2002a.

9. See Georgia Technical University 1999.

10. A literature search on various combinations of "Internet," "computers," "information technology," "technology," "rejection," "drop out," "non-use," "barriers," and "have-nots" yielded very little. "Barriers" yielded the most hits, but much of that was about national level adoption or education. "Drop outs" also yielded quite a few hits, including some interesting material about young people who dropped out of school or university as a result of spending too much time online.

11. Winner (2000) discusses a survey of more than 1,500 adults and 600 children conducted on behalf of National Public Radio, the Kaiser Family Foundation, and the Kennedy School of Government during November and December 1999. Winner does not provide the details, but he claims that "a small but not insignificant minority" do not have a computer nor any plans to acquire one. Three-fourths of this unspecified minority do not feel this as a lack.

12. "Multi-user domains" is a generic term for the huge variety of online, usually text-based role-playing games. Much early Internet research, of which Turkle's is perhaps the best known, focused on text-based applications such as these.

Chapter 4

1. This is done, e.g., in Brosveet and Sørensen's 2000 study of Norwegian society's domestication of ICT, where the concept is used on the macro level, and in

278 Notes to pages 82–111

Levold's 2001 study of domestication of an information technological research position, which uses the concept on an individual level.

2. Between 1934 and 1960 a license was needed to import and to sell cars for private use in Norway. The reason was mainly related to the balance of payment, as private cars were considered too luxurious to be freely imported (Østby 1994, 1995).

3. See the following policy papers: Samferdselsdepartementet 1996; Stortinget 1997; Stortinget 1998.

4. The public library also provides access, but is closed afternoons and evenings.

5. The råners described earlier are around 20 years old. Earlier research, as well as the experiences of my informants, reveals that the distinctions in interests and orientation in the youth culture starts in the upper primary school years (see Jørgensen 1994).

Chapter 5

1. A dramatic example of the kind of issue we have in mind was provided by the contaminated blood scandals of the early 1980s, when many people dependent on blood transfusions and products (including people with hemophilia) succumbed to HIV-related infections. Perhaps the trauma was greatest in France, precisely because of that country's public commitment to voluntary donation of blood as an expression of social solidarity. The anthropologist Paul Rabinow (1999: 71–72) has suggested that this showed the French that their system "could not operate on benevolently donated supplies of blood alone, nor was it as exempt from the laws of the international market as the citizenry had been led to believe."

2. Current US strategic missile defense initiatives notwithstanding.

3. Some defense technologies are apropos here, as are "orphan" drugs (which lack a large enough commercial market to be developed in the private sector).

4. For an excellent discussion of this and related issues, see Lewinsohn-Zamir 1998.

5. Law and Akrich (1994) discuss the analogous notion of configuring "good customers."

6. Problems do arise in conflict-ridden areas of the globe, which may lack a recognized, legitimate government.

7. "Early-onset group B streptococcal disease—United States, 1998–1999," *Morbidity and Mortality Weekly Report* 49 (2000), no. 35: 793–796; "Decreasing incidence of perinatal group B streptococcal disease—United States, 1993–1995." *Morbidity and Mortality Weekly Report* 46 (1997), no. 21: 473–477.

8. The data in this section come in part from interviews carried out by D. A. Rose at the National Institute of Allergy and Infectious Diseases in autumn 1997.

9. "Early-onset group B streptococcal disease—United States, 1998–1999," p. 793.

10. Source: www.cdc.gov/ncidod/dbmd/gbs.

11. "Early-onset group B streptococcal disease—United States, 1998–1999," p. 793.

12. The extent to which, or whether, this constitutes configuring by "the state" is a matter of some debate, as Brown (2001: 57) has pointed out. However, we are reasonably satisfied that scientific work conducted primarily in government laboratories is sufficient, for our purposes, to talk about states configuring users.

13. Interview with Ingrid Geesink in Bilthoven, The Netherlands, 1998. Quoted in Blume and Geesink 2000.

14. Interview with Stuart Blume in London, 1997. Quoted in Blume and Geesink 2000.

15. Source: www.cdc.gov. Under the auspices of the CDC, the Vaccines for Children program (VFC) provides funding for the purchase of vaccines by the government at discounted prices, to be provided free of charge to eligible children both in these public clinics as well as by private providers. For those children who do not qualify to receive free immunizations under this program, a long-standing program, known as section 317 or the Immunization Grant Program, offers an additional opportunity to obtain free vaccinations by providing states with supplementary funds to carry out immunization-related activities, including purchases (again at a substantially discounted rate). Outside of VFC and section 317, additional funds are appropriated to states in the form of Maternal and Child Health Block grants, Medicaid funding, and other smaller programs for the purchase of vaccines.

16. The Expanded Programme on Immunization (EPI), begun by the WHO in 1974, initially included six different vaccines for incorporation: measles, polio, BCG, diphtheria, pertussis, and tetanus (Wright 1995: 610). Other routine pediatric vaccines include mumps, rubella, hepatitis B, hemophilus influenza b, and varicella. These latter, however, have been only slowly, if at all incorporated into the schedules of poorer countries.

17. Source: www.cdc.gov.

18. Ibid.

19. Though these authors are referring to immunization practices in developing countries, we feel that, by and large, much the same can be said of more industrialized nations.

20. One should also consider what it means to be "informed." Information campaigns are typically initiated by public health establishments, public health clinics, health maintenance organizations and private physicians, and are heavily weighted toward the promotion of vaccination.

21. This seems to be an excellent example of the Foucauldian notion of "governmentality," in which various strategies are implemented by the state and its institutions as a way to "urge on individuals for their own benefit [and direct their] 'free will'" (Higgs 1998).

22. See also Dew 1999, which incorporates more explicit Foucauldian analysis than this earlier work.

23. The concept behind herd immunity is simple: if enough people receive enough vaccine in a given population, the opportunity for infectious disease agents to transmit and spread from one person to another decreases. With most of a population immunized, the disease-causing organism is confined to fewer and fewer hosts (thereby lessening the opportunity to spread and reproduce). Moreover, weakened pathogens excreted by persons already immunized find their way into water supplies and the surrounding environment. It is through this and other mechanisms that non-immunized persons become "passively" immunized, which also contributes to herd immunity. The much sought after outcome of all this is a drastic reduction in the incidence of the disease, or ideally, the eradication of the antigen entirely.

24. These risks, however small, have been a constant source of concern to the vaccines industry. In the US these risks were felt to be an important restraint on the commitments of pharmaceutical firms to develop future vaccines (recall our GBS example). In that country, the result was the National Vaccines Injury Compensation Program, which was designed to create a "no-fault" environment for developers and manufacturers (and compensation for victims and their families, where deemed appropriate) in instances where a vaccine harmed someone.

25. The IOM advises the federal government in matters pertaining to medical care, research, and education. Under the charter, the IOM has contributed a number of decisive reports and recommendations with regard to vaccine research, development, and delivery issues.

26. Source: www.909shot.com.

27. Ibid.

28. Understanding, of course, that "the marketplace," across national contexts, is one that is often regulated, at least to the extent that prices (and sometimes demand) are impacted by legislation.

29. We use this term somewhat loosely, recognizing that vaccines are not collective, or public, goods as defined by economists. They are, however, akin to them in certain fundamental ways relating both to their protective properties as well as the types of markets in which they are often exchanged, perhaps illustrating an example of a "mixed" good as defined by Green (1992).

30. After all, what would be the point of a campaign of mass immunization predicated on herd immunity if the state could not pursue policies to ensure that herd immunity was achieved?

Chapter 6

1. In the United States, one woman in eight is reported to have breast cancer during her lifetime. The same statistic in Britain is one in twelve.

2. Epstein (1996) briefly mentions this phenomenon in his analysis of AIDS activists, quoting one activist as follows: "I *never* represented 'people with AIDS.' I represented *activists*. And those are different people, you know. They are a subset of people with AIDS."

3. In fact, not only have British citizens resisted any efforts to dismantle and privatize the NHS, but surveys have shown that British respondents are particularly proud of their national health-care system and do not want it to change in any significant way.

Chapter 7

1. This change of policy focus can largely be attributed to the work of women's health advocates who raised these issues during the 3-year preparatory phase.

2. These practices, and the appraisal of women's health advocates, have been comprehensively reported and analyzed by Hartmann (1987).

3. There is no similar history of men and contraceptives. Accordingly, men's health advocacy groups in the area of reproduction have not developed.

4. See also Talwar et al. 1994; Jones 1994; Jones 1996.

5. Beatrijs Stemerding was the coordinator of the campaign to call for a stop to the research on anti-fertility "vaccines."

6. An impressive amount of social scientific research data is available on mostly women's use of contraceptives. A wealth of data on patterns of contraceptive use involving a number of demographic variables has been generated by USAID-funded Demographic and Health Surveys. Shah (1995) analyzed the data from the DHS, which entailed findings based on nationally representative samples of 360,000 women of reproductive age in 44 developing countries of Africa, Asia, Latin America, and the Caribbean, in addition to a literature survey. The pharmaceutical industry has also done research on consumer preferences. For example, Ortho-McNeil Pharmaceuticals initiated an annual survey on contraceptive use patterns in 1969 among approximately 8,000 women (quoted in Report of Workshop, 1995). At the Population Council, social scientific research was carried out by its Research Division, which published the results in its own bimonthly journal, *Studies in Family Planning*.

7. For instance, Hardon (1997a) reviewed 20 articles published in scientific journals such as *Contraception* and *Fertility and Sterility* within the past decade which present results of clinical studies on anti-fertility vaccines. She reports: "While making explicit what they aim to develop, the researchers do not explain where

these requirements originate. Nowhere do they refer to empirical research on the perceived needs of users and providers who actually live in the diverse socio-cultural settings in which the new contraceptive technology is eventually to be used."

8. See also Ravindran and Berer 1994; Heise 1997.

9. Of course this is one of the central insights of action-research approaches, and the comment has been made by other critical traditions in sociology as well.

10. See also the Gender Advisory Panel's Terms of Reference.

11. On the distinction between "context of testing" and "context of use," see also Pinch 1993.

12. See also Hardon 1990; Richter 1993; Schrater 1995.

13. See also Talwar 1996: 397.

14. See also Call for a Stop to Research on Anti-Fertility "Vaccines" (Immunological Contraceptives); Wieringa 1994; Forum for Women's Health 1995.

15. See also Richter 1993; Wieringa 1994.

16. See also WHO/HRP 1993: 21.

17. See also Stevens 1990: 563; Griffin 1990: 521, quoted in Richter 1993: 35.

18. The researchers never mentioned the lag period or the unpredictable time span of protective immune response in discussing the acceptability of the proposed method.

19. All the recommendations of the SERG meeting were directed toward the performance of clinical research and the provision of methods. The meeting recommended that the WHO/HRP should review the ethical guidelines for doing clinical research and formulate guidelines for the provision of fertility regulation methods. In addition, recommendations were made for implementing these guidelines, such as the organization of seminars and workshops and the installation of monitoring groups in the clinical trials and introductory stages of technology development (SERG 1994).

20. Griffin (1996: 143) wrote: "The intended performance profile of fertility-regulating vaccines, in particular anti-hCG vaccines, is that they would not cause endocrine and metabolic disturbances associated with contraceptive steroids, they would not require daily pill-taking, they would not present the storage and disposal problems of barrier methods, they would not require specialized insertion and removal procedures as with implants and IUDs, they would not depend on the strict self-discipline demanded by 'natural' family planning, they would be naturally reversible unlike sterilization, and they would offer the woman or man personal confidentiality of use." See also Griffin 1994: 88; Griffin, Jones, and Stevens 1994: 108.

Chapter 8

1. See e.g. Brown 1992; Epstein 1996; Kleinman 2000b; Myhre 2001.

2. Here I mean to suggest the virtues of using the sociological literature on social movements in studying lay participation in science. Few scholars have attempted to do this—but see Petersen and Markle 1981; Indyk and Rier 1993; Moore 1993; Epstein 1996; 1997b; Myhre 2001.

3. On the politics of representation "in all senses of the word," see Bourdieu 1985 (quote from p. 727); Bourdieu 1991, chapters 8–10; Bourdieu 1998, chapter 3. For a critical review of uses of the concept of "representation" in science studies, see Lynch 1994. On the various meanings collapsed into the English word "representation," see Spivak 1988: 275–280.

4. NIH and FDA sources tend to refer to "gender differences" even though they typically appear to have biological or anatomical differences in mind. Some members of these agencies have described a deliberate avoidance of the term "sex" for fear of confusion with "sexuality."

5. The Centers for Disease Control and Prevention (CDC), another federal agency that, like the NIH and the FDA, is part of the HHS Department, has also implemented policies on inclusion of women and minorities (Jones, Snider, and Warren 1996). (Inclusion was mandated for extramural research in 1995 and intramural research in 1996.) The CDC's research portfolio (which includes preventive vaccine trials, for example) is quite small in comparison to that of the NIH. For that reason, and because little to no public attention focused on the CDC's policies on inclusion, I ignore that agency for the purposes of this discussion.

6. Apparently, the earliest such study was Kinney et al. 1981. For a dissenting view on the underrepresentation of women, see Kadar 1994. In what is probably the best quantitative analysis of gender representation to date, Bird and Flood (1997) examined all clinical research published in the *New England Journal of Medicine* and the *Journal of the American Medical Association* in 1980, 1985, 1990, and 1995. Bird and Flood found that women were less likely than men to be underrepresented or excluded from studies of gender-specific diseases, but more likely to be underrepresented or excluded from studies of diseases affecting both men and women.

7. In response to perceptions of an epidemic of heart disease in men in the United States, much cardiovascular research launched in the 1970s and the 1980s focused on men, who do tend to get heart attacks at earlier ages than do women (see Healy 1991). Nevertheless, heart disease remains the number one killer of women in the United States. In the case of the Physicians' Health Study, what the congressional uproar obscured was that investigators intended to study women in a subsequent clinical trial (now in progress); they chose to exclude women from the original study because the available population of older women physicians was too small to yield a reliable answer about the protective effect of aspirin in women (interview with Buring).

8. Source: Jaschik 1990. A subsequent study by the GAO of the FDA, presented to Congress in 1992, added fuel to the controversy surrounding that agency's policies on inclusion of women. After surveying manufacturers of drugs that had been approved by the agency in recent years, the investigators concluded that "for more than 60 percent of the drugs, the representation of women in the test population was less than the representation of women in the population with the corresponding disease" (Nadel 1992: 2–3).

9. Here I pass over the complex details of the legislative history from 1990 to 1993 (see Auerbach and Figert 1995; Narrigan et al. 1997; Primmer 1997). The first version of the act died in the House of Representatives in 1990. After its reintroduction, the act was passed by Congress in 1991 but vetoed by President Bush, largely because of his opposition to a separate section of the bill that would have overturned the ban on fetal tissue research. Once President Clinton, soon after his inauguration, issued an executive order throwing out the fetal tissue ban, the act moved quickly through Congress and thence to Clinton's desk. However, these changes to NIH research policy were only one dimension of the women's health agenda promoted by the Congressional Caucus on Women's Issues in the form of an omnibus "Women's Health Equity Act," and many of the other measures included therein never were enacted.

10. More generally, on the transformation of the politics of women's health in recent decades, see Weisman 1998, 2000; Treichler, Cartwright, and Penley 1998; Eckman 1998.

11. The revised directive, which took effect with the 2000 census, identifies "five minimum categories for data on race: American Indian or Alaska Native, Asian, Black or African American, Native Hawaiian or other Pacific Islander, and White." It also specifies "two categories for data on ethnicity: 'Hispanic or Latino' and 'Not Hispanic or Latino.'" Finally, the revision allows respondents to identify themselves as belonging to more than one category.

12. As I have argued elsewhere, the fusion of these two kinds of claims—ethical and epistemological—in support of diverse, inclusive, and heterogeneous trials endowed the arguments of AIDS activists with particular force and credibility and helped them to travel back and forth between lay and expert arenas (Epstein 1995, 1996).

13. Clearly there are many potential implications of the new inclusionary policies, most of which are simply beyond the scope of this chapter. For an evaluative approach that focuses on questions of justice, see Baird 1999.

Chapter 9

1. In their studies of the design process, Walsh et al. (1992: 124) and Cockburn and Fürst Dilic (1994: 10) found only a few female designers.

2. Bram Porrey has created an extensive database and a catalogue with illustrations of all shaving devices that Philips produced since 1939. He is an active mem-

ber of the Philishave Collector Club. This club of hobbyists was established in 1990, one year after the Philips invited diverse collectors to celebrate the fiftieth anniversary of the Philishave. In 1996 the Club had about 100 members. In 1996 the Collectors Club, which is independent of the Philips company, published *Van Klapmes tot Philishave*, a concise history of shaving devices (Derks et al. 1996).

3. The idea of electric dry shaving was generated in the late 1910s, but not until the end of the 1920s did electric motors became small enough to be used in shavers (Derks et al. 1996).

4. In the first half of the 1930s, more than 1.2 million electric dry shavers were sold in the United States (Derks et al. 1996: 16).

5. Other producers though did use the word "razor." Two examples were the "Women"s Razor" and the "Laurel Lady's Boudoir Safety Razor" (Derks et al. 1996: 37–38).

6. The content of gender is diverse, and not all masculinities have technological competence as a core issue. However, I agree with Connell (1987) that technological competence is central to the dominant, hegemonic form of masculinity.

Chapter 10

1. For a similar criticism, see Silverstone and Haddon 1996.

2. Although my argument is very likely to be valid for the relationship between technology and identities in general, I will restrict my analysis to gender identities.

3. Actually, gender theorists adopting ethnomethodological and interactionist approaches have been the first to describe the ongoing processes of gender presentation and gender attribution involved in categorizing people as belonging to a specific gender as "doing gender" (Garfinkel 1967; Goffman 1976; West and Zimmerman 1987). An early example of this sociological work on gender as process is Kessler and McKenna's *Gender: An Ethnomethodological Approach* (1978), which describes how we come to impute sex identity to people on the basis of how they perform gender. Butler's theory of performativity has enriched the view of gender as doing by incorporating theories of speech acts, psychoanalysis, and poststructural philosophy.

4. Butler's work represents a multi-disciplinary approach which combines theoretical perspectives developed in Austin's theory of speech acts, the poststructuralist philosophy of Derrida and Foucault, and psychoanalytical theory (Austin 1962; Butler 1990, 1993).

5. Although Butler (1993) emphasizes the importance of language in producing gendered bodies, she does not argue that bodies are merely linguistic constructs. Butler's argument must be considered as an epistemological rather than an ontological claim (Vasterling 1999: 19).

6. The UK seems to present an exception here. The historian Kate Fisher has described how, in the early decades of the twentieth century, working-class communities in the UK considered the use of contraceptives predominantly as a man's job. In the 1920s and the 1930s, the first birth-control clinics in the UK failed to convince potential female clients of the attractiveness of cervical contraceptive caps and sponges. Fisher has described this resistance as the result of "conflicting cultures of contraception." The idea that women should control their own fertility, first put forward by Mary Stopes in the UK and other feminist birth-control campaigners in Europe and the US, and reiterated during the second wave of feminism in the 1960s and the 1970s, clashed with the existing culture of contraception in working-class communities where both women and men held men responsible for birth control. Fisher described these contraceptive practices in the early decades of this century as "a clearly male culture." In those days, men "played a very significant role in all aspects of contraceptive use: in initiating discussions about birth control, in determining which methods to use, in making sexual advances and in deciding how frequently contraception would be employed, in finding out about methods, and in obtaining any appliances used" (Fisher: 8). Today, traces of similar "male cultures" still exist in several African countries (Stokes 1980).

7. With the term "technosociality" I paraphrase the concept of "biosociality" introduced by Paul Rabinow to describe social movements that focus on health conditions of specific groups. Rabinow (1992) defined this term as: "persons having specific conditions (illnesses) who are organized, coordinated, and who feel a kinship based on their shared experience." I prefer the term "technosociality" to explain the emergence of the women's health movement because it was the emergence of contraceptive technologies rather than the condition of pregnancy that urged women to organize themselves in women's health groups.

8. This does not imply that there are no men who actually practice contraception: many men use condoms and a minority have chosen vasectomy. As Connell has suggested, the cultural ideal of masculinity does not necessarily correspond to the actual activities of the majority of men. Hegemonic masculinity does not mean the total cultural dominance of one specific form of masculinity. Alternatives may exist, but they are subordinated in the dominant cultural narratives (Connell 1987, 1995). Actually, the past decade has shown a substantial increase in the use of condoms, not as contraceptives but to prevent HIV and sexually transmitted diseases (STDs). In the US, condom use among heterosexual men and women at risk of HIV and STDs increased from 10 percent in 1990 to 23 percent in 1992 and then leveled off again to 20 percent in 1996 (Catania et al. 2001: 179). Another US survey shows how condom use among women increased significantly between 1988 and 1995 (Bankole et al. 1999: 264). This survey also indicates that condoms cannot be considered as strictly male contraceptives because women buy them and take care of their use as well. Although the increase in condom use reflects important changes in the attitudes and behavior toward the use of condoms among men to prevent diseases, it is not yet clear whether and how this will affect men's attitudes and behavior toward condoms as contraceptives thus challenging hegemonic views of masculinity and taking

responsibility for contraception. For an analysis of male responsibility in birth-control decisions, see Tone 2001.

9. Connell introduced his theory as a critique on previous gender theories, par-ticularly sex role theory, which largely neglected questions of power. For a more detailed description of Connell's theoretical approach, see Demetriou 2001.

10. Although Connell's structuralist approach to gender and Butler's poststruc-turalist theory of the performativity of gender represent two different intellectual traditions, I have chosen to combine their conceptual vocabulary because both theories go beyond essentialism and avoid a voluntaristic position. Although Butler's early work (e.g. 1990) was identified with voluntarism, her later work (1993, 1995) addressed the limits of the performativity of gender and under-scored that her view of gender implied substantial stability. According to Butler (1993: 95), performativity "cannot be understood outside of a process of iter-ability, a regularized and constrained repetition of norms." Connell (1987, 1995) also explicitly addresses the constraints on the performances of particular forms of gender. In the last decade, a large body of research has emerged that seeks a common ground between structuralist and poststructuralist account of gender. For a more detailed discussion of this literature, see Saetnan et al. 2000: 5–7. Recently, Butler's and Connell's work show even more convergence. In "Restaging the Universal: Hegemony and the Limits of Formalism," Butler dis-cussed the merits of the Gramscian notion of hegemony, a notion which is, as we have seen, central to Connell's approach, to understand social transformations. Reflecting on the different ways to understand social transformations, she con-cluded that "the theory of performativity is not far from the theory of hegemony in this respect: both emphasize the way in which the social world is made—and new possibilities emerge—at various levels of social action through a collabora-tive relation with power" (Butler 2000: 14).

11. See e.g. Connell 1987: 186, where Connell discusses Cynthia Cockburn's 1983 study of the printing industry.

12. Connell developed his "three-fold model of the structure of gender relations" in his first three books (1987: 90–118; 1995; 73–76; 1996: 161–162). In his most recent book, Connell identified linguistic practices, which he called "the structure of symbolism," as a fourth structuring principle of gender (2000: 26, 42–43, 150–155). For a detailed discussion of Connell's work, see Demetriou 2001.

13. For exceptions, see Moore and Schmidt 1999 and Mamo and Fishman 1999. Moore and Schmidt have analyzed how discursive practices used by semen banks in the US construct differences among men and simultaneously maintain hege-monic forms of masculinities. Mamo and Fishman have analyzed discourses on Viagra. Both studies are, however, largely restricted to the use of these technologies.

14. Other locations in which this renegotiation of masculinities takes place are family planning policies and clinics, women's health organizations, and the new media. For an extended analysis of how actors in these worlds articulate identi-ties of the future use of new male contraceptives, see Oudshoorn 2003.

15. My research is based on an analysis of publications in scientific journals and press bulletins reporting the results of these clinical trials; protocols, and other written materials of these trials; reports of interviews with men who participated in these trials; and interviews with leading scientists and clinicians involved in these trials, including William Bremner, Alvin Matsumoto, and Alvin Paulsen of the Division of Endocrinology of the Department of Medicine at the University of Washington in Seattle (October 18, 1994) and Fred Wu of the Department of Medicine at the University of Manchester (April 22, 1994).

16. Cockburn and Ormrod (1993) defined the distinction between subjective identity (the gendered sense of self as created and experienced by the individual) and projected identity (the potential, actual, or desired gender identity as others perceive or portray it).

17. Clarke and Montini (1993) have made the important point that users can be configured in their absence thus creating what they have called "implicated actors" who experience the consequences of being configured as users. This has been very common in the history of female contraceptives.

18. Since the late 1970s, noncompliance has become an important concern in the medical community (Epstein 1996: 205).

19. Actually, the first publication that reported the suppression of sperm production by hormones was published as early as 1939 (Heckel 1939). In the 1950s, several small-scale trials were organized by Gregory Pincus, who eventually became one of the "fathers" of the first hormonal contraceptive pill for women. Pincus tested the same hormonal compound he used for his clinical trials with women (Enovid) on eight psychotic mental patients. He reported it to have a "definite sterilizing effect" (Vaughan 1972: 40). This was the first and the last time that Pincus included men in his trials. A more continuous testing practice emerged in the early 1970s.

20. This analysis is based on the Medline database.

21. A similar role of clinical trials as a means to have access to medical care has been described by Jessika van Kammen for women in Third World countries who participate in the testing of contraceptive vaccines (van Kammen 2000b).

22. For a further analysis of the skepticism of journalists, see Oudshoorn 1999.

23. The major source for this part of my analysis consists of the results of so-called acceptability studies based on questionnaires and focus group discussions among male contraceptive trial participant carried out by Karin Ringheim, a social scientist affiliated with the US Agency for International Development, as part of the two large-scale multi-center clinical trials organized by the WHO in the late 1980s and the early 1990s. My original plan to interview male trial participants failed because researchers were reluctant to cooperate in facilitating contacts with these men because they expected a negative interference with their own research.

24. To be sure, not all participants of the group discussions in which these self-images were expressed considered themselves as different from other men. Men in Singapore and Bangkok considered themselves as "any other man on the street" or "more or less the same" as other men (Ringheim 1993: 22).

25. A poll carried out among women in Europe for the European Society of Human reproduction and Embryology in 1997 reported that almost half of married women were dissatisfied with their current contraceptive method (Arlidge 1997).

26. For a more detailed analysis of the attitude of feminists toward male contraceptive development, see chapter 2 of Oudshoorn 2003.

27. The articulation of non-hegemonic identities of men met severe resistance among other relevant actors in constructing the cultural feasibility of new male contraceptives, most notably the news media. See Oudshoorn 1999, 2003.

Chapter 11

1. See Marx 1994.

2. See also Cowan 1983.

3. Nye's work complements the work of many others, including Thomas P. Hughes, who concentrated on the production of electricity. See Hughes 1983.

4. See also Fischer 1992.

5. See e.g. Silverstone and Haddon 1996; Lie and Sørenson 1996; Kline and Pinch 1996.

6. See e.g. Cockburn and Ormrod 1993; de Grazia and Furlough 1996.

7. In history, the development of a consumption perspective has led to a reinterpretation of the nature of the Industrial Revolution. It was not only a revolution in production methods, but also witnessed a major revolution in consumption. A now-classic study is Neil McKendrick, John Brewer, and J. H. Plumb, *The Birth of a Consumer Society* (Indiana University Press, 1982).

8. We have been inspired by Ruth Oldenziel and others to develop this idea. See e.g. Lerman, Mohun, and Oldenziel 1997.

9. This definition emphasizes that consumer culture is not a late consequence of industrialization or modernization, something that followed after these processes were accomplished. Instead it was part of the very making of the modern societies.

10. Rosenberg also provides a good introduction to Schumpeter.

11. The process of mediation can be characterized as a learning process. Following Hoogma and Schot (2001), we would like to specify this concept by

introducing here a distinction between single-loop and double-loop learning. Single-loop learning entails learning about the effectiveness of a certain technology for achieving a specific goal, i.e. learning aimed at verification. Double-loop learning consists of learning about the assumptions (or scripts; see below) built into the technology. Double-loop learning implies allowing room in introduction process for interaction between specific design options and possible user preferences. Double-loop learning is not implicated in the "learning by doing-using-interacting" approaches discussed here. In these approaches users are mainly perceived as knowledge providers for manufacturers who consequently learn to make better products. In this chapter we explore the dynamics and the outcomes of both learning/mediation processes.

12. Kemp, Schot, and Hoogma (1998: 178) write: "The new technologies have not proven what they are worth yet, so the consumers are not sure what to expect. The meaning and implications of the new technologies have yet to be specified by their applications in practice." See also Teubal 1987.

13. For histories of these mediators, see Fine and Leopold 1993; Goldstein 1997; Blaszczyk 2002; Kline 2000; Horowitz and Mohun 1998; de Wit, Albert de la Bruheze, and Berendsen 2001.

14. On the concept of technology junction, see Schot 1992. On forums, arenas, agendas, agenda building, and technological development, see Albert de la Bruheze 1992. This study addresses issues of access, inclusion and exclusion, or restated: power and interests. Technology junctions cannot be entered freely, and (especially institutional) mediators are of course neither neutral nor interest free. For the flow of our central arguments these issues are not addressed here.

15. We can refer to many studies, including some in this volume. For their broad coverage and innovative work we like to refer to Lie and Sørensen 1996 and to Lerman, Mohun, and Oldenziel 1997.

16. For a similar perspective see Woolgar 1991.

17. For a discussion of methods used by designers in this process, see Akrich 1995.

18. This section is based on den Hartog and Albert de la Bruheze 2000.

19. In the 1920s and the 1930s the Dutch dairy industry started experimenting with carton packages for whipped cream, grated cheese, and ice cream. The carton was coated with a paraffin layer on the inside to make it waterproof and to prevent the packaged food from getting a carton taste and smell.

20. *Misset's Zuivelbereiding en -Handel* 63 (1957), no. 33: 713.

21. After the Second World War, the Dutch government and parliament agreed that more account should be taken of the consumer. This resulted, among other things, in the realization of the Publiekrechtelijke Bedrijfsorganisatie (Statutory Industrial Organization), in which employees and employers were represented.

The Act on Industrial Organization establishing the PBO provided that the Dutch Industry had to consult consumers and their organizations regularly (SER 1991).

22. The diffusion of the glass milk bottle in the Netherlands had a long history. Introduced at the end of the nineteenth century by urban dairy shops and propagated by physicians and other "hygienists," it remained an article of luxury for ordinary people till the 1950s.

23. "Moderne melkverpakking in discussie," *Misset's Zuivelbereiding en -Handel* 73 (1967), no. 14: 274.

24. The disposable milk carton became a niche product complementary to the then-advancing glass milk bottle. Among the niche markets were schools, sport manifestations, military barracks, exhibitions, music festivals, outdoor recreation, and travel and transport. These niche markets were very important "learning environments" for accustoming milk consumers to the disposable milk carton.

25. Centrale Verpakkingscommissie, Notitie inzake (melk)verpakking en milieu verontreiniging, 1971: 8.

26. Based on Albert de la Bruheze 2000.

27. Unilever Historical Archive (UHA), Box AHK 1679, File 038.664, Memorandum of C.J. van Buuren for Food Coordination II, January 29, 1965.

28. UHA, Box 1841, File 038.664—Snacks, 1966–1970, Memorandum of G. J, van Leeuwen, Van den Berg & Jurgens, to G. Kellam, Technology Division, Unilever Research Laboratory, Colworth House, Bedford, November 9, 1967.

29. UHA, Box AHK 2418, File 10256/16, "Review of Calve Borrelnootjes Snacks," Calve Corn/Foods Marketing Division, October 1974; UHA, Box AHK 2418, File 10256/16, letter from B. H. Arnstedt, Unilever Marketing Rotterdam, to Food & Drink Coordination, November 14, 1974.

30. UHA, Reports 880, "Wholesome Snacks Working Group Report IV—A wholesome snack," 1985; UHA, Reports 880, "Wholesome Snacks Working Group Report V—The Competitive Environment for snacks," 1985.

Chapter 12

1. Ruth Schwartz Cowan (1987) was one of the first historians of technology to draw attention to these mediating boundaries between consumers and manufacturers in her work on the "consumption junction." For other research that focuses upon mediators and the process of mediation, see Goldstein 1997. Kline (2000) also points out the mediating role played by home economists in rural electrification in the US, and Bijker and Bijsterveld (2000) emphasize the role of the "Vak women" in mediating between architects and consumers in house design in the Netherlands. Cockburn and Ormrod (1993) deal with how

microwave ovens are sold. The general point about the role of mediators in developing technology was first made by Latour (1986). On the role of mediators in music, see Hennion 1989.

2. For general work on the sociology of selling, see Prus 1989 and Clark and Pinch 1996. For a study of the selling technological components, see Darr 2002.

3. Repair technicians too can tell us about users. See Orr 1996.

4. On the domestication of technology, see Lie and Sørenson 1996 and Silverstone and Hirsch 1992.

5. Christina Lindsay makes good use of such material in her chapter in this volume.

6. For an approach to markets influenced by science and technology studies, see Callon 1998. For an attempt to extend sociological ideas into economics that focuses on the synthesizer, see Pinch 2001.

7. NAMM is now known as the International Music Products Association.

8. The Doors used it on *Strange Days* (1967), the Byrds on *The Notorious Byrd Brothers* (1968), and the Beatles on *Abbey Road* (1969).

9. Interview with Jim Scott, October 26, 1997.

10. Ibid.

11. Interview with Bob Moog, November 15, 1997.

12. Interview with Bob Moog, June 5, 1996.

13. Jon S. Wilson, "Hyman, at Museum, Gives Moog Synthesizer Concert," *New York Times,* August 22, 1970.

14. Ibid.

15. See e.g. Doerscuk 1986.

16. Interview with David Van Koevering, January 30, 1999. All subsequent quotations of Van Koevering are from this interview.

17. Interview with Keith Emerson, January 30, 1999. All subsequent quotations of Emerson are from this interview.

18. Market pitchers routinely use the technique of creating demand by claiming that they have only a limited number of items for sale and saying that people who do not buy immediately will "miss out." For more details, see Clark and Pinch 1996.

19. See Clark and Pinch 1996.

20. In the nineteenth century these traveling salesmen were known as "drummers" (Zunz 1990). For accounts of early salespeople in the electrical industry and the telephone industry, see Goldstein 1997 and Fischer 1992.

21. Quoted on p. 124 of Vail 2000.

22. Interview with Jim Scott, October 26, 1997.

23. Interview with Bob Moog, November 15, 1997.

24. However, Moog kept 15% of the company, and that was worth $300,000 by the time Norlin bought Moog Music in 1973.

25. Interview with Moog, November 15, 1997.

26. Interview with Moog, November 15, 1997.

27. Ibid.

28. Ibid.

29. The data are from Théberge 1997: 74.

30. See Clark and Pinch 1995.

31. Salesmen struggling to bring about a new market is reminiscent of scientists struggling to produce and stabilize new phenomena. See Pickering 1995.

32. On developing new markets for science and technology, see Callon 1998.

33. See e.g. Cockburn and Ormrod 1993.

34. Van Koevering's strategy seems to fit with Latour's analysis of Pasteur's moving back between field and lab as he enlisted new groups to his work on the anthrax virus—see Latour 1983.

35. Latour (1992) has suggested that natural actants are the missing masses of technology studies.

References

Adams, J. 1993. "Resistance to 'modernity': Southern Illinois farm women and the cult of domesticity." *American Ethnologist* 20: 89–113.

Adams, J. 1994. *The Transformation of Rural Life: Southern Illinois, 1890–1990.* University of North Carolina Press.

Adorno, T. W. 1991. *The Culture Industry: Selected Essays on Mass Culture.* Verso.

Akrich, M. 1992. "The de-scription of technical objects." In *Shaping Technology/Building Society,* ed. W. Bijker and J. Law. MIT Press.

Akrich, M. 1995. "User representations: Practices, methods and sociology." In *Managing Technology in Society: The Approach of Constructive Technology Assessment,* ed. A. Rip, T. J. Misa, and J. Schot. Pinter.

Akrich, M., and Latour, B. 1992. "A summary of a convenient vocabulary for the semiotics of human and nonhuman assemblies." In *Shaping Technology/Building Society,* ed. W. Bijker and J. Law. MIT Press.

Albert de la Bruheze, A. 1992. *Political Construction of Technology: Nuclear Waste Disposal in the United States, 1945–1972.* Eburon.

Albert de la Bruheze, A. 2000. "Snacks." In *Techniek in Nederland in de twintigste eeuw—Deel II,* ed. J. Schot et al. Walburg.

Altevie. 2001. European Internet Stats. http://www.altevie.net (August 19).

Angier, N. 1994a. "Fierce competition marked fervid race for cancer gene." *New York Times,* September 20.

Angier, N. 1994b. "Scientists identify a mutant gene tied to hereditary breast cancer." *New York Times,* September 15.

Annadale, E., and Clark, J. 1996. "What is gender? Feminist theory and the sociology of human reproduction." *Sociology of Health and Illness* 18, no. 1: 17–45.

Appadurai, A., ed. 1986. *The Social Life of Things: Commodities in Cultural Perspective.* Cambridge University Press.

Argersinger, P. H., and Argersinger, J. A. E. 1984. "The machine breakers: Farmworkers and social change in the rural midwest of the 1870s." *Agricultural History* 58: 393–410.

Arlidge, J. 1997. "'First man on the pill' can't wait to start." *Manchester Evening News,* October 6.

Arnold, E., ed. 1984. *Party Lines, Pumps, and Privies: Memories of Hoosier Homemakers.* Indiana Extension Homemakers Association.

Arnold, E., ed. 1985. *Voices of American Homemakers.* National Extension Homemakers Council.

Ashmore, M, Mulkay, M., and Pinch, T. 1989. *Health and Efficiency: A Sociology of Health Economics.* Open University Press.

Auerbach, J. D., and Figert, A. E. 1995. "Women's health research: Public policy and sociology." *Journal of Health and Social Behavior,* extra issue: 115–131.

Austin, J. 1962. *How to Do Things with Words.* Harvard University Press.

Bain, J., Rachlis, V., Robert, E., and Khait, Z. 1980. "The combined use of oral medroxyprogesterone acetate and methyltestosterone in a male contraceptive trial programme." *Contraception* 21, no. 4: 365–379.

Baird, K. L. 1999. "The new NIH and FDA medical research policies: Targeting gender, promoting justice." *Journal of Health Politics, Policy and Law* 24, no. 3: 531–565.

Bangemann, M. 1994. Europe and the Global Information Society: Recommendations to the European Council. Internal report, European Council.

Bankole, A., Darroc, J. E., and Singh, S. 1999. "Determinants of trends in condom use in the United States, 1988–1995." *Family Planning Perspectives* 31, no. 6: 264–271.

Bardini, T. 2000. *Bootstrapping: Douglas Engelbart, Coevolution, and the Origins of Personal Computing.* Stanford University Press.

Bardini, T., and Horvath, A. 1995. "The social construction of the personal computer user." *Journal of Communication* 45, no. 3: 40–66.

Barfield, A., Melo, J., Coutinho, E., Alvarez-Sanchez, F., Faundes, A., Brache, V., Leon, P., Frick, J., Bartsch, W. H., Weiske, P., Mishell, D., Bernstein, G., and Ortiz, A. 1979. "Pregnancies associated with sperm concentrations below 10 million/ml in clinical studies of a potential male contraceptive method, monthly depot medroxyprogesterone acetate and testosterone esters." *Contraception* 20, no. 2: 121–127.

Baudet, H. 1986. *Een Vertrouwde Wereld: 100 Jaar Innovatie in Nederland.* Bert Bakker.

Baudrillard, J. 1988. "Consumer society." In *Selected Writings.* Polity.

Bauer, M. 1995a. "Resistance to new technology: Nuclear power, information technology and biotechnology." In *Resistance to New Technology*, ed. M. Bauer. Cambridge University Press.

Bauer, M. (ed.). 1995b. *Resistance to New Technology: Nuclear Power, Information Technology, and Biotechnology*. Cambridge University Press.

Bauman, Z. 1996. "On communitarians and human freedom: or, how to square the circle." *Theory, Culture and Society* 13, no. 2: 79–90.

Beecher, H. K. 1966. "Ethics and clinical research." *New England Journal of Medicine* 74: 1354–1360.

Benangiano, G. 1995. Letter to N. Wieringa, August 4. Archive Women's Health Action Foundation, Amsterdam.

Benangiano, G., and Cottingham, J. 1997. "Contraceptive methods: Potential for abuse." *International Journal of Gynecology and Obstetrics* 56: 39–46.

Bereano, P., Bose, C., and Arnold, E. 1985. "Kitchen technology and the liberation of women from housework." In *Smothered by Invention*, ed. W. Faulkner and E. Arnold. Pluto.

Berg, A. 1996. Digital Feminism. Ph.D. thesis, Norwegian University of Science and Technology.

Berg, A. J. 1994. "Technological flexibility: Bringing gender into technology (or was it the other way round?)." In *Bringing Technology Home*, ed. C. Cockburn and R. Fürst Dilic. Open University Press.

Berg, A. J., and Lie, M. 1993. "Feminism and constructivism: Do artifacts have gender?" *Science, Technology and Human Values* 20: 332–351.

Berger, P. L., and Luckmann, T. 1966. *The Social Construction of Reality: A Treatise in the Sociology of Knowledge*. Doubleday.

Bernardes, J., and Cameron, E. 1998 "Gender and disadvantage in health: Men's health for a change." *Sociology of Health and Illness* 20, no. 5: 73–93.

Berner, B., and Mellström, U. 1997. "Looking for mister engineer: Understanding masculinity and technology at two Fin de Siécles." In *Gendered Practices*, ed. B. Berner. University of Linkoping.

Bijker, W. E. 1987. "The social construction of Bakelite: Toward a theory of invention." In *The Social Construction of Technological Systems*, ed. W. Bijker et al. MIT Press.

Bijker, W. E. 1995a. *Of Bicycles, Bakelites and Bulbs: Toward a Theory of Sociotechnical Change*. MIT Press.

Bijker, W. E. 1995b. "Sociohistorical technology studies." In *Handbook of Science and Technology Studies*, ed. S. Jasanoff et al. Sage.

Bijker, W. E., and Bijstersveld, K. 2000. "Women walking through plans: Technology, obduracy and gender identity." *Technology and Culture* 41: 485–515.

Bijker, W. E., Hughes, T. P., and Pinch, T. J., eds. 1987. *The Social Construction of Technological Systems: New Directions in the Sociology and History of Technology.* MIT Press.

Bijker, W. E., and Law, J. 1992. *Shaping Technology/Building Society: Studies in Sociotechnical Change.* MIT Press.

Bijker, W. E., and Pinch, T. J. 1987." The social construction of facts and artifacts: or how the sociology of science and the sociology of technology might benefit each other." In *The Social Construction of Technological Systems*, ed. W. Bijker et al. MIT Press.

Bijsterveld, K. 1991. "The nature of aging. Some problems of an 'Insider's Perspective' illustrated by Dutch debates about aging (1945–1982)." Presented at annual conference of Society for Social Studies of Science, Cambridge, Massachusetts.

Bird, C. E., and Flood, A. B. 1997. Women's representation as subjects in clinical research: A systematic review of original research published in *JAMA* and *NEJM* in 1980, 1985, 1990, 1995. Brown University Center for Gerontology and Health Care Research, grant report submitted to Office of Research on Women's Health, National Institutes of Health.

Bjurström, E. 1990. "Raggare. En tolkning av en stils uppkomst och utveckling." In *Spelrum*, ed. P. Dahlen and M. Rönnberg. Filmförlaget.

Blaszczyk, R. L. 2002. *Imagining Consumers: Design and Innovation from Wedgwood to Corning.* Johns Hopkins University Press.

Blume, S. 1992. *Insight and Industry: On the Dynamics of Technological Change in Medicine.* MIT Press.

Blume, S. 1999. "Histories of cochlear implantation." *Social Science and Medicine* 49, no. 9: 1257–1268.

Blume, S., and Geesink, I. 2000. "Vaccinology: An industrial science?" *Science as Culture* 9, no. 1: 41–72.

BMJ. 1964. "Immunization against poliomyelitis." *British Medical Journal* 1: 132–133.

Bocock, R. 1993. *Consumption.* Routledge.

Boston Women's Health Book Collective. 1973. *Our Bodies, Ourselves.* Simon & Schuster.

Bourdieu, P. 1984. *Distinction: A Social Critique of the Judgement of Taste.* Routledge.

Bourdieu, P. 1985. "The social space and the genesis of groups." *Theory and Society* 14, no. 6: 723–744.

Bourdieu, P. 1991. *Language and Symbolic Power.* Harvard University Press.

Bourdieu, P. 1995. *Distinksjonen.* Oslo: Pax forlag.

Bourdieu, P. 1998. *Practical Reason: On the Theory of Action.* Stanford University Press.

Brache, V., Alvarez, F., Croxatto, H., Salvatierra, A. M., et al. 1992. "Whole beta-hCG: Tetanus toxoid." Presented at Scientific Session on Immunological Aspects of Reproductive Health, Moscow.

Brackett, E. E., and Lewis, E. B. 1934. Use of Electricity on Nebraska Farms, 1920–1934. Bulletin no. 289, University of Nebraska College of Agriculture Experiment Station.

Brawley, O. W. 1995. "Response to 'Inclusion of women and minorities in clinical trials and the NIH Revitalization Act of 1993—The perspective of NIH clinical trialists.'" *Controlled Clinical Trials* 16: 293–295.

Brawley, O. W. 1998. "The study of untreated syphilis in the Negro male." *International Journal of Radiation Oncology* 40, no. 1: 5–8.

Breast Cancer Action 1996. *Policy on Genetic Testing for Breast Cancer Susceptibility.* USA.

Breast Cancer Action. 1997. *Taking Action, Creating Change: Breast Cancer Action Annual Report.*

Brenner, B. 1996. "Off with their breasts!" *Breast Cancer Action Newsletter* 36, June: 2.

Brokaw, T. 1994. "There's an important breakthrough. . . ." NBC Nightly News, September 13.

Brosveet, J., and Sørensen, K. H. 2000. "Fishing for fun and for profit? National domestication of multimedia: The case of Norway." *The Information Society* 16: 263–276.

Brown, D. 1980. *Electricity for Rural America: The Fight for the REA.* Greenwood.

Brown, D. 1994. "Gene for an inherited form of breast cancer is located." *Washington Post*, September 20.

Brown, M. 2001. "The civic shaping of technology: California's Electric Vehicle Program." *Science, Technology and Human Values* 26, no. 1: 56–81.

Brown, P. 1992. "Popular epidemiology and toxic waste contamination: Lay and professional ways of knowing." *Journal of Health and Social Behavior* 33: 267–281.

Bruce, J. 1987. "Users' perspectives on contraceptive technology and delivery systems: Highlighting some feminist issues." *Technology in Society* 9: 359–383.

Bryder, L. 1999. "'We shall not find salvation in inoculation': BCG vaccination in Scandinavia, Britain and the USA, 1921–1960." *Social Science and Medicine* 49: 1157–1167.

Butler, J. 1990. *Gender Trouble: Feminism and the Subversion of Identity.* Routledge.

Butler, J. 1993. *Bodies That Matter: On the Discursive Limits of "Sex."* Routledge.

Butler, J. 1995. "Melancholy gender/refused identification." In *Constructing Masculinity*, ed. M. Berger et al. Routledge.

Butler, J. 2000. "Restaging the universal: Hegemony and the limits of formalism." In *Contingency, Hegemony, Universality*, ed. J. Butler et al. Verso.

Callon, M., ed. 1998. *The Laws of the Markets.* Blackwell.

Callon, M., and Rabehariso. 1999. "La Lecon d'Humanite de Gino." *Reseaux* 17: 189–233.

Caro, R. A. 1982. *The Years of Lyndon Johnson*, vol. 1: *The Path to Power.* Random House.

Casper, M. 1994. "At the margins of humanity: Fetal positions in science and medicine." *Science, Technology and Human Values* 19, no. 3: 307–323.

Casper, M., and Clarke, A. 1998. "Making the pap smear into the right tool for the job: Cervical cancer screening in the United States, c. 1940–1995." *Social Studies of Science* 28 (2/3): 255–290.

Castells, M. 2001. *The Internet Galaxy.* Oxford University Press.

Catania, J. A., Canchola, J., Binson, D., Dolcini, M. M., Paul, P. P., Fisher, L., Choi, K. -H., Pollcak, L., Chang, J., Yarber, W. L., Heima, J. R., and Coates, T. 2001. "National trends in condom use among at-risk heterosexuals in the United States." *Journal of Acquired Immune Deficiency Syndromes* 27: 176–182.

Chabaud-Rychter, D. 1994. "Women users in the design process of a food robot: Innovation in a French domestic appliance company." In *Bringing Technology Home*, ed. C. Cockburn and R. Fürst-Dilic. Open University Press.

Chambers, I. 1985. *Urban Rhythms.* Macmillan.

Chessher, A. H. 1964. *Let There Be Light: A History of Guadalupe Valley Electric Cooperative.* Naylor.

Christie, B. 1996. "Genetic research gave clue to risk of breast cancer." *Scotsman,* May 29.

Clark, C., and Pinch, T. 1996. *The Hard Sell: The Language and Lessons of Street-Wise Marketing.* HarperCollins.

Clarke, A. 1998. *Disciplining Reproduction: Modernity, American Life and 'The Problem of Sex."* University of Chicago Press.

Clarke, A. Forthcoming. *Grounded Theory after the Postmodern Turn.*

Clarke, A., and Montini, T. 1993. "The many faces of RU486: Tales of situated knowledges and technological contestations." *Science, Technology and Human Values* 18, no. 1: 42–78.

Clarke, A. E., and Olesen, V. L., eds. 1998. *Revisioning Women, Health, and Healing: Feminist, Cultural, and Technoscience Perspectives.* Routledge.

Cockburn, C. 1983. *Brothers: Male Dominance and Technological Change.* Pluto.

Cockburn, C. 1985. *Machinery of Dominance: Women, Men and Technical Know-How.* Pluto.

Cockburn, C. 1992. "The circuit of technology: Gender, dentity and power." In *Consuming Technology,* ed. S. Silverstone and E. Hirsch. Routledge.

Cockburn, C., and Fürst Dilic, R., eds. 1994. *Bringing Technology Home: Gender and Technology in a Changing Europe.* Open University Press.

Cockburn, C., and Ormrod, S. 1993. *Gender and Technology in the Making.* Sage.

Cohen, L. 1990. *Making a New Deal: Industrial Workers in Chicago.* Cambridge University Press.

Connell, R. W. 1987. *Gender and Power.* Polity.

Connell, R. W. 1995. *Masculinities.* Polity.

Connell, R. W. 2000. *The Men and the Boys.* Polity.

Coombs, R., Green, K., Richards, A., and Walsh, V., eds. 2001. *Technology and the Market: Demand, Users and Innovation.* Elgar.

Cooper, D., and Lanza, R. 2000. *Xeno: The Promise of Transplanting Animal Organs in Humans.* Oxford University Press.

Corea, G. 1992. *The Invisible Epidemic: The Story of Women and AIDS.* HarperCollins.

Cottingham, J. 1995. Interview with the author, October 20.

Cottingham, J. 1997. "Introduction." In *Beyond Acceptability.* Published by *Reproductive Health Matters* for World Health Organization.

Cottingham, J., and Benangiano, G. 1997. "Contraceptive methods: Potential for abuse." *International Journal of Gynecology and Obstetrics* 56: 39–46.

Cotton, P. 1990a. "Is there still too much extrapolation from data on middle-aged white men?" *Journal of the American Medical Association* 263, no. 8: 1049–1050.

Cotton, P. 1990b. "Examples abound of gaps in medical knowledge because of groups excluded from scientific study." *Journal of the American Medical Association* 263, no. 8: 1051–1055.

Coutard, O., ed. 1999. *The Governance of Large Technical Systems.* Routledge.

Cowan, R. S. 1976. "The industrial revolution in the home: Household technology and social change in the twentieth century." *Technology and Culture* 17: 1–23.

Cowan, R. S. 1983. *More Work for Mother: The Ironies of Household Technology from the Open Hearth to the Microwave.* Basic Books.

Cowan, R. S. 1987. "The consumption junction: A proposal for research strategies in the sociology of technology." In *The Social Construction of Technological Systems*, ed. W. Bijker et al. MIT Press.

Cussins, C. 1998. "Ontological choreography. Agency for women patients in an infertility clinic." In *Differences in Medicine*, ed. A. Mol and M. Berg. Duke University Press.

CyberAtlas. 2002a. "Men Still Dominate World Wide Internet Use." http://cyberatlas. internet.com (August 19).

CyberAtlas. 2002b. "World's Online Populations." http://cyberatlas.internet.com (August 19).

Cyber Dialogue. 2000. Cyber Dialogue Study Shows US Internet Audience Growth Slowing. http://www.cyberdialogue.com (March 30).

Dale, B. 1995. "Service og velferd i uttynningssamfunnet." In *Livskraftige uttynningssamfunn: en nordisk antologi*, ed. K. Aasbrenn NordREFO.

Danbom, D. B. 1979. *The Resisted Revolution: Urban America and the Industrialization of Agriculture, 1900–1930*. Iowa State University Press.

Darr, A. 2002. "The technicization of sales work: An ethnographic study in the US electronics industry." *Work, Employment and Society* 16: 47–65.

Davidson, A. R., Ahn, K. C., Chandra, S., Diaz-Guerro, R., Dubey, D. C., and Mehryar, A. 1985. "Contraceptive choices for men: Existing and potential male methods." Prepared for presentation at Seminar on Determinants of Contraceptive Method Choice, East West Population Institute, Honolulu.

Davies, K., and White, M. 1995. *Breakthrough: The Race to Find the Breast Cancer Gene*. Wiley.

Davis, K., ed. 1997. *Embodied Practices: Feminist Perspectives on the Body*. Sage.

de Grazia, V. 1996a. "Empowering women as citizens-consumers." In *The Sex of Things*, ed. V. de Grazia with E. Furlough. University of Califormia Press.

de Grazia, V. 1996b. "Introduction." In *The Sex of Things*, ed. V. de Grazia with E. Furlough. University of Califormia Press.

de Grazia, V., and Furlough, E., eds. 1996. *The Sex of Things: Gender and Consumption in Historical Perspective*. University of California Press.

Demetriou, D. Z. 2001. "Connell's concept of hegemonic masculinity: A critique." *Theory and Society* 30: 337–361.

den Hartog, A. P., and Albert de la Bruheze, A. 2000. "Verpakking." In *Techniek in Nederland in de twintigste eeuw—Deel III*, ed. J. Schot et al. Walburg.

Derks, S., Nuijen, W. Spierings, P., and de Weijer, P. 1996. *Van Klapmes tot Philishave*. Philishave Verzamelaarsclub.

Dew, C. B. 1994. *Bond of Iron: Master and Slave at Buffalo Forge.* Norton.

Dew, K. 1995. The measles vaccination campaign in New Zealand, 1985 and 1991: The issues behind the panic. Working paper 10, Department of Sociology and Social Policy, Victoria University of Wellington.

Dew, K. 1999. "Epidemics, panic and power: Representations of measles and measles vaccines." *Health* 3, no. 4: 379–398.

de Wit, O., Albert de la Bruheze, A., and Berendsen, M. 2001. "Ausgehandelter Konsum: Die Verbreitung der modernen Küche, des Kofferradios und des Snack Food in den Niederlanden." *Technikgeschichte* 86, no. 2: 133–155.

de Wit, O., Berendsen, M., and Albert de la Bruheze, A. 2000. "The mediation of consumption: The negotiated diffusion of the modern kitchen, the transistor radio and fast food in the Netherlands." Presented at annual congress of Society for the History of Technology, Munich.

Djerassi, C. 1970. "Birth control after 1984." *Science* 169: 941–951.

Doerscuk, B., ed. 1986. *Rock Keyboard.* Quill/Keyboard.

Douglas, M., and Isherwood, B. 1979. *The World of Goods.* Allen Lane.

Douglas, S. J. 1987. *Inventing American Broadcasting, 1899–1922.* Johns Hopkins University Press.

Downey, G. L., and Dumit, J., eds. 1997. *Cyborgs and Citadels: Anthropological Interventions in Emerging Sciences and Technologies.* School of American Research Press.

Dresser, R. 1992. "Wanted: Single white male for medical research." *Hastings Center Report,* January–February: 24–29.

Dresser, R. 1999. "Public advocacy and allocation of federal funds for biomedical research." *Milbank Quarterly* 77, no. 2: 257–274.

Dreyfus, C., ed. 1977. *Seizing Our Bodies: The Politics of Women's Health.* Vintage.

Dreyfus, H., and Rabinow, P. 1982. *Michel Foucault: Beyond Structuralism and Hermeneutics,* second edition. University of Chicago Press.

Duden, B. 1991. *The Woman beneath the Skin: A Doctor's Patients in Eighteenth-Century Germany.* Harvard University Press.

du Gay, P., Hall, S., Janes, L., Mackay, H., and Negus, K. 1997. *Doing Cultural Studies: The Story of the Sony Walkman.* Sage.

Dugdale, A. 1998. "Materiality: Juggling sameness and difference." In *Actor Network Theory and After,* ed. J. Law and J. Hassard. Blackwell and *Sociological Review.*

Duster, T. 1990. *Backdoor to Eugenics.* Routledge.

Eckman, A. K. 1998. "Beyond 'the Yentl syndrome': Making women visible in post-1990 women's health discourse." In *The Visible Woman*, ed. P. Treichler et al. New York University Press.

Edgar, H., and Rothman, D. J. 1990. "New rules for new drugs: The challenge of AIDS to the regulatory process." *Milbank Quarterly* 68, suppl. 1: 111–142.

Epstein, J. S. 1998. "Introduction: Generation X, youth, culture and identity." In *Youth Culture*, ed. J. Epstein. Blackwell.

Epstein, S. 1991. "Democratic science? AIDS activism and the contested construction of knowledge." *Socialist Review* 91, no. 2: 35–64.

Epstein, S. 1995. "The construction of lay expertise: AIDS activism and the forging of credibility in the reform of clinical trials." *Science, Technology and Human Values* 20, no. 4: 408–437.

Epstein, S. 1996. *Impure Science: AIDS, Activism, and the Politics of Knowledge*. University of California Press.

Epstein, S. 1997a. "Activism, drug regulation, and the politics of therapeutic evaluation in the AIDS era: A case study of ddC and the 'surrogate markers' debate." *Social Studies of Science* 27, no. 5: 691–726.

Epstein, S. 1997b. "AIDS activism and the retreat from the genocide frame." *Social Identities* 3, no. 3: 415–438.

European Commission-DG XIII-C/E. 1998. "Telematics applications programme 'Design for All' for an inclusive information society." In *Design for All and ICT Business Practice*. EC reference no. 98.70.022.

Ewen, S. 1976. *Captains of Consciousness: Advertising and the Social Roots of the Consumer Culture*. McGraw-Hill.

Fagen, M. D., ed. 1975. *A History of Engineering and Science in the Bell System: The Early Years, 1875–1925*. Bell Telephone Laboratories.

Farman, I. 1992. *Tandy's Money Machine*. Mobium.

Faulkner, W. 2000. "The power and the pleasure? A research agenda for making gender stick to engineers." *Science, Technology and Human Values* 25, no. 1: 87–119.

Faulkner, W. 2001. "'I'm No Athlete [but] I Can Make This Thing Sing!' Men's Pleasures in Technology." Unpublished.

Faulkner, W., and Arnold, E., eds. 1985. *Smothered by Invention: Technology in Women's Lives*. Pluto.

Fauske, H. 1993. "Lokal forankring og global orientering? Holdningar til heimstaden blant unge i Oppland." In *Ungdom i lokalmiljø*, ed. K. Heggen et al. Samlaget.

Fausto-Sterling, A. [1985] 1992. *Myths of Gender: Biological Theories about Women and Men,* second edition. Basic Books.

Feenberg, A. 1992. "On being a human subject: Interest and obligation in the experimental treatment of incurable disease." *Philosophical Forum* 23, spring: 213–230.

Feinstein, A. R. 1983. "An additional basic science for clinical medicine: II. The limitations of randomized trials." *Annals of Internal Medicine* 99, no. 4: 544–550.

Ferguson, K. J. 1998. "'Caught in 'no-man's land': The negro cooperative extension service and the ideology of Booker T. Washington, 1900–1918." *Agricultural History* 72: 33–54.

Fine, B., and Leopold, E. 1993. *The World of Consumption.* Routledge.

Fischer, C. S. 1987a. "The revolution in rural telephony, 1900–1920." *Journal of Social History* 21: 5–26.

Fischer, C. S. 1987b. "Technology's retreat: The decline of rural relephony in the United States." *Social Science History* 11: 295–327.

Fischer, C. S. 1992. *America Calling: A Social History of the Telephone to 1940.* University of California Press.

Fisher, K. 1998. "Conflicting cultures of contraception: Birth control clinics and the working classes in Britain between the wars." Presented at Workshop on Cultures of Medicine, Wellcome Institute for the History of Medicine, London.

Fleck, J. 1994. "Learning by trying: The implementation of configurational technology." *Research Policy* 23: 637–652.

FNZ. 1957. *De toepassingsmogelijkheden van papierverpakking voor melk: Een onderzoek van de commissie voor melk in papierverpakking.* Technisch Wetenschappelijke Serie, no. 11.

Foegh, M. 1983. "Evaluation of steroids as contraceptives in men." *Acta endocrinologica* suppl. 260: 7–46.

Foegh, M., Nicol, K., Bruunshuus, I., Petersen, and Schou, G. 1980. "Clinical evaluation of long-term treatment with levo-norgestel and testosterone enanthate in normal men." *Contraception* 21, no. 6: 631–640.

Forum for Women's Health. 1995. "The case against anti-fertility vaccines." *Medico Friend Circle Bulletin* 218: 1–7.

Fosso, E. J. 1997. Industristeders arbeidstilbud og generasjoners forhold til utdanning, arbeid og sted—eksemplet Årdal. Ph.D. thesis, Norwegian School of Economics and Business Administration and University of Bergen.

Foucault, M. 1979. *Discipline and Punish: The Birth of the Prison.* Vintage.

Foucault, M. 1980. *Power/Knowledge,* ed. C. Gordon. Pantheon.

Franklin, S., and Ragone, H. 1998. *Reproducing Reproduction: Kinship, Power, and Technological Innovation.* University of Pennsylvania Press.

Freeman, H. P. 1998. "The meaning of race in science—Considerations for cancer research: Concerns of special populations in the National Cancer Program." *Cancer* 82, no. 1: 219–225.

Freeman, P., and Robbins, A. 1991. "The elusive promise of vaccines." *American Prospect,* winter: 80–90.

French National League for Liberty in Vaccination. 1998. Court Hearing against the Hepatitis B Vaccine, July 1998. Excerpted from www.ctanet.fr/vaccination-information/plainte.htm.

Friedman, A. 1989. *Computer System Development: History, Organisation and Implementation.* Wiley.

Gabe, J., ed. 1995. *Medicine, Health and Risk: Sociological Approaches.* Blackwell.

Gangarosa, E., Galazka, A., Wolfe, C., Phillips, L., Gangarosa, R., Miller, E., and Chen, R. 1998. "Impact of anti-vaccine movements on pertussis control: The untold story." *Lancet* 351: 356–361.

Garfinkel, H. 1967. *Studies in Ethnomethodology.* Prentice-Hall.

Garschagen, O. 1998. "Amerikanen in de rij voor 'penispil.'" *Volkskrant,* May 1.

Garvey, P. 2001. "Driving, drinking and daring in Norway." In *Car Cultures,* ed. D. Miller. Berg.

Gender Advisory Panel. 1996. Report of the first meeting of the HRP Gender Advisory Panel, Geneva.

Genetic Interest Group. 2000. About GIG, Genetic Interest Group.

Georgia Technical University. 1999. Internet User Surveys. http://www.cc.gatech.edu/gvu/user_surveys (March 28, 2000).

Gevas, P. C. 1995. Letter to Judy Norsigian, May 31.

Gibbons, M., Limoges, C., Nowotny, H., Schwartzman, S., Scott, P., and Trow, M. 1994. *The New Production of Knowledge: The Dynamics of Science and Research in Contemporary Societies.* Sage.

Giddens, A. 1990. *The Consequences of Modernity.* Polity.

Gieryn, T. F. 1983. "Boundary-work and the demarcation of science from non-science: strains and interests in professional ideologies of scientists." *American Sociological Review* 48: 781–795.

Gieryn, T. F. 1995. "Boundaries of Science." In *Handbook of Science and Technology Studies,* ed. S. Jasanoff et al. Sage.

Gilbert, G. N., and Mulkay, M., eds. 1984. *Opening Pandora's Box: A Sociological Analysis of Scientific Discourse.* Cambridge University Press.

Gilboa, N. 1996. "Elites, lamers, narcs and whores." In *Wired Women,* ed. L. Cherny and E. Weise. Seal.

Gilman, S. L. 1985. *Difference and Pathology: Stereotypes of Sexuality, Race, and Madness.* Cornell University Press.

Ginsburg, F. D., and Rapp, R., eds. 1995. *Conceiving the New World Order: The Global Politics of Reproduction.* University of California Press.

Glander, H-J. 1987. "Bemerkungen zur nichthormonalen reversibelen Kontrazeption beim mann." *Hautartzt* 38, no. 6: 321–326.

Goffman, E. 1976. *Gender Advertisements.* Harvard University Press.

Goldberg, D. T. 1997. *Racial Subjects: Writing on Race in America.* Routledge.

Goldklang. 1998. Interview: TRS-80.

Goldstein, C. M. 1997. "From service to sales: Home economics in light and power, 1920–1940." *Technology and Culture* 38, no. 1: 121–152.

Gomart, E., and Hennion, A. 1999. "A sociology of attachment: music amateurs, drug users." In *Actor Network Theory and After,* ed. J. Law and J. Hassard. Blackwell and *Sociological Review.*

Gray, C. H., ed. 1995. *The Cyborg Handbook.* Routledge.

Green, D. 1992. "The price elasticity of mass preferences." *American Political Science Review* 86, no. 1: 128–148.

Greenough, P. 1995a. "Global immunization and culture: Compliance and resistance in large-scale public health campaigns." *Social Science and Medicine* 41, no. 5: 605–607.

Greenough, P. 1995b. "Intimidation, coercion and resistance in the final stages of the South Asian smallpox eradication campaign, 1973–1975." *Social Science and Medicine* 41, no. 5: 633–645.

Griffin, P. D. 1988. "Vaccines for fertility regulation." In *Research in human reproduction: biennal report 1986–1987.* World Health Organization.

Griffin, P. D. 1992. "Options for immunocontraception and issues to be addressed in the development of birth control vaccines." *Scandinavian Journal of Immunology* 36, suppl. 11: 111–117.

Griffin, P. D. 1994. "Immunization against hCG." *Human Reproduction* 9, suppl. 2: 88–95.

Griffin, P. D. 1995. Interview with the author, October 19.

Griffin, P. D. 1996. "The immunoregulation of fertility—changes in perspectives." *American Journal of Reproductive Immunology* 35: 140–147.

Griffin, P. D., and Jones, W. R. 1991. "The preliminary clinical evaluation of the safety and efficacy of a fertility regulating vaccine." *Statistics in Medicine* 10: 177–190.

Griffin, P. D., Jones, W. R., and Stevens, V. C. 1994. "Anti-fertility vaccines: current status and implications for family planning programmes." *Reproductive Health Matters* 3: 70–75.

Griffith, L. 1998. "Humour as resistance to professional dominance in community mental health teams." *Sociology of Health and Illness* 20, no. 6: 874–895.

Grint, K., and Gill, R. 1995. *The Gender-Technology Relation: Contemporary Theory and Research.* Taylor & Francis.

Grobe. 1998. Interview.

Guerin, J. Rollet. 1988. "Inhibition of spermatogenesis in men using various combinations of oral progestagens and percutaneous or oral androgens." *International Journal of Andrology* 11: 187–199.

Gupta, S. K., Talwar, G. P., Rose, N. R., and Burek, C. J. 1991. "Preliminary studies on the indium slide immunoassay for estimation of human chorionic gonadotropin and antihuman chorionic gonadotropin antibody." *Proceedings of the Society for Experimental Biology and Medicine* 196: 106–109.

Gusfield, J. R. 1981. *The Culture of Public Problems: Drinking-Driving and the Symbolic Order.* University of Chicago Press.

Habermeyer, K. F. 1990. "Product use and product improvement." *Research Policy* 19: 271–283.

Haddon, L. 1988. "The home computer: The making of a consumer electronic." *Science as Culture* 2: 7–51.

Haddon, L. 1992. "Explaining ICT technologies: The case of the home computer owner." In *Consuming Technologies*, ed. R. Silverstone and E. Hirsch. Routledge.

Hafner, K., and Markoff, J. 1991. *Cyberpunks: Outlaws and Hackers on the Computer Frontier.* Simon & Schuster.

Hall, S. 1973. "Encoding and decoding in the TV discourse." Reprinted in *Culture, Media, Language*, ed. S. Hall et al. (Hutchinson, 1981).

Hall, S. 1995. "New cultures for old." In *A Place in the World*, ed. D. Massey and P. Jess. Oxford University Press.

Hamilton, J. A. 1996. "Women and health policy: On the inclusion of females in clinical trials." In *Gender and Health*, ed. C. Sargent and C. Brettell. Prentice-Hall.

Handelsman, D. J. 1991. "Bridging the gender gap in contraception: Another hurdle cleared." *Medical Journal of Australia* 154, no. 4: 230–233.

Hanhart, J. 1995. "Norplant moving from South to North." In *A Healthy Balance*, ed. E. Hayes. Women's Health Action Foundation.

Hanson, B. 1997. *Social Assumptions, Medical Categories*. JAI.

Håpnes, T. 1996. "Not in their machines: How hackers transform computers into subcultural artefacts." In *Making Technology Our Own*, ed. M. Lie and K. Sørensen. Scandinavian University Press.

Haraway, D. 1985. "Manifesto for cyborgs: Science, technology and socialist feminism in the 1980s." *Socialist Review* 80: 65–108. Reprinted as "A cyborg manifesto: Science, technology and socialist feminism in the late twentieth century" in D. Haraway, *Symians, Cyborgs and Women* (FAB, 1991).

Haraway, D. 1997. *Modest_Witness@Second_Millenium: FemaleMan©_Meets_OncoMouse™*. Routledge.

Harding, S. 1986. *The Science Question in Feminism*. Cornell University Press.

Hardon, A. P. 1990. "An analysis of research on new contraceptive vaccines." *Women and Pharmaceuticals Bulletin* 4: 22–24.

Hardon A. P. 1992. "The needs of women versus the interests of family planning personnel, policy-makers and researchers: Conflicting views on safety and acceptability of contraceptives." *Social Science and Medicine* 6, no. 35: 753–766.

Hardon, A. P. 1995. "User-studies on hormonal contraceptives: Towards gender-aware and experience-near approaches." Presented at WHO/HRP meeting on User Perspectives on Fertility Regulation Technologies, Geneva.

Hardon, A. P. 1997a. "Contesting claims on the safety and acceptability of anti-fertility vaccines." *Reproductive Health Matters* 10: 68–80.

Hardon, A. P. 1997b. "Women's views and experiences of hormonal contraceptives: What we know and what we need to find out." In *Beyond Acceptability*. Published by *Reproductive Health Matters* for World Health Organization.

Hardon, A. P. 1998. Interview with the author, April 28.

Hartmann. B. 1987. *Reproductive Rights and Wrongs: The Global Politics of Population Control and Contraceptive Choice*. Harper and Row.

Hartouni, V. 1991. "Containing women: Reproductive discourse in the 1980s." In *Technoculture*, ed. C. Penley and A. Ross. University of Minnesota Press.

Harvey, D. 1990. *The Condition of Post-Modernity*. Blackwell.

Haseltine, F. B. 1997. "Conclusion." In *Women's Health Research*, ed. F. Haseltine and B. Greenberg Jacobson. Health Press International.

Healy, B. 1991. "The Yentl syndrome." *New England Journal of Medicine* 325, no. 4: 274–276.

Hebdige, D. 1979. *Subculture: The Meaning of Style.* Methuen.

Heckel, N. J. 1939. *Proceedings of the Society of Experimental Biology and Medicine* 40: 658.

Heggen, K. 1996. "Ungdom og lokalmiljø—forankring eller fråkopling?" In *Ung på 90-tallet,* ed. T. Øia. Cappelen.

Heise, L. L. 1997. "Beyond acceptability: Reorienting research on contraceptive choice." In *Beyond Acceptability.* Published by *Reproductive Health Matters* for World Health Organization.

Hennion, A. 1989. "An intermediary between production and consumption: The producer of popular music." *Science Technology and Human Values* 14: 400–424.

Henwood, F., Hughes, G. Kennedy, H., Miller, N., and Wyatt, S. 2001. "Cyborg lives in context: Writing women's technobiographies." In *Cyborg Lives?* ed. F. Henwood et al. Raw Nerve Books.

Hesen, J. C. 1971. "De verwerking van aardappelen tot chips." *Voedingsmiddelen-technologie* 2: 13, 15.

Hetland, P. 1996. Exploring Hybrid Communities. IMK report 29, Department of Media and Communication, University of Oslo.

Hetland, P. 1999. "The mediation of expertise." Presented at annual meeting of Society for Social Studies of Science, San Diego.

Hetland, P., Knutzen, P., Meissner, R., and Olsen, O. E. 1989. Nært men likevel fjernt: Telematikk og lokal utvikling. Report RF 123/89, Rogalandsforskning.

Higgs, P. 1998. "Risk, governmentality and the reconceptualization of citizenship." In *Modernity, Medicine and Health,* ed. G. Scambler and P. Higgs. Routledge.

Highfield, R. 1994. "Breast cancer test may cost too much for NHS: Checks for 30,000 women at risk pose unprecedented problems." *Daily Telegraph* (London), September 16.

Hjort, T., and Griffin, P. D. 1985. "The identification of candidate antigens for the development of birth control vaccines: An international multi-center study on antibodies to reproductive tract antigens, using clinically defined sera." *Journal of Reproductive Immunology* 8: 271–278.

Hoffman, D. L., and Novak, T. P. 1998. Bridging the Digital Divide: The Impact of Race on Computer Access and Internet Use. Working paper, Vanderbilt University.

Hofmann, J. 1996. "Schrijvers, teksten en schrijfhandelingen. Geconstrueerde werkelijkheden in tekstverwerkingsprogrammatuur." *Kennis en Methode* 20: 9–37.

Hogle, L. 1999. *Recovering the Nation's Body: Cultural Memory, Medicine, and the Politics of Redemption.* Rutgers University Press.

Hollingsworth, J., Hage, J., and Hanneman, R. 1990. *State Intervention in Medical Care: Consequences for Britain, France, Sweden, and the United States, 1890–1970.* Cornell University Press.

Hoogma, R., and Schot, J. W. 2001. "How innovative are users? A critique of learning by doing and using." In *Technology and the Market,* ed. R. Coombs et al. Elgar.

Horkheimer, M., and Adorno, T. W. 1979. *Dialectic of Enlightenment.* Verso. First published in 1947.

Horowitz, R., and Mohun, A., eds. 1998. *His and Hers: Gender, Consumption and Technology.* University Press of Virginia.

Hoskins, K. F., et al. 1995. "Assessment and counseling for women with a family history of breast cancer: A guide for clinicians." *Journal of the American Medical Association* 273, no. 7: 577–585.

Hubak, M. 1996. "The car as cultural statement: Car advertising as gendered socio-technical scripts." In *Making Technology our Own?* ed. M. Lie and K. Sørensen. Scandinavian University Press.

Hughes, T. P. 1983. *Networks of Power: Electrification in Western Society, 1880–1930.* Johns Hopkins University Press.

Hughes, T. P. 1989. *American Genesis: A Century of Invention and Technological Enthusiasm.* Viking Penguin.

Iannotti, J., and Williams, G. 1998. "Total shoulder arthroplasty—Factors influencing prosthetic design." *Orthopedic Clinics of North America* 29, no. 3: 377–385.

ICPD. 1994. International Conference on Population and Development, Program of Action, New York: UN International Conference on Population and Development Secretariat.

Indyk, D., and Rier. D. 1993. "Grassroots AIDS knowledge: Implications for the boundaries of science and collective action." *Knowledge: Creation, Diffusion, Utilization* 15: 3–43.

Institute of Medicine. 1985. *Vaccine Supply and Innovation.* National Academy Press.

International Conference on Population and Development, United Nations. 1994. "Program of Action."

Internet Society. 2001. All About the Internet. http://www.isoc.org/internet/stats/ (June 11).

Ismail, S. 1998. Vintage computer festival. www.siconic.com/vcf1998.

Jacobus, M., Keller, E. F., and Shuttleworth, S., eds. 1990. *Body/Politics: Women and the Discourses of Science.* Routledge.

Jasanoff, S. S. 1987. "Contested boundaries in policy-relevant science." *Social Studies of Science* 17: 195–230.

Jaschik, S. 1990. "Report says NIH ignores own rules on including women in its research." *Chronicle of Higher Education* 27, June: A-27.

Jellison, K. 1993. *Entitled to Power: Farm Women and Technology, 1913–1963.* University of North Carolina Press.

Jenkins, R. V. 1975. "Technology and the market: George Eastman and the origins of mass amateur photography." *Technology and Culture* 16: 1–19.

Jones, J. H. [1981] 1993. *Bad Blood: The Tuskegee Syphilis Experiment.* Free Press.

Jones, W. K., Snider, D. E., and. Warren, R. C. 1996. "Deciphering the data: Race, ethnicity, and gender as critical variables." *Journal of the American Medical Women's Association* 51, no. 4: 137–138.

Jones, W. R. 1982. *Immunological Fertility Regulation.* Blackwell.

Jones, W. R., et al. 1988. "Phase I clinical trial of World Health Organization's birth control vaccine." *Lancet,* June 11: 1295–1298.

Jones, W. R. 1994. "Vaccination for contraception." *Australian and New Zealand Journal of Obstetrics and Gynecology* 34, no. 3: 320–329.

Jones, W. R. 1996. "Contraceptive vaccines." *Bailliere's Clinical Obstetrics and Gynecology* 10, no. 1: 69–86.

Jordan, T. 2001. "Measuring the Internet: Host counts versus business plans." *Information, Communication and Society* 4, no. 1: 34–53.

Jordanova, L. 1989. *Sexual Visions: Images of Gender in Science and Medicine between the Eighteenth and Twentieth Centuries.* University of Wisconsin Press.

Jørgensen, G. 1994. To Ungdomskulturer: Om vedlikehold av sosiale og kulturelle ulikheter i Bygdeby. Report 1/1994. Vestlandsforskning.

Kadar, A. G. 1994. "The sex-bias myth in medicine." *Atlantic Monthly,* August: 66–70.

Katz, J. E., and Aspden, P. 1998. "Internet dropouts in the USA: The Invisible Group." *Telecommunications Policy* 22, no. 4/5: 327–339.

Kaufert, P. A. 1998. "Women, resistance, and the breast cancer movement." In *Pragmatic Women and Body Politics,* ed. M. Lock and P. Kaufert. Cambridge University Press.

Kemp, R., Schot, J. W., and Hoogma, R. 1998. "Regime shifts through processes of niche information." *Technology Analysis and Strategic Management* 10, no. 2: 175–195.

Kessler, A. 1992. "Establishment and early development of the programme." In *Reproductive Health,* ed. J. Khanna et al. World Health Organization.

Kinney, E. L., Trautmann, J., Gold, J. A., Vesell, E. S., and Zelis, R. 1981. "Underrepresentation of women in new drug trials: Ramifications and remedies." *Annals of Internal Medicine* 95: 495–499.

Kirejczyk, M. 1993. "Shifting the burden onto women: The gender character of in vitro fertilization." *Science as Culture* 3, no. 17: 507–522.

Kirkham, P., ed. 1996. *The Gendered Object.* Manchester University Press.

Klein, R. 2001. *The New Politics of the National Health Service.* Prentice-Hall.

Kleinman, D. L. 2000a. "Democratizations of science and technology." In *Science, Technology, and Democracy,* ed. D. Kleinman. State University of New York Press.

Kleinman, D. L., ed. 2000b. *Science, Technology, and Democracy.* State University of New York Press.

Kline, R. 1997a. "Agents of Modernity: Home Economists and Rural Electrification, 1925–1950." In *Rethinking Home Economics,* ed. S. Stage and V. Vincenti. Cornell University Press.

Kline, R. 1997b. "Ideology and the New Deal: 'Fact Film' Power and the Land." *Public Understanding of Science* 6: 19–30.

Kline, R. 2000. *Consumers in the Country: Technology and Social Change in Rural America.* Johns Hopkins University Press.

Kline, R., and Pinch, T. 1996. "Users as agents of technological change: The social construction of the automobile in the rural United States." *Technology and Culture* 37, no. 4: 763–795.

Koch, L., and Stemerding, D. 1994. "The sociology of entrenchment: A cystic fibrosis test for everyone?" *Social Science and Medicine* 39: 1211–1220.

Kramarae, C., ed. 1988. *Technology and Women's Voices.* Routledge & Kegan Paul.

Kraybill, D. B. 1989. *The Riddle of Amish Culture.* Johns Hopkins University Press.

Krieger, N., and Fee, E. 1996a. "Man-made medicine and women's health: The biopolitics of sex/gender and race/ethnicity." In *Man-Made Medicine,* ed. K. Moss. Duke University Press.

Krieger, N., and Fee, E. 1996b. "Measuring social inequalities in health in the United States: A historical review, 1900–1950." *International Journal of Health Services* 26, no. 3: 391–418.

Kyberd, P., Evans, M., and Winkel, S. 1998. "An intelligent anthropomorphic hand, with automatic grasp." *Robotica* 16: 531–536.

Lacquer, T. 1987. "Orgasm, generation, and the politics of reproductive biology." In *The Making of the Modern Body,* ed. C. Gallegher and T. Lacquer. University of California Press.

Lamvik, G. 1996. "A fairy tale on wheels: The car as a vehicle for meaning within a Norwegian subculture." In *Making Technology Our Own?* ed. M. Lie and K. Sørensen. Scandinavian University Press.

Latour, B. 1983. "Give me a laboratory and I will raise the world." In *Science Observed*, ed. K. Knorr-Cetina and M. Mulkay. Sage.

Latour, B. 1986. *Science in Action: How to Follow Scientists and Engineers through Society*. Harvard University Press.

Latour, B. 1988. *The Pasteurization of France*. Harvard University Press.

Latour, B. 1992. "Where are the missing masses? Sociology of a few mundane artifacts." In *Shaping Technology/Building Society*, ed. W. Bijker and J. Law. MIT Press.

Latour, B. 1997. *De Berlijnse Sleutel*. Van Gennep.

Latour, B., and Woolgar, S. 1979. *Laboratory Life: The Social Construction of Scientific Facts*. Sage.

Laurance, J. 1996. "Women at risk of cancer agonise over mastectomy." *Times* (London), July 11.

Law, J., and Akrich, M. 1994. "On customers and costs: A story from public sector science." *Science in Context* 7, no. 3: 539–561.

Lee, C. 1989. Wired Help for the Farm: Individual Electric Generating Sets for Farms, 1880–1930. Ph.D. dissertation, Pennsylvania State University.

Leonard, D. 1998. *Wellsprings of Knowledge: Building and Sustaining the Source of Innovation*. Harvard Business School Press.

Leonard-Barton, D. 1988. "Implementation as mutual adoption of technology and organization." *Research Policy* 17: 251–267.

Lerman, N. E., Mohun, A. P., and Oldenziel, R. 1997. "The shoulders we stand on and the view from here: Historiography and directions for research." *Technology and Culture* 38: 9–30.

Lerman, N. E., Mohun, A. P., and Oldenziel, R., eds. 1997. "Gender analysis and the history of technology." *Technology and Culture* 38, no. 1: 1–213.

Leung, L., and Wei, R. 1999 "Who are the mobile phone have-nots? Influences and consequences." *New Media and Society* 1, no. 2: 209–226.

Levering, R., Katz, M., and Moskowitz, M. 1984. *The Computer Entrepreneurs: Who's Making It Big and How in America's Upstart Industry*. New American Library.

Levold, N. 2001. "Doing gender" in academia: The domestication of an information-technological researcher-position." In *The Social Production of Technology*, ed. H. Glimmel and O. Juhlin. Gothenburg: Business and Administration Studies.

Levy, R. A. 1993. *Ethnic and Racial Differences in Response to Medicines: Preserving Individualized Therapy in Managed Pharmaceutical Programs.* National Pharmaceutical Council.

Lewinsohn-Zamir, D. 1998. "Consumer preferences, citizen preferences, and the provision of public goods." *Yale Law Journal* 108, no. 2: 377–398.

Lewis, D. L., and Goldstein, L., eds. 1983. *The Automobile and American Culture.* University of Michigan Press.

Lie, M. 1995. "Technology and masculinity: The case of the computer." *European Journal of Women's Studies* 2: 379–394.

Lie, M. 1996. "'Excavating' the present: The computer as material culture." *Knowledge and Society* 10: 51–68.

Lie, M., and Sørenson, K. H., eds. 1996. *Making Technology Our Own? Domesticating Technology into Everyday Life.* Scandinavian University Press.

Liff, S. 1999. Cybercafés and Telecottages: Increasing Public Access to Computers and the Internet. Unpublished.

Lissner, E. A. 1992. "Frontiers in nonhormone male contraceptive research." In *Issues in Reproductive Technology I*, ed. H. Holmes. Garland.

Lobel, B., Olivo, J. F., Guille, F., and Le Lannou. D. 1989. "Contraception in men: Efficacy and immediate toxicity. A study of 18 cases." *Acta Urologica Belgica* 57, no. 1: 117–124.

Lock, M. 1993. "The concept of race: An ideological construct." *Transcultural Psychiatric Research Review* 30: 203–227.

Lock, M. 2000. "On dying twice: Culture, technology and the determination of death." In *Living and Working with the New Medical Technologies*, ed. M. Lock et al. Cambridge University Press.

Lock, M., and Kaufert, P. A. 1998. *Pragmatic Women and Body Politics.* Cambridge University Press.

Lohan, M. 2000. "Constructive tensions in feminist technology studies." *Social Studies of Science* 30, no. 6: 895–916.

Love, S. (with K. Lindsay). 1995. *Dr. Susan Love's Breast Book.* Addison-Wesley.

Lubar, S. 1998. "Men/women/production/consumption." In *His and Hers*, ed. R. Horowitz and A. Mohun. University Press of Virginia.

Lundvall, B. 1985. *Product Innovation and User-Producer Interaction.* Aalborg University Press.

Lundvall, B. 1988. "Innovation as an interactive process: From user-producer interaction to the national system of innovation." In *Technical Change and Economic Theory*, ed. G. Dosi et al. Pinter.

Lury, C. 1996. *Consumer Culture.* Polity.

Lyles, F. 1996. Preferred technology, Inc.: Aphton's vaccines. http://www.aphton.com/preferre.htm.

Lynch, M. 1994. "Representation is overrated: Some critical remarks about the use of the concept of representation in science studies." *Configurations* 1: 137–149.

Mackay, H. 1997. *Consumption and Everyday Life.* Sage.

Mackay, H., and Gillespie, G. 1992. "Extending the social shaping of technology approach: Ideology and appropriation." *Social Studies of Science* 22, no. 4: 685–716.

Mackay, H., Crane, C., Beynon-Davies, P., and Tudhope, D. 2000. "Reconfiguring the user: using rapid application development." *Social Studies of Science* 30, no. 5: 737–759.

Mamo, L., and Fishman, J. R. 1999. "Potency in all the right places: Viagra as a technology of the gendered body." *Body and Society* 7, no. 4: 13–37.

Marcuse, H. 1964. *One Dimensional Man.* Routledge.

Marks, J. 1995. *Human Biodiversity: Genes, Race, and History.* Aldine de Gruyter.

Marlin, E. 1998. "But will I grow tits?" *New Woman*, August: 226–228.

Martin, B., and Scott, P. 1992. "Automatic vehicle identification: A test of theories of technology." *Science, Technology and Human Values* 17, no. 4: 485–505.

Martin, E. 1989. *The Woman in the Body.* Open University Press.

Martin, E. 1994. *Flexible Bodies: Tracking Immunity in American Culture—From the Days of Polio to the Age of AIDS.* Beacon.

Martin, M. 1991. *"Hello Central?" Gender, Technology and the Culture in the Formation of Telephone Systems.* McGill–Queens University Press.

Marvin, C. 1988. *When Old Technologies Were New: Thinking about Electric Communication in the Late Nineteenth Century.* Oxford University Press.

Marx, K. 1980. *Marx's Grundrisse*, ed. D. McLellan. Paladin.

Marx, L. 1994. "The idea of 'technology' and postmodern pessimism." In *Technology, Pessimism, and Postmodernism*, ed. Y. Ezrahi et al. Kluwer.

Mastroianni, A. C., Faden, R., and Federman. D., eds. 1994. *Women and Health Research: Ethical and Legal Issues of Including Women in Clinical Studies.* National Academy Press.

Mayntz, R., and Hughes, T. P., eds. 1988. *The Development of Large Technical Systems.* Westview.

McBride, D. 1991. *From TB to AIDS: Epidemics among Urban Blacks since 1900.* State University of New York Press.

McCracken, G. 1988. *Culture and Consumption: New Approaches to the Symbolic Character of Consumer Goods and Activities.* Indiana University Press.

McGaw, J. 1982. "Women and the history of American technology." *Signs* 7: 789–828.

McGaw, J. 1989. "No passive victims, no separate spheres: A feminist perspective on technology's history." In *In Context,* ed. H. Cutcliffe and R. Post. Bethlehem.

Meinert, C. L. 1995a. "The inclusion of women in clinical trials." *Science* 269: 795–796.

Meinert, C. L. 1995b. "Comments on NIH clinical trials valid analysis requirement." *Controlled Clinical Trials* 16: 304–306.

Miles, I., and Thomas, G. 1995. "User resistance to new interactive media: Participants, processes and paradigms." In *Resistance to New Technology,* ed. M. Bauer. Cambridge University Press.

Millard, E. 1997. "Gender identity and the construction of the developing reader." *Gender and Education* 9, no. 1: 31–48.

Miller, D. 1995. *Acknowledging Consumption: A Review of New Studies.* Routledge.

Miller, K. B. 1905. *American Telephone Practice,* fourth edition. McGraw-Hill.

Mitchison, N. A. 1990. "Lessons learned and future needs." In *Gamete interaction,* ed. N. Alexander et al. Wiley-Liss.

Moore, J. J., and Schmidt, M. A. 1999. "On the construction of male differences: Marketing variations in technosemen." *Men and Masculinities* 1, no. 4: 331–351.

Moore, K. 1993. Doing Good While Doing Science: The Origins and Consequences of Public Interest Science Organizations in America, 1945–1990. Ph.D. dissertation, University of Arizona.

Morgan, L. M., and Michaels, M. W., eds. 1999. *Fetal Subjects, Feminist Positions.* University of Pennsylvania Press.

Morley, D. 1992. *Television, Audiences and Cultural Studies.* Routledge.

Morley, D. 1995. "Theories of consumption in media studies." In *Acknowledging Consumption,* ed. D. Miller. Routledge.

Moser, I. 2000. "Against normalisation: subverting norms of ability and disability." *Science as Culture* 9, no. 2: 201–240.

Moser, I., and Law, J. 1998. "Materiality, textuality, subjectivity: Notes on desire, complexity and inclusion." *Concepts and Transformations: International Journal of Action Research and Organizational Renewal* 3: 207–227.

Moser, I., and Law, J. 2001. "Making voices": New Media Technologies, Disabilities, and Articulation. Centre for Science Studies and the Department of Sociology, Lancaster University.

Mowery, D., and Mitchell, V. 1995. "Improving the reliability of the US vaccine supply: An evaluation of alternatives." *Journal of Health Politics, Policy and Law* 20, no. 4: 973–1000.

Mueller, M.-R. 1998. "'Women and minorities' in federal research for AIDS." *Race, Gender and Class* 5, no. 2: 79–98.

Muraskin, W. 1998. *The Politics of International Health: The Children's Vaccine Initiative and the Struggle to Develop Vaccines for the Third World.* State University of New York Press.

Myhre, J. 2001. Medical Mavens: Gender, Science, and the Consensus Politics of Breast Cancer Activism. Ph.D. dissertation, University of California, Davis.

Myriad Genetics, Inc. 1996a. Annual Report.

Myriad Genetics, Inc. 1996b. Myriad Genetics Introduces the First Comprehensive Breast/Ovarian Cancer Susceptibility Test. Press release.

Myriad Genetics, Inc. 1998. Myriad Genetics Signs Agreement with Aetna US Healthcare to Provide Cancer Testing to Its Members. Press release.

Myriad Genetics, Inc. 1999. *Testing for Hereditary Risk of Breast and Ovarian Cancer Is It Right for You? An Informational Program for People Considering Genetic Susceptibility Testing.* Videotape.

Myriad Genetics, Inc. 2000. *Breast and Ovarian Cancer: Given the Choice, Wouldn't You Rather Deal with the Known Than the Unknown?*

Nadel, M. V. 1990. Statement before Subcommittee on Health and the Environment, Committee on Energy and Commerce, US House of Representatives, June 18. General Accounting Office.

Nadel, M. V. 1992. Women's Health: FDA Needs to Ensure More Study of Gender Differences in Prescription Drug Testing. General Accounting Office.

Nagelsmit, L. 1975. *Melk en melkverpakking,* Den Haag: Centrale Verpakkings-commissie Zuivelindustrie.

Nahon, D., and Lander, N. 1993. "The masculine mystique." *Canadian Pharmacological Journal* 126: 458–459.

Nairn, K., McCormack J., and Liepins R. 2000. "Having a place or not? Young peoples experiences of rural and urban environments." Presented at Making and Breaking Borders, Nordic Youth Research Symposium, University of Helsinki.

Narrigan, D., Sprague Zones, J., Worcester, N., and Grad, M. J. 1997. "Research to improve women's health: An agenda for equity." In *Women's Health,* ed. S. Burt Ruzek et al. Ohio State University Press.

National Action Plan on Breast Cancer. 1996. Position Paper: Hereditary Susceptibility Testing for Breast Cancer. US Department of Health and Human Services.

National Breast Cancer Coalition. 1996a. Commentary on the ASCO Statement on Genetic Testing for Cancer Susceptibility.

National Breast Cancer Coalition. 1996b. Genetic Testing for Heritable Breast Cancer Risk. National Breast Cancer Coalition.

National Commission for the Protection of Human Subjects of Biomedical and Behavioral Research. 1979. The Belmont Report: Ethical Principles and Guidelines for the Protection of Human Subjects of Research. Office for Protection from Research Risks, National Institutes of Health, US Department of Health, Education, and Welfare.

National Telecommunications and Information Administration (NTIA). 2000. Falling Through the Net: Toward Digital Inclusion.

Nechas, E., and Foley, D. 1994. *Unequal Treatment: What You Don't Know About How Women Are Mistreated by the Medical Community.* Simon & Schuster.

Nederlands Verpakkingscentrum (NVC). 1959. Commissie 'Huisvrouw en Verpakking'—Subcommissie Melkverpakking. *Melkverpakking ter Tafel,* Den Haag: NVC, Verpakkingsreeks, no. 9.

Neice, D. C. 2002. "Cyberspace and social distinctions: Two metaphors and a theory." In *Inside the Communication Revolution,* ed. R. Mansell. Oxford University Press.

Neth, M. 1996. *Preserving the Family Farm: Women, Community, and the Foundations of Agribusiness in the Midwest, 1900–1940.* Johns Hopkins University Press.

NIAID. 2000. *The Jordan Report: Accelerated Development of Vaccines.* US Government Printing Office.

Nichter, M. 1995. "Vaccinations in the third world: A consideration of community demand." *Social Science and Medicine* 41, no. 5: 617–622.

Nieschlag E., Lerch, A., and Nieschlag, S. 1994. *Institut für Reproduktionsmedizin der Westfälischen Wilhelms-Universität Münster.* Westfälischen Wilhelms-Universität.

Nightingale, E. 1977. "Recommendations for a national policy on poliomyelitis vaccination." *New England Journal of Medicine* 297: 249–253.

Nissen, J. 1993. *Pojkarna vid datorn.* Symposium Graduale.

NOP Research Group. 1999. Internet User Profile Study, Wave 8 Core Data. http://www.nopres.co.uk (March 29, 2000).

Norsk Gallup Institutt, 1999. Intertrack November. http://www.gallup.no/menu/internett/default.htm.

Nua. 2002a. How Many Online? http://www.nua.com (August 19).

Nua. 2002b. More European Women Online. http://www.nua.com (August 19).

320 *References*

NVAC. 1991. "The measles epidemic: The problems, barriers, and recommendations." *JAMA* 266, no. 11: 1547–1582.

NVIC. 1999. Hearing of the criminal justice, drug policy and human resources subcommittee of the House Government Reform Committee on the Hepatitis B Vaccine, May 18.

Nye, D. 1990. *Electrifying America: Social Meanings of a New Technology, 1880–1940.* MIT Press.

Nye, D. E. 1998. *Consuming Power: A Social History of American Energies.* MIT Press.

Oakley, A. 1984. *The Captured Womb: A History of the Medical Care of Pregnant Women.* Blackwell.

O'Connor, J., Orloff, A., and Shaver, S. 1999. *States, Markets, Families: Gender, Liberalism, and Social Policy in Australia, Canada, Great Britain, and the United States.* Cambridge University Press.

O'Dell, T. 2001. "Raggare and the panic of mobility: Modernity and hybridity in Sweden." In *Car Cultures,* ed. D. Miller. Berg.

Oldenziel, R. 1997. "Boys and their toys: The Fisher Body Craftsmen's Guild, 1930–1968, and the making of a male technical domain." *Technology and Culture* 38: 60–98.

Oldenziel, R. 1999. *Making Technology Masculine: Men, Women and Modern Machines in America 1870–1945.* Amsterdam University Press.

Oldenziel, R. 2001. "Consumption-junction revisited." In *Science, Medicine, and Technology in the 20th Century,* ed. L. Schiebinger. University of Pennsylvania Press.

Om Singh et al. 1989. "Antibody response and characteristics of antibodies in women immunized with three contraceptive vaccines inducing antibodies against human chorionic gonadotropin." *Fertility and Sterility* 52, no. 5: 739–744.

Orient, J. 1999. Statement of the American Association of Physicians and Surgeons on Vaccines: Public Safety and Personal Choice. www.aapsonline.org/aaps/testimony/hepbstatement.htm.

Orr, J. 1996. *Talking About Machines.* Cornell University Press.

Østby, P. 1994. "Escape from Detroit: The Norwegian conquest of an alien artifact." In *The Car and Its Environments,* ed. K. Sørensen. European Commission.

Østby, P. 1995. Flukten fra Detroit: Bilens inntregrasjon i det norske samfunnet. STS-report 24/95, University of Trondheim.

Oudshoorn, N. 1994. *Beyond the Natural Body: An Archeology of Sex Hormones.* London and Routledge.

Oudshoorn, N. 1996. *Gender Scripts in Technologie: Noodlot of Uitdaging?* Oratie Universiteit Twente.

Oudshoorn, N. 1998. "Representatie of script? Over gender, de woorden en de dingen." *Tijdschrift voor Genderstudies* 1, no. 3: 5–13.

Oudshoorn, N. 1999. "On masculinities, technologies and pain: The testing of male contraceptive technologies in the clinic and the media." *Science, Technology and Human Values* 24, no. 2: 265–290.

Oudshoorn, N. 2000. "Imagined men: Representations of masculinities in discourses on male contraceptive technology." In *Bodies of Technology*, ed. A. Saetnan et al. Ohio State University Press.

Oudshoorn, N. 2003. *The Male Pill: A Biography of a Technology in the Making.* Duke University Press.

Oudshoorn, N., Brouns, M., and van Oost, E. 2005. "Diversity and agency in the design and use of medical video-communication technologies." In *Inside the Politics of Technology*, ed. H. Harbers. Amsterdam University Press.

Oudshoorn, N., Rommes, E., and Stienstra, M. 2003. "Configuring the user as everybody: Gender and design cultures in information and communication technologies." *Science, Technology and Human Values,* forthcoming.

Oudshoorn, N., Saetnan, A. R., and Lie, M. 2002. "On gender and things: Reflections on an exhibition on gendered artifacts." *Women's Studies International Forum* 25, no. 4: 471–483.

Parens, E., and Asch, A. 1999. "The disability rights critique of prenatal genetic testing: Reflections and recommendations." *Hastings Center Reports* 29, no. 5: U1–S22.

Paulsen, C. A., Bremner, W. J., and Matsumoto, A. M. 1994. Male Contraceptive Development. Population Center for Research in Reproduction. Department of Medicine, University of Washington and Veterans Administration Medical Center, Seattle.

Person, H. S. 1950. "The rural electrification administration in perspective." *Agricultural History* 24: 70–89.

Petchesky, R. 1987. "Fetal Images: The power of visual culture in the politics of reproduction." *Feminist Studies* 13, no. 2: 263–292.

Petchesky, R. P., and Judd, K., eds. 1998. *Negotiating Reproductive Rights: Women's Perspectives across Countries and Cultures.* International Reproductive Rights Research Action Group. Zed Books.

Petersen, J. C., and Markle, G. E. 1981. "Expansion of conflict in cancer controversies." *Research in Social Movements, Conflict and Change* 4: 151–169.

Pew Internet Project. 2001. Internet and American Life.

Pew Internet Project. 2002. Getting Serious Online.

Pfeffer, N. 1985. "The hidden pathology of the male reproductive system." In *The Sexual Politics of Reproduction*, ed. H. Homans. Gower.

Piantadosi, S. 1995. "Commentary regarding 'inclusion of women and minorities in clinical trials and the NIH Revitalization Act of 1993—The perspective of NIH clinical trialists." *Controlled Clinical Trials* 16: 307–309.

Piantadosi, S., and Wittes, J. 1993. "Politically correct trials." *Controlled Clinical Trials* 14: 562–567.

Pickering, A. 1995. *The Mangle of Practice: Time, Agency, and Science*. University of Chicago Press.

Pile, S., and Thrift, N., eds. 1995. *Mapping the Subject: Geographies of Cultural Transformation*. Routledge.

Pinch, T. 1993. "Testing—one, two three . . . testing! Toward a sociology of testing." *Science, Technology, and Human Values* 18, no 1: 25–42.

Pinch, T. 2001. "Why you go to a piano store to buy a synthesizer: Path dependence and the social construction of technology." In *Path Dependence and Creation*, ed. R. Garud and P. Karnoe. LEA.

Pinch, T. J., and Bijker, W. E. 1984. "The social construction of facts and artifacts: Or how the sociology of science and the sociology of technology might benefit each other." *Social Studies of Science* 14: 399–431.

Pinch, T. J., and Bijker, W. E. 1987. "The social construction of facts and artifacts: or how the sociology of science and the sociology of technology might benefit each other." In *The Social Construction of Technological Systems*, ed. W. Bijker et al. MIT Press.

Pinch, T., and Trocco, F. 2002. *Analog Days: The Invention and Impact of the Moog Synthesizer*. Harvard University Press.

Pinn, V. W., and Jackson, D. M. 1996. "Advisory committee to NIH Office passes resolutions." *Journal of Women's Health* 5, no. 6: 549–553.

Plotkin, S., and Mortimer, E., eds. 1994. *Vaccines*, second edition. Saunders.

Plotkin, S., and Orenstein, W., eds. 1999. *Vaccines*, third edition. Saunders.

Polednak, A. P. 1989. *Racial and Ethnic Differences in Disease*. Oxford University Press.

Population Council. 1990. Annual Report. The Population Council.

Porrey, B. 1998. De Geschiedenis van het Scheren. Personal notes.

Prentice, T. 1990. "Are we ready for the male pill?" *Times* (London), October 2.

Primmer, L. 1997. "Women's health research: Congressional action and legislative gains: 1990–1994." In *Women's Health Research*, ed. F. Haseltine and B. Greenberg Jacobson. Health Press International.

Prins, B. 1995. "The ethics of hybrid subjects: Feminist constructivism according to Donna Haraway." *Science, Technology and Human Values* 20, no. 3: 352–367.

Proctor, R. N. 1988. *Racial Hygiene: Medicine under the Nazis.* Harvard University Press.

Progress in Human Reproduction Research. 1997. Newsletter of the WHO/HRP, No. 44.

Prus, R. 1989. *Making Sales: Influence as a Personal Accomplishment.* Sage.

Pursell. C. 2001. "Feminism and the rethinking of the history of technology." In *Science, Medicine, and Technology in the 20th Century,* ed. L. Schiebinger. Princeton University Press.

Rabinow, P. 1992. "Artificiality and enlightenment: From sociobiology to biosociality." In *Incorporations,* ed. J. Crary and S. Kwinter. Zone.

Rabinow, P. 1999. *French DNA.* University of Chicago Press.

Rakow, L. 1992. *Gender on the Line: Women, the Telephone and Community Life.* University of Illinois Press.

Randall, A. J. 1986. "The philosophy of Luddism: The case of the West England woolen workers, ca. 1790–1809." *Technology and Culture* 27: 1–17.

Rapp, M. 1930. Fuels Used for Cooking Purposes in Indiana Rural Homes. Bulletin no. 339, Purdue University Agricultural Experiment Station, May.

Rapp, R. 1998. "Refusing prenatal diagnosis: The meanings of bioscience in a multicultural world." *Science, Technology and Human Values* 23, no. 1: 45–71.

Rasmussen, B., and Hapnes, T. 1991. "Excluding women from the technologies of the future? A case study of the culture of computer science." *Futures,* December: 1107–1119.

Ravindran, T. K. S., and Berer, M. 1994. "Contraceptive safety and effectiveness: Re-evaluating women's needs and professional criteria." *Reproductive Health Matters* 3: 6–11.

Ray, C., and Talbot, H. 1999. "Rural telematics: The Information society and rural development." In *Virtual Geographies,* ed. M. Crang et al. Routledge.

Reclaim the Streets! 2001. http://www.reclaimthestreets.net (June 7).

Report of a workshop. 1995. Investigating Women's Preferences for Contraceptive Technology. Harvard Center for Population and Development Studies, Harvard School of Public Health.

Richardson, L. K. 1961. *Wisconsin REA: The Struggle to Extend Electricity to Rural Wisconsin, 1935–1955.* University of Wisconsin Agricultural Experiment Station.

Richter, J. 1994. *Vaccination against Pregnancy: Miracle or Menace?* BUKO Pharma-Kampagne, Bielefeld.

Richter, J. 1996. *Vaccination against Pregnancy: Miracle or Menace?* Zed Books.

Ringheim, K. 1993. Guidance for Future Social Science Research on Hormonal Methods for Men. Findings from Followup Questionnaires and Focus Groups Discussions with Former Clinical Trial Participants. Report to steering committee of Task Force on Methods for the Regulation of Male Fertility.

Ringheim, K. 1995. "Evidence for the acceptability of an injectable hormonal method for men." *International Family Planning Perspectives* 21, no. 2: 75–80.

Ringheim, K. 1996a. "Male involvement and contraceptive methods for men, present and future." Presented at APHA session Towards Gender Partnership in Reproductive Health, New York.

Ringheim, K. 1996b. "Wither methods for men? Emerging gender issues in contraception." *Reproductive Health Matters* 7, May: 79–89.

Roach, Mack, III. 1998. "Is race an independent prognostic factor for survival from prostate cancer?" *Journal of the National Medical Association* 90, no. 11 (supplement): S713–S719.

Robey, B., Rutstein, S. O., Morris, L., and Blackburn, R. 1992. "The reproductive revolution: New survey findings." *Population Reports*, Series M, 11: 1–2.

Rogers, L. 1994. "Breast test may save daughter from surgery." *Sunday Times* (London), September 16.

Rommes, E., van Oost, E., and Oudshoorn, N. 1999. "Gender and the design of a digital city." *Information Technology, Communication and Society* 2, no. 4: 476–495.

Rommes, E. 2002. *Gender Scripts and the Internet: The Design and Use of Amsterdam's Digital City.* Twente University Press.

Rose, D. 1998. Moving toward a Comprehensive System of Vaccine Research and Development and Delivery in the United States (1979–1995). Master's thesis, University of Amsterdam.

Rose, D. 1999. A history of diphtheria antitoxin production around the turn of the century: Manuscript, Department of Social and Behavioral Sciences, University of California, San Francisco.

Rosenberg, N. 1976. *Perspectives on Technology.* Cambridge University Press.

Rosenberg, N. 1982. *Inside the Black Box: Technology and Economics.* Cambridge University Press.

Rosengren, A. 1994. "Some notes on the male motoring world in a Swedish community." In *The Car and Its Environments*, ed. K. Sørensen. European Commission.

Rosser, Sue V. 1994. *Women's Health—Missing from US Medicine.* Indiana University Press.

Rothman, David J. 1991. *Strangers at the Bedside.* Basic Books.

Rothschild, J., ed. 1999. *Design and Feminism: Re-Visioning Spaces, Places and Everyday Things.* Rutgers University Press.

Russell, Stewart. 1986. "The social construction of artefacts: A response to Pinch and Bijker." *Social Studies of Science* 16: 331–346.

Ruzek, S. B. 1978. *The Women's Health Movement: Feminist Alternatives to Medical Control.* Praeger.

Sabo, D., and Gordon, D. 1995. *Men's Health and Illness: Gender, Power, and the Body.* Sage.

Saetnan, A. 2000. "Women's involvement with reproductive medicine: Introducing shared concepts." In *Bodies of Technology*, ed. A. Saetnan et al. Ohio State University Press.

Saetnan, A., Oudshoorn, N., and Kirejczyk, M., eds. 2000. *Bodies of Technology: Women's Involvement with Reproductive Medicine.* Ohio State University Press.

Saltus, R. 1994. "Mutated gene tied too early breast cancer is located." *Boston Globe*, September 20.

Samferdselsdepartementet, 1996. Den Norske IT-veien bit for bit. Statssekretærutvalget for ITs utredning om norsk IT politikk. http://odin.dep.no/html/nofovalt/offpub/utredninger/it/it-veien/

Scale, D. 2002. The Male Pill and the Science Museum: Creating Consumers for Male Hormonal Contraceptives. Master's thesis, Department of History and Philosophy of Science. University of Cambridge.

Schiebinger, L. 1987. "Skeletons in the closet: The first illustrations of the female skeleton in eighteenth-century anatomy." In *The Making of the Modern Body*, ed. C. Gallegher and T. Lacquer. University of California Press.

Schiebinger, L. 1993. *Nature's Body: Gender in the Making of Modern Science.* Beacon.

Schmidt, M., and Moore, L. S. 1998. "Constructing a good catch, picking a winner: The development of technosemen and the deconstruction of the monolithic male." In *Cyborg Babies*, ed. R. Davis-Floyd and J. Dumit. Routledge.

Schot, J. W. 1992. "Constructive technology assesment and technology dynamics: The case of clean technologies." *Science, Technology and Human Values* 17, no. 1: 36–56.

Schrater, A. F. 1992. "Contraceptive vaccines: promises and problems." In *Issues in Reproductive Technology*, ed. H. Bequart Holmes. Garland.

Schrater, A. F. 1995. "Immunization to regulate fertility: Biological and cultural frameworks." *Social Science and Medicine* 41, no. 5: 657–671.

Schrater, A. F. 1996. Interview with the author, October 22.

Schroeder, P. 1998. *24 Years of House Work and the Place Is Still a Mess: My Life in Politics.* Andrews McNeel.

Schroeder, P., and Snowe, O. 1994. "The politics of women's health." In *The American Woman, 1994–95,* ed. C. Costello and A. Stone. Norton.

Schürmeyer, T., Belkien, L., Knuth, U. A., and Nieschlag, E. 1984. "Reversible azoospermia induced by the anaboloic steroid 19–nortestosterone." *Lancet,* Februrary 25: 417–420.

Scientific and Ethical Review Group. 1994. Discussion on ethical aspects of research, development and introduction of fertility regulating methods. Geneva, June 1–2.

Scott, J. C. 1985. *Weapons of the Weak: Everyday Forms of Peasant Resistance.* Yale University Press.

Scott, J. W. 1986. "Gender: A useful category of historical analysis." *American Historical Review* 91: 1053–1075.

SER. 1991. *Van consument en consumptie: 25 jaar advies en overleg door bedrijfsleven en consument.* Den Haag: Staatsuitgeverij.

SERG. 1994. Discussion on ethical aspects of research, development, and introduction of fertility regulating methods. Geneva, 1–2 June.

Severson, H. 1962. *The Night They Turned On the Lights: The Story of the Electric Power Revolution in the North Star State.* Midwest Historical Features.

Severson, H. 1964. *Determination Turned on the Power: A History of the Eastern Iowa Light and Power Cooperative*

Severson, H. 1965. *Architects of Rural Progress: A Dynamic Story of the Electric Cooperatives as Service Organizations in Illinois.* Association of Illinois Electric Cooperatives.

Severson, H. 1972. *Corn Belt: A Pioneer in Cooperative Power Production.* Corn Belt Power Cooperative.

Shah, I. 1995. Perspectives on Methods of Fertility Regulation: Setting a Research Agenda. Unpublished.

Shankland, L. 1993. "Putting the male 'pill' to the test." *Western Mail* (Edinburgh), July 15.

Shuldener, A. 1998. "From automatic restaurants to *automatieks*: A comparison of the American and Dutch use of the automat." Presented at SHOT conference.

Silverstone, R. 1989. Families, Technologies, and Consumption: The Household and Information and Communication Technologies. CRICT discussion paper, Brunel University.

Silverstone, R. 2000. "Under construction: New media and information technologies in the societies of Europe." Framework paper for European Media Technology and Everyday Life Network (EMTEL 2).

Silverstone, R., and Haddon, L. 1996. "Design and the domestication of information and communication technologies: Technical change and everyday life." In *Communication by Design*, ed. R. Silverstone and R. Mansell. Oxford University Press.

Silverstone, R., and Hirsch, E., eds. 1992. *Consuming Technologies: Media and Information in Domestic Spaces*. Routledge.

Silverstone, R., Hirsch E., and Morley, D. 1992. "Information and communication technologies and the moral economy of the household." In *Consuming Technologies*, ed. R. Silverstone and E. Hirsch. Routledge.

Singleton, V. 1996. "Feminism, sociology of scientific knowledge and postmodernism: Politics, theory and me." *Social Studies of Science* 26: 445–468.

Skelton, T., and Valentine, G., eds. 1998. *Cool Places: Geographies of Youth Cultures*. Routledge.

Skocpol, T. 1996. *Boomerang: Clinton's Health Security Effort and the Turn against Government in US Politics*. Norton.

Skoglund, R. D., and Paulsen, A. A. 1973. "Danazol B testosterone combination: A potentially effective means for reversible male contraception, A preliminary report." *Contraception* 7, no. 5: 357–365.

Slater, D. 1997. *Consumer Culture and Modernity*. Cambridge University Press.

Slaughter, S. 1993. "Innovation and learning during implementation: A comparison of user and manufacturer innovations." *Research Policy* 22: 81–95.

Smith, G. 1997. "Male pill is off the menu." *Daily Record* (Edinburgh), December 26.

Smith, Mark D. 1991. "Zidovudine—Does it work for everyone?" *Journal of the American Medical Association* 266: 2750–2751.

Sørenson, C. A. 1944. "Rural electrification: A story of social pioneering." *Nebraska History* 25: 257–270.

Sørenson, K. H. 1994. "Adieu Adorno: The moral emancipation of the household." In *Technology in Use*. STS-arbeidsnotat, Center for Teknologi og Samfunn, Trondheim.

Sørenson, K. H., Aune, M., and Hatling, M. 2000. "Against linearity: On the cultural appropriation of science and technology." In *Between Understanding and Trust*, ed. M. Dierkes and C. von Grote. Harwood.

Sørensen, K. H., and Sørgård, J. 1994. "Mobility and modernity: Towards a sociology of cars." In *The Car and Its Environments*, ed. K. Sørensen. European Commission.

Soufir, J. C., Jouannet, P., Marson, J., and Soumach, A. 1983. "A reversible inhibition of sperm production and gonadotrophin secretion in men following combined oral medroxyprogesterone acetate and percutaneous testosterone treatment." *Acta Endocrinologica* 102, no. 4: 625–632.

Sparke, P. 1996. *As Long as It's Pink: The Sexual Politics of Taste.* New York University Press.

Spilkner, H., and Sørenson, K. H. 2000. "A ROM of one's own or a home for sharing? Designing the inclusion of women in multimedia." *New Media and Society* 3: 268–285.

Spivak, G. C. 1988. "Can the subaltern speak?" In *Marxism and the Interpretation of Culture*, ed. C. Nelson and L. Grossberg. University of Illinois Press.

Stabiner, K. 1997. *To Dance with the Devil: The New War on Breast Cancer.* Delacorte.

Star, S. L. 1991. "Power, technology and the phenomenology of conventions: On being allergic to onions." In *A Sociology of Monsters*, ed. J. Law. Routledge.

Starr, D. 1998. *Blood: An Epic History of Medicine and Commerce.* Knopf.

Starr, P. 1982. *The Social Transformation of American Medicine.* Basic Books.

Stemerding, B. 1995. "International campaign to stop the research on antifertility 'vaccines.'" In *A Healthy Balance?* ed. E. Hayes. Women's Health Action Foundation.

Stemerding, B. 1998. Interview with the author, February 23.

Stepan, N. 1982. *The Idea of Race in Science: Great Britain, 1800–1960.* Macmillan.

Stevens, V. C. 1990. "Birth control vaccines and the immunological approaches to the therapy of noninfectious diseases." *Infectious Disease Clinics of North America* 4, no. 2: 343–354.

Stevens, V. C. 1992. "Future perspective for vaccine development." *Scandinavian Journal of Immunology* 36, suppl. 11: 137–143.

Stevens, V. C. 1995. Interview with the author, August 9.

Stevens, V. C. 1996. "Progress in the development of human chorionic gonadotropin anti-fertility vaccines." *American Journal of Reproductive Immunology* 35: 148–155

Stewart, J. 1999. Cybercafés: Computers in the community, not communities in the computers. Unpublished.

Stokes, B. 1980. Men and Family Planning. Paper 41, Worldwatch Institute.

Storey, J. 1999. *Cultural Consumption and Everyday Life.* Arnold.

Stortinget, 1997. Stortingsmelding 31 (1996–97) Om distrikts og regionalpolitikken.

Stortinget, 1998. Stortingsmelding 37 (1997–1998) IT-kompetanse i et regionalt perspektiv.

Strasser, S. 1990. *Satisfaction Guaranteed: The Making of the American Mass Market.* Pantheon.

Streefland, P., Chowdhury, A., and Ramos-Jimenez, P. 1999. "Patterns of vaccination acceptance." *Social Science and Medicine* 49: 1705–1716.

Strickland, Stephen P. 1972. *Politics, Science and Dread Disease: A Short History of United States Medical Research Policy.* Harvard University Press.

Suchman, L. 1994. "Working relations of technology production and use." *Computer Supported Cooperative Work (CSCW)* 2: 21–39.

Suchman, L. 2001. "Human/machine reconsidered." In Suchman, *Plans and Situated Actions,* second revised edition. Cambridge University Press.

Summerton, J., ed. 1994. *Changing Large Technical Systems.* Westview.

Sutter, P., Tangermann, R., Aylward, P., and Cochi, S. 2001. "Poliomyelitis eradication: Progress, challenges for the end game, and preparation for the post-eradication era." *Infectious Disease Clinics of North America* 15, no. 1: 41ff.

Sweetenham, E. 1994. "Good health: Male pill made me more virile." *Daily Mail* (London), April 26.

Talwar, G. P. 1994. "Immunocontraception." *Current Opinion in Immunology* 6: 698–704.

Talwar, G. P. 1996. "Molecular endocrinology to new vaccines." Lecture delivered at tenth-anniversary celebration of Molecular Biology Unit, Banara Hindu University, Varanasi.

Talwar, G. P., Om Singh, R. Pal, N. Chatterjee, et al. 1993. "A birth control vaccine is on the horizon for family planning." *Annals of Medicine* 25: 207–212.

Talwar, G. P., Om Singh, R. Pal, N. Chatterjee, et al. 1994. "A vaccine that prevents pregancy in women." *Proceedings of the National Academy of Science* 91: 8532–8536.

Tapper, M. 1999. *In the Blood: Sickle Cell Anemia and the Politics of Race.* University of Pennsylvania Press.

Tapscott, D. 1998. *Growing Up Digital: The Rise of the Net Generation.* McGraw-Hill.

Tavris, Carol. 1992. *The Mismeasure of Woman.* Simon & Schuster.

Tejeda, H. A., Green, S. B., Trimble, E. L., Ford, L., High, J. L., Ungerleider, R. S., Friedman, M. A., and Brawley, O. W. 1996. "Representation of African-Americans, Hispanics, and whites in National Cancer Institute cancer treatment trials." *Journal of the National Cancer Institute* 88: 812–816.

ten Bruggencate, T. 1969. *Enkele aspecten rondom de melkverpakking.* Den Haag: Centrale Verpakkingscommissie Zuivelindustrie.

Terry, J., and Urla, J., eds. 1995. *Deviant Bodies: Critical Perspectives on Difference in Science and Popular Culture.* Indiana University Press.

Teubal, M. 1987. *Innovation Performance, Learning and Government Policy.* University of Wisconsin Press.

Thau, R. 1992. Anti-LHRH and anti-pituitary gonadotropin vaccines: their development and clinical applications." *Scandinavian Journal of Immunology* 36, suppl. 11: 127–130.

Thau, R., Croxatto, H., Luukkainen. T., Alvarez, F., et al. 1989. "Advances in the development of an anti-fertility vaccine." In *Reproductive Immunology*, ed. L. Mettler and W. Billington. Elsevier.

Théberge, P. 1997. *Any Sound You Can Imagine: Making Music Consuming Technology.* Wesleyan University Press.

Thomas, G., and Wyatt, S. 2000. "Access is not the only problem: Using and controlling the Internet." In *Technology and In/Equality*, ed. S. Wyatt et al. Routledge.

Thomas, S. B., and Quinn, S. C. 1991. "The Tuskegee Syphilis Study, 1932 to 1972: Implications for HIV education and AIDS risk education programs in the black community." *American Journal of Public Health* 81, no. 11: 1498–1505.

Thornton, S. 1997. "General introduction." In *The Subcultures Reader*, ed. K. Gelder and S. Thornton. Routledge.

Timmermans, S., and Leiter, V. 2000. "The redemption of thalidomine: Standardizing the risk of birth defects." *Social Studies of Science* 30, no. 1: 41–73.

Tisdall, S. 2000. "Will Europe's third way be the dotcom way?" *Guardian*, March 22.

Titmuss, R. 1971. *The Gift Relationship: From Human Blood to Social Policy.* Pantheon.

Tone, A. 2001. *Devices and Desires: A History of Contraception in America.* Hill and Wang.

Treichler, P. A. 1991. "How to have theory in an epidemic: The evolution of AIDS treatment activism." In *Technoculture*, ed. C. Penley and A. Ross. University of Minnesota Press.

Treichler, P. A. 1999. *How to Have Theory in an Epidemic: Cultural Chronicles of AIDS.* Duke University Press.

Treichler, P. A., Cartwright, L., and Penley, C. 1998. "Introduction: Paradoxes of visibility." In *The Visible Woman*, ed. P. Treichler et al. New York University Press.

Trescott, M. M., ed. 1979. *Dynamos and Virgins Revisited: Women and Technological Change in History.* Scarecrow.

Tucker, W. H. 1994. *The Science and Politics of Racial Research.* University of Illinois Press.

Turkle, S. 1984. *The Second Self. Computers and the Human Spirit.* Simon & Schuster.

Turkle, S. 1988. "Computational reticence: Why women fear the intimate machine." In *Technology and Women's Voices,* ed. C. Kramarae. Routledge & Kegan Paul.

Turkle, S. 1995. *Life on the Screen: Identity in the Age of the Internet.* Simon & Schuster.

Tweton, D. J. 1988. *The New Deal at the Grass Roots: Programs for the People in Otter Tail County, Minnesota.* Minnesota Historical Society Press.

United Nations. 1994. Report on the International Conference on Population and Development, Cairo A/Conf. 171/13 New York.

United Nations Development Program (UNDP). 1999. Human Development Report. http://www.undp.org/hdro/report.html (October 19).

Upadhyay, S. N., Kaushic, C., and Talwar, G. P. 1990. "Antifertility effects of Neem oil by a single intrauterine administration: A novel method for contraception." *Proceedings of the Royal Society London* B242: 175–179.

US Congress. 1993. Public Law. National Institutes of Health Revitalization Amendment.

US Congress. 1997. Public Law. Food and Drug Administration Modernization Act of 1997.

US Department of Agriculture. 1966. *Yearbook of Agriculture, 1965.* Government Printing Office.

US Department of Commerce, Bureau of the Census. 1952. *US Census of Agriculture, 1950: General Report,* vol. 2, *Statistics by Subject.* Government Printing Office.

Vail, M., ed. 2000. *Vintage Synthesizers,* second edition. Miller Freeman.

van Kammen, J. 1999. "Representing users' bodies: The gendered development of anti-fertility vaccines." *Science, Technology and Human Values* 24, no. 3: 307–337.

van Kammen, J. 2000a. "Do users matter?" In *Bodies of Technology,* ed. A. Saetnan et al. Ohio State University Press.

van Kammen, J. 2000b. Conceiving Contraceptives: The Involvement of Users in Anti-Fertility Vaccines Development. Ph.D. thesis, University of Amsterdam.

van Oost, E. 1995. "Over 'vrouwelijke' en 'mannelijke' dingen." In *Vrouwenstudies in de Jaren Negentig,* ed. M. Brouns et al. Coutinho.

van Oost, E. 2000. "Making the computer masculine: The historical roots of gendered representations." In *Women, Work and Computerization.* Kluwer.

Vasterling, V. 1999. "Butler's sophisticated constructivism: A critical assessment." *Hypatia* 14, no. 3: 17–38.

Vaughan, P. 1972. *The Pill on Trial.* Penguin.

Veenman, J., and Jansma, L. 1980. "The 1978 Dutch polio epidemic: A sociological study of the motives for accepting or refusing vaccination." *Netherlands Journal of Sociology* 16, no. 1: 21–48.

Veljkovic, V., Metlas, R., Kohler, H., Urnovitz, H., Prljic, J., Veljkovic, N., Johnson. E., and Muller, S. 2001. "AIDS epidemic at the beginning of the third millennium: Time for a new AIDS vaccine strategy." *Vaccine* 19, no. 15–16: 1855–1862.

Verbeek, P. P. 2000. *De Daadkracht der Dingen.* Boom.

Verenigde Glasfabrieken. 1965. *Melkverpakking en Consument.* Verenigde Glasfabrieken.

Volland, V. 1994. "Scientists isolate gene causing type of breast cancer." *St. Louis Post-Dispatch,* September 16.

Von Hippel, E. 1976. "The dominant role of users in the scientific instrument innovation process." *Research Policy* 5: 212–239.

Von Hippel, E. 1988. *The Sources of Innovation.* Oxford University Press.

Von Hippel, E., and Tyre, M. 1995. "How learning by doing is done: Problem identification in novel process equipment." *Research Policy* 24: 1–12.

Waara, P. 1996. *Ungdom i Gränsland.* Boréa.

Wailoo, K. 1997. *Drawing Blood: Technology and Disease Identity in Twentieth-Century America.* Johns Hopkins University Press.

Wajcman, J. 1991. *Feminism Confronts Technology.* Polity.

Wakeford, N. 1999. "Gender and the landscapes of computing in an Internet café." In *Virtual Geographies,* ed. M. Crang et al. Routledge.

Waldby, C. 1996. *AIDS and the Body Politic: Biomedicine and Sexual Difference.* Routledge.

Walker, L. P. 1998. *Advertising Progress: American Business and the Rise of Consumer Marketing.* Blackwell.

Walker, M. 1996. "Home Extension work among African-American farm women in East Tennessee, 1920–1939." *Agricultural History* 70: 487–502.

Walsh, V., Roy, R., Bruce, M., and Potter, S. 1992. *Winning by Design: Technology, Product Design and International Competitiveness.* Blackwell.

Ward, M. 2000. "Some Britons still refuse to surf." BBC News Online, July 11.

Weisman, C. S. 1998. *Women's Health Care: Activist Traditions and Institutional Change.* Johns Hopkins University Press.

Weisman, C. S. 2000. "Breast cancer policymaking." In *Breast Cancer: Society Shapes and Epidemic,* ed. A. Kasper and S. Ferguson. St. Martin's.

West, C., and Zimmerman, D. H. 1987. "Doing gender." *Gender and Society* 1: 125–151.

WHO (World Health Organization). 1992a. *WHO Laboratory Manual for the Examination of Human Semen and Sperm-cervical Mucus Interaction,* third edition. Cambridge University Press.

WHO. 1992b. *Guidelines for the Use of Androgens in Men.* World Health Organization.

WHO. 1996. WHO Completes International Trial of a Hormonal Contraceptive for Men. Press release, Geneva, April 2.

WHO/HRP. 1992. Reproductive health: A key to a brighter future. Biennial report 1990–1991, World Health Organization.

WHO/HRP. 1993. *Fertility Regulating Vaccines.* World Health Organization.

WHO/HRP. 1996. *Biennial Report 1994–1995.* World Health Organization.

WHO/HRP/ATR. 1995. *Annual Technical Report.* World Health Organization.

WHO/HRP/ITT. 1991. *Creating Common Ground: Women's Perspectives on the Selection and Introduction of Fertility Regulating Technologies.* World Health Organization.

Wieringa, N. 1994. "Anti-fertility vaccines: The wrong road?" *International Journal of Risk and Safety in Medicine* 6: 1–9.

Winner, L. 2000. "Enthusiasm and concern: Results of a new technology poll." *Tech Knowledge Revue,* February 29.

Wittes, B., and Wittes, J. 1993. "Group therapy." *New Republic* 5, April: 15–16.

Wollerich, H. 2000. "Van eitje tot Quadra. Fabriek in Drachten produceert al vijftig jaar Philishaves." *Dagblad Tubantia,* November 7.

Wolpe, P. 1998. "The triumph of autonomy in American bioethics: A sociological view." *Bioethics and Society: Constructing the Ethical Enterprise.* Prentice-Hall.

Woolgar, S. 1991. "Configuring the user: The case of usability trials." In *A Sociology of Monsters,* ed. J. Law. Routledge.

Working Group for the Chief Medical Officer. 1996. Genetics and Cancer Services, Department of Health.

World Health Organization Task Force on Methods for the Regulation of Male Fertility. 1996. "Contraceptive efficacy of testosterone-induced azoospermia and oligospermia in normal men." *Fertility and Sterility* 65, no. 4: 821–829.

Wright, B. D., et al., eds. 1987. *Women, Work, and Technology.* University of Michigan Press.

Wright, Lawrence. 1994. "One drop of blood." *New Yorker,* July 25: 46–55.

Wright, P. 1995. "Global immunization—A medical perspective." *Social Science and Medicine* 41, no. 5: 609–616.

Wyatt, S., Thomas, G., and Terranova, T. 2002. "They came, they surfed, they went back to the beach: Conceptualising use and non-use of the Internet." In *Virtual Society?* ed. S. Woolgar. Oxford University Press.

Ziehe, T. 1989. *Kulturanalyser: Ungdom, utbildning, modernitet.* Symposium bibliotek.

Zunz, O. 1990. *Making America Corporate, 1870–1920.* University of Chicago Press.

Zussman, R. 1992. *Intensive Care: Medical Ethics and the Medical Profession.* University of Chicago Press.

Contributors

Adri Albert de la Bruheze is a lecturer at the Center for Studies of Science, Technology and Society at the University of Twente. Since 1994 he has been a researcher and editorial secretary for the seven-volume research project Technology in the Netherlands in the Twentieth Century (TIN-20). His current research interest is the mutual construction of products, consumption, and consumers in twentieth-century Holland.

Stuart Blume is a professor of science dynamics at the University of Amsterdam. His books include *Toward a Political Sociology of Science* (Free Press, 1974) and *Insight and Industry* (MIT Press, 1992). His current research interests and publications focus on the dynamics of the development of technologies for deaf people and the development of vaccines.

Steven Epstein is an associate professor in the sociology department at the University of California, San Diego. He is also affiliated with UCSD's interdisciplinary Science Studies Program. His book *Impure Science: AIDS, Activism, and the Politics of Knowledge* (University of California Press, 1996) received three major awards. His current research examines the politics of inclusion and the management of difference in US biomedical research.

Ronald Kline is a professor in the school of electrical engineering and the department of science and technology studies at Cornell University. His current research interest is the history of information theory in the United States and Britain and its application, adaptation, and modification in the physical and social sciences. His major publications are *Steinmetz: Engineer and Socialist* (Johns Hopkins University Press, 1992) and *Consumers in the Country: Technology and Social Change in Rural America* (Johns Hopkins University Press, 2000).

Anne Sofie Laegran is a research fellow in the department of geography at the Norwegian University of Science and Technology, where she is also affiliated with the Center for Technology and Society. She is interested in how technologies are integrated into everyday life spaces, and in how spaces and places are produced in the intersection between technology and social practice. Her dissertation is on Internet cafés in rural and urban parts of Norway.

Christina Lindsay is a senior user research consultant at Philips Design. Her primary field of research is strategic design, particularly innovation projects concerning sustainable solutions. She is researching and developing new methodologies and tools to bring users into the design process as partners.

Nelly Oudshoorn is a professor of science and technology studies. She is the author of *Beyond the Natural Body: An Archeology of Sex Hormones* (Routledge, 1994) and *The Male Pill: A Biography of a Technology in the Making* (Duke University Press, 2003) and a co-editor of *Bodies of Technology: Women's Involvement with Reproductive Technologies* (Ohio State University Press, 2000). Her current research focuses on the construction of trust and the emergence of new intermediary organizations in tele-monitoring technologies for heart patients.

Shobita Parthasarathy is an assistant professor of Public Policy at the University of Michigan. She is completing a book comparing the development of genetic testing for breast and ovarian cancer in the United States and Britain (to be published by The MIT Press). She is currently researching two projects, investigating the European politics regarding patenting medical genetics and biotechnology and the role and success of patient advocacy groups in influencing government policy and corporate strategy.

Trevor Pinch is a professor of science and technology studies and of sociology at Cornell University. He is a co-author of *The Golem: What You Should Know about Science* (second edition: Canto, 1998) and *The Golem at Large: What You Should Know about Technology* (Canto, 1998). His most recent research has been on the connections between STS and sound, noise, and music. A co-author of *Analog Days: The Invention and Impact of the Moog Synthesizer* (Harvard University Press, 2002), he is currently completing a new book for the Golem series on the topic of medicine.

Dale Rose is a doctoral candidate at the University of California, San Francisco. His primary academic interests are in the study of the innovation and use of medical technologies, the construction of knowledge and the claims of knowledge makers, public health policies and practices, and the work life of professionals. Currently he is researching the evolution of emergency medical services in the United States in terms of its organization at various levels of government (federal, state, local) and with respect to the actual practices that constitute the relatively new field of emergency medicine.

Johan Schot is a professor of social history of technology at the Eindhoven University of Technology. He is a co-editor of *Managing Technology in Society: The Approach of Constructive Technology Assessment* (Pinter, 1995) and a co-author of *Experimenting for Sustainable Transport* (SPON, 2002).

Jessika van Kammen has conducted a study on the future of technology and integrated health care in the Netherlands for the Dutch Study Center for Technology Trends (STT). She currently works as an adviser on the implementation of research results in health-care practices at the Organization for Health Research and Development (ZonMw). She is the editor of *Zorgtechnologie: Kansen voor Innovatie en Gebruik* (STT, 2002).

Ellen van Oost is an associate professor at the University of Twente. In her research she combines theoretical perspectives of script analyses and domestication analyses to gain insight into processes that shape both digital services and its gendered users. She is a co-author (with Wim Smit) of *De wederzijdse beïnvloeding van technologie en samenleving. Een Technology Assessment benadering* (Coutinho, 1999) and a co-editor of *De opkomst van de informatietechnologie in Nederland* (Ten Hagen and Stam, 1998).

Sally Wyatt works in the department of communication studies at the University of Amsterdam. She is currently doing research about the ways in which women and men draw on different information sources, including the Internet, to assess the risks associated with hormone replacement therapy and Viagra. A co-editor of *Technology and In/Equality: Questioning the Information Society* (Routledge, 2000), she also contributed three chapters to *Cyborg Lives? Women's Technobiographies*, edited by Flis Henwood, Helen Kennedy, and Nod Miller (Raw Nerve Press, 2001). She was president of the European Association for the Study of Science and Technology between 2001–2004.

Index

Printed in the United States
by Baker & Taylor Publisher Services